Magnetism

Carmen-Gabriela Stefanita

Magnetism

Basics and Applications

 Springer

Dr. Carmen-Gabriela Stefanita
NanoDotTek
01803 Burlington
Massachusetts
USA

ISBN 978-3-642-22976-3   e-ISBN 978-3-642-22977-0
DOI 10.1007/978-3-642-22977-0
Springer Heidelberg Dordrecht London New York

Library of Congress Control Number: 2011941396

Printed on acid-free paper

Springer is part of Springer Science+Business Media (www.springer.com)

To Alex and Chris
  for their unconditional support

Preface

This book discusses the basics of contemporary magnetism as encountered in today's familiar applications, such as the magnetic tape on a credit card or the read head in a personal computer. Nevertheless, the discussion is not particularly centered on immediate products, as emphasis is placed on revealing key concepts and magnetic phenomena on which such practical applications rest. Therefore, this book does not have a commercial focus, nor is it meant to provide an up-to-date reference on the latest research accomplishments of the different magnetics subjects. On the contrary, it is up to the reader to uncover the connection between these concepts and everyday life where magnetism reveals itself, and as such continue to explore further. After all, the main goal is to entice people to recognize on their own the myriad of subtle associations developed in time between disparate and seemingly unrelated areas of magnetics. For this reason, it should not come as a surprise when the discussion takes the reader over a variety of combined physical manifestations or different length scales of phenomena, revealing aspects not always immediately recognizable as belonging to magnetism. Moreover, the questions/problems at the end of each chapter are meant to help further understanding, and at the same time test the reader's broad knowledge of magnetics. Quite often, these problems are not necessarily a continuation of concepts discussed in that particular chapter, but rather unexpected questions about topics that require further study. This is to motivate the reader to look into unanticipated aspects of magnetics. Clearly, the goal is to

---

[1] *Schopenhauer. Brevier*, herausgegeben von Raymund Schmidt, Dieterich'schen Verlagsbuchhandlung zu Leipzig, S. 262, 1938.

deviate from the "spoonfeeding" that many students find comfortable, but is not appropriate for an academic education. However, this book is still an accessible starting point for someone interested in the broader area of magnetics, and can serve as an introductory course either at the late undergraduate or the early graduate level. After all, academic education has to keep up with today's increased complexity of technological developments.

The book begins with a first and second chapter introduction to some historical concepts that should make parts of the text more accessible. Some basic ideas of magnetics go back roughly fifty years or even a century in time, having experienced mainly bulk practical demonstrations. Nonetheless, these ideas find nowadays new avenues of expression as applications move to smaller scales of fabrication. For instance, magnetic domains and their wall movement have started to preoccupy once more those interested in new magnetic storage techniques, while the long researched spin of the electron is finally put to more ubiquitous use through new devices. Similarly, nondestructive techniques relying on micromagnetics and magnetic flux leakage have preoccupied the minds of many researchers for many decades, and for this reason their successful application to engineering applications is reviewed as we move on to Chap. 3. In view of their lasting endurance in the magnetic recording industry, the familiar magneto-optical recording and readout techniques are discussed in Chap. 4. On the other hand, some new magneto-optical applications mainly used for thin film characterization are mentioned as well, given that magnetic and optical interactions allow us to gain some valuable information about magnetization orientation or magnetization reversal. The many developments of the last century have revealed aspects of magnetism that some scientists and engineers may not be aware of. The treatment in Chap. 5 of the combined effects of electrical, thermal, and magnetic phenomena is trying to unveil precisely this lesser known aspect of magnetics, since it is easy for magnetics principles to be eclipsed by better known physical properties encountered in these applications. As the discussion in the book progresses, the number of new concepts increases, and more and more unfamiliar ideas may surface, such as the notion of a spin valve, elaborated on in Chap. 6. The best way to keep up is to maintain a broad interdisciplinary base of knowledge, whether through other academic courses or research. This adds a level of difficulty not always anticipated when reading the book, and will perhaps lead to some bumps in the road to understanding. Some knowledge of semiconductor device fabrication will come in handy, as well as familiarity with more laboratory-oriented growth techniques, in particular when reading Chap. 7 which deals with the newly emerged field of spin electronics. Prior exposure to at least basic quantum mechanics is a key requirement, as many quantum mechanical concepts, such as for instance, the Pauli exclusion principle, Fermi wavelength, and superposition of quantum states, are assumed to be familiar to the reader. But rest assured, all this is worth the trouble when the reader finally reaches Chap. 8 where the latest magnetic recording media developments are uncovered. Nevertheless, the discussion of the many fascinating subfields of magnetics is not coming to an end, even when the book does. As such, it continues to write many chapters through many other applications that are being developed every day.

Some of the discussed technologies such as quantum devices exist almost exclusively in the most advanced laboratories. There is hope that these technologies may replace one day the current type of semiconductor electronic devices. Unfortunately, until now the ever increasing speed of computers has thrown more and more obstacles in the way of the semiconductor industry. With nearly a billion transistors on an Intel Dual Core chip, following the familiar paths of optical lithography may become troublesome sooner than expected. This is because optical lithography depends on projecting light through stencil-like masks creating layer after layer of infinitesimal structures that become part of a transistor. Nevertheless, the wavelengths of lights currently used are merely too large to print the required dense patterns. Of course, tremendous research efforts have been spent and continue to be spent on developing systems with radiation beyond the visible spectrum, but regrettably they do not seem commercially viable. Alternative techniques such as double patterning lithography (i.e., doubling up the printed layers) have also emerged; however, it is not yet known how long they can keep up with the miniaturization of transistors. Advances in materials research combined with promising new areas in magnetics may offer ingenious solutions for overcoming these miniaturization barriers. Regrettably, for conventional semiconductor devices it may also mean replacing them, as new devices emerge with diversified functions that could take over some of the old, familiar tasks. However, many new device developments have encountered challenges of their own, because the actual construction of quantum devices that could perform at least the same tasks as electronic devices has proven to have its own difficulties. For the time being, the existence of such devices gives an indication of the direction of future technological developments that do not necessarily follow Moore's law that experts say is likely to hit a brick wall. If nothing else, these novel devices may provide spin-offs into further areas of magnetics.

The book is the work of Carmen-Gabriela Stefanita. A deep appreciation and gratitude go to Dr. Min Namkung of the NASA Goddard Space Flight Center, Greenbelt, Maryland, who has lent his valuable expertise to the discussion of some fundamental micromagnetics concepts.

Burlington, Massachusetts   *Carmen-Gabriela Stefanita*

Contents

List of Symbols

The units contained in this table represent the most commonly used and do not belong to only one system of units. Proper conversion between systems is required. It is assumed the meaning of symbols such as V (volt), K (kelvin), and A (ampere) is known to the reader.

Symbol	Definition	Commonly used units
X	Magnetic susceptibility	H/m
M	Magnetization	A/m or Wb m
H_0	Applied magnetic field intensity	A/m or Oe
μ_0	Permeability of vacuum $\mu_0 = 4\pi 10^{-7}$ H/m	H/m
χ_0	Relative susceptibility	
T	Absolute temperature	K
Θ_N	Néel temperature	K
p	Magnetic pole strength	Wb
r	Distance between magnetic poles	cm
F	Magnetic force	N
l	Length of bar magnet	cm
L	Torque	N m
θ	Angle between the directions of magnetic field and bar magnet magnetization	
U	Potential energy	J
m	Magnetic moment	A m^2
dV	Unit volume	cm^3
L	Total orbital angular momentum	Units of $\hbar$
S	Total spin angular momentum	Units of $\hbar$
J	Total atomic angular momentum	Units of $\hbar$

Symbol	Definition	Commonly used units
J	Exchange integral for two atoms with spins S_i and S_j	J
$\hbar$	Reduced Planck constant	J s or N m s
n	Number of electrons in an atom	
$\vec{\mu}_L$	Magnetic moment associated with $\mathbf{L}$	Bohr magnetons
$\vec{\mu}_S$	Magnetic moment associated with $\mathbf{S}$	Bohr magnetons
$\vec{\mu}$	Total magnetic moment	Bohr magnetons
β	Bohr magneton	$\frac{|e|\hbar}{2mc} = 10^{-24}\ J/T$
μ_J	Component of μ parallel to J	Bohr magnetons
g	Landé g-factor	
J_z	z component of J	Units of $\hbar$
z	Axis in the x,y,z Cartesian system	
U_{ij}	Potential energy between two atoms with spins S_i and S_j	J
m_J	Eigenvalues of J_z	
μ_{J_z}	Projection of μ_J along z	Bohr magnetons
N	Number of atoms in the sample (e.g., Avogadro number)	
$\hat{H}$	Hamiltonian	
E_m	Eigenvalues of $\hat{H}$ (also known as magnetic energy)	J
Z	Partition function	
kT	Thermal energy	J
x	Ratio of magnetic and thermal energies	
$\langle \mu_{J_z} \rangle$	Expectation value of the magnetic moment μ_{J_z}	Bohr magnetons
$\langle J_z \rangle$	Expectation value of J_z	Units of $\hbar$
B_J	Brillouin function	
bcc	Body centered cubic	
λ	Magnetostriction	
λ_s	Isotropic saturation magnetostriction	
λ_{100}	Saturation longitudinal magnetostriction along [100]	
λ_{111}	Saturation longitudinal magnetostriction along [111]	
E_{exchange}	Exchange energy (density)	J/cm^3
$E_{\text{magnetostatic}}$	Magnetostatic energy (density)	J/cm^3

Symbol	Definition	Commonly used units
$E_{\text{magnetocrystalline}}$	Magnetocrystalline (anisotropic) energy density	J/cm^3
$E_{\text{magneotelastic}}$	Magnetoelastic energy density	J/cm^3
E_{wall}	Wall energy density	J/cm^3
δ	Domain width	nm
δ	Domain wall thickness (not to be confused with domain width)	nm
J	Exchange coupling constant	J
S	Spin quantum number	
K	Magnetic anisotropy constant	J/m^3
a	Lattice constant	Å
H	Magnetic field intensity	A/m or Oe
U	Potential energy density	N/m^2
$\vec{M}_s$	Domain magnetization vector	A/m or Wb m
$\vec{B}$	Magnetic field density vector	T or Gs
x	Domain wall position	M
σ	Stress	N/m^2
ε	Strain	
$V(x)$	Potential due to lattice defects	
C_{11}		
α	Energy (density) contribution responsible for an easy axis	J/cm^3
β	Energy (density) contribution from the isotropic background	J/cm^3
$\text{MBN}_{\text{energy}}$	See text for description	$\text{mV}^2\ \text{s}$
V_{rms}^2	Squared voltage rms signal of MBN	V^2
Θ	Angle at which a magnetic field is applied	Degrees or radians
Φ	Magnetic easy axis direction	
φ	Magnetic scalar potential	V s m^{-1}
$\ln a$	Natural logarithm of amplitude attenuation factor	
Γ	Phase of electromagnetic field	Degrees or radians
Φ	Phase change of electromagnetic field	Degrees or radians
μ_0	Magnetic permeability of free space	H/m
μ	Magnetic permeability of the material	H/m
F	Frequency of excitation	Hz

Symbol	Definition	Commonly used units
Ω	Angular frequency of excitation	rad/s
Σ	Electrical conductivity	S/m
r_i	Inside pipe diameter	m
r_o	Outside pipe diameter	m
T	Thickness of pipe wall	m
P	Electrical resistivity	Ω m
V	Velocity of the MFL "pig"	m/s
A	Magnetic vector potential	V s m^{-1}
J_s	Surface current density	A/m^2
T	Time	s
kT	Thermal activation energy	J
C	Crystallographic axis	
S	Direction of polarization (German: senkrecht)	
P	Direction of polarization (German: parallel)	
P	Recording wavelength for magnetic mark	nm
Λ	Readout light wavelength	nm
NA	Numerical aperture of the objective lens	
D	Diameter of first Airy ring	nm
v_w	Linear velocity of the domain wall	m/s
μ_w	Domain wall mobility	cm^2/V s
H_σ	Magnetic field (intensity) due to Bloch wall energy gradient	Oe
H_M	Demagnetizing field (intensity)	Oe
H_{ext}	External magnetic field (intensity)	Oe
H_{cw}	Domain wall coercivity	Oe
L	Minimum mark length	nm
V	Linear velocity of the rotating disk	m/s
D	Displacement of the domain wall	nm
T	Wall displacement time	ns
Θ	Complex Kerr rotation angle	
Θ	Kerr rotation	
E	Ellipticity	
ZT	Figure-of-merit	
A	Seebeck coefficient	V/K
α_S	Perpendicular Seebeck coefficient	V/K
α_P	Parallel Seebeck coefficient	V/K

Symbol	Definition	Commonly used units
A	Magnetoelectric voltage coefficient	V/cm Oe
Σ	Electrical conductivity	$\Omega^{-1}\mathrm{m}^{-1}$
K	Thermal conductivity	W/(m K)
P	Electrical (contact) resistivity	Ω m
E_0	Potential barrier	J
E	Electron charge	1.6022×10^{-19} C
H	Planck constant	6.6261×10^{-34} J s
D	Barrier thickness	m
P_t	Tunneling probability	
$m*$	Effective mass	kg
x	Thickness	m
ΔQ	Heat	J
Q_{tr}	Energy flux density	W/m^2
Δt	Time interval	s
A	Surface area	m^2
T	Temperature	K
T_j	Junction temperature	K
T_a	Ambient temperature	K
P	Power dissipation	W
P	Polarization density	C/m^2
R_{ja}	Thermal resistance	°C/W
$\Delta \vec{q}_{td}$	Change in thermal diffusion flux	W/m
$\prod$	Peltier coefficient	V m
J	Electric current density	A/m^2
$\vec{q}$	Total thermal diffusion flux	W/m
$\tilde{\phi}$	Electrochemical potential	V or J/C
k_B	Boltzmann constant	1.38065×10^{-23} J/K
E_a	Energy gap	eV
σ_0	Residual conductivity	$\Omega^{-1}\mathrm{m}^{-1}$
Λ	Phonon mean free path	m
P	Scattering parameter	
Φ	Angle between principal thermoelectric axis and length of crystal	
A	Length of crystal	m
B	Width of crystal	m
N	Nernst coefficient	m^2/(K s)
E_{tt}	Ettingshausen coefficient per unit volume	K/J
R_L	Righi–Leduc coefficient	m^2/(V s)

Symbol	Definition	Commonly used units
k_c	Coupling factor	
R_H	Hall coefficient	m^3/C
B	Magnetic field density	T
E	Electric field intensity	V/m
V	Voltage	V
T	Sample thickness	cm
ε_e	Piezoelectric coefficient	C/m^2
$\in_m$	Piezomagnetic coefficient	Oe^{-1}
$\rho_\uparrow$	"Spin-up" resistivity	Ω m
$\rho_\downarrow$	"Spin-down" resistivity	Ω m
ρ_+	Majority spin resistivity	Ω m
ρ_-	Minority spin resistivity	Ω m
P	Resistivity	Ω m
ρ_F	Resistivity in case of a ferromagnetic alignment	Ω m
ρ_{AF}	Resistivity in case of an antiferromagnetic alignment	Ω m
$\Delta\rho$	Change in resistivity	Ω m
MR	Magnetoresistance	
N	Total number of interfaces	
$R(0)$	Resistance of the system at zero magnetic field	Ω
$R(H)$	Resistance of the system at a magnetic field value H	Ω
A	Parameter measuring spin transport asymmetry	
M	Electrochemical potential	eV
E	Energy level	eV
Γ	Coupling to the source or drain	eV
Γ	Half-width at half-maximum	nm
N	Number of energy levels	
E	Energy	eV
$G(E)$	Green's function	
B_z	Magnetic field density along the z direction	T
I	Identity matrix	
μ_B	Bohr magneton	5.7883×10^{-5} eV/T
X	Axis direction, but also distance along the x direction (in nm)	
T_2	Transverse relaxation time	ns

Symbol	Definition	Commonly used units
j_M	Net magnetization current (density)	A/cm^3
j_e	Net electric current (density)	A/cm^3
μ_B	Bohr magneton $\mu_B = 9.274\,009$ $49(80) \times 10^{-24}$ J/T	J/T
η_M	Spin injection efficiency for a single heterojunction	
η'_M	Spin injection efficiency for a double heterojunction	
$G_\uparrow$	"Spin-up" conductance	S
$G_\downarrow$	"Spin-down" conductance	S
$G_\uparrow^{tot}$	Total "spin-up" conductance	S
$G_\downarrow^{tot}$	Total "spin-down" conductance	S
$T_\uparrow$	"Spin-up" transmission probability	
$T_\downarrow$	"Spin-down" transmission probability	
G	Electron spin g-factor	
l_{sd}	Spin diffusion length	nm
v_F	Fermi velocity	m/s
$\tau_{\uparrow\downarrow}$	Spin-flip time	ns
Λ	Mean free path	nm
N	Spin density at distance x from the interface	cm^{-3} or J^{-1}
n_0	Spin density at the interface	cm^{-3} or J^{-1}
X	Distance from the interface	nm
I	Sensing current	A
H	External magnetic field intensity	Oe
B	External magnetic field density	T
M	Specimen magnetization	T
S	Electron spin	
g_0	Free electron g-factor	$g_0 = 2.0023$
Hcp	Hexagonal closed packed	
κ_α	Magnetic anisotropy constant	J/cm^3
Y	Magnetic switching volume	cm^3
λ	Light wavelength used in lithography	nm
NA	System numerical aperture	
p	Pattern period	nm
θ	Half-angle between the two beams in interference lithography	

Chapter 1
Traditional Magnetism

Abstract A pointed piece of a mineral called magnetite was noticed to turn north–south when supported in air or placed on the surface of water. As such, magnetite was the first magnetic material to be discovered, and people wasted no time exploiting it. From the dawn of the human race, when magnetic phenomena began to fascinate us, we have, thus, learned how to creatively exploit magnetic properties. Since then, magnetism has become a highly diversified discipline that has opened up new possibilities for scientific and technological developments. Novel magnetic applications emerge, roughly every year, in their scientific move toward unexplored natural boundaries that push those of human imagination. Furthermore, new technologies that are currently being developed incorporate more and more magnetic structures, spanning a range of applications, often replacing others in more established fields. Interestingly, a significant number of modern conveniences are based on magnetic phenomena that were already studied half a century ago, but are only now benefiting from increased appreciation. And yet, aside from natural magnets, most magnetic materials are not really known to the public at large, whether they are naturally occurring substances, or artificially created compounds.

Varieties of magnetic properties have atomic origins, and have already become observable on length scales of the order of a few nanometers. For this reason, and given the possibilities these days of advanced fabrication technologies, magnetic structures can be nanoengineered to the extent that entirely new materials are obtained. In this manner, artificial metamaterials can be successfully created, relying on combined magnetic phenomena not observable in the original individual compounds. For instance, when a soft phase is incorporated into an amorphous matrix, a completely different magnetic material can be obtained, such as embedded magnetic nanoclusters, or granular nanosized polymer materials. Similarly, many modern magnetic devices that are presently enjoying widespread use are functional on a nanoscale. The increasingly recognizable structures known as nanodots, where quantum–mechanical effects are no longer negligible, have become a common research topic in laboratories preoccupied with the search for improved magnetic storage. It is therefore no surprise that concepts such as quantum well states and

C.-G. Stefanita, *Magnetism*, DOI 10.1007/978-3-642-22977-0_1,
© Springer-Verlag Berlin Heidelberg 2012

spin degrees of freedom come into play, expanding nanodot areas of applications to quantum computing and spin electronics.

This introductory chapter is meant to provide the reader with minimal information on some traditional aspects of magnetism. However, the task of revealing a comprehensive view on the historical concepts of magnetics is best left to previously published, classic treatises. Only a few key concepts are mentioned, with the selection of topics left to the discretion of the author. How can we develop advanced nanostructured devices, if we do not even know when we can classify a material as magnetic? Which magnetic properties have served as historical uses of magnetic materials, and what new ideas are actually old concepts presented in a new light? We cannot move forward toward advanced magnetics topics without trying to answer a few critical questions, and in the process of doing so, discover that not all hot subjects in today's magnetics are actually that new. Nevertheless, in order to facilitate a minimal understanding of fundamental ideas on which new technologies are based, we need to first become familiar with basic concepts in magnetics. This way, we can facilitate a transition toward more modern applications discussed in later chapters.

1.1 Magnetic Concepts

1.1.1 Magnetic Dipoles

Bearing on similarities with electrostatics, magnetism has been traditionally viewed as an interaction between magnetic poles of strengths p_1 and p_2 separated by a distance r. This is analogous to the Coulomb interaction between electrically charged particles, and can be written as [1]

$$F = \frac{p_1 p_2}{4\pi\mu_0 r^2},$$ (1.1)

where F is the force acting on a magnetic pole and μ_0 is the permeability of vacuum. In a complementary approach, it can also be said that a magnetic field producing electric current or another magnetic pole, will exert a force F on the initial magnetic pole of strength p

$$F = pH_0,$$ (1.2)

where H_0 is the applied magnetic field due to the electric current or the other magnetic pole. We know from Maxwell's electromagnetic theory that magnetic poles occur in pairs. As such, when a magnet is cut into pieces, each piece will have a pair of poles. Magnetic poles exert a force on each other such that like poles repel each other with force F described by (1.1), while north and south poles attract. Equation (1.2) implies that if a magnetic material is brought near a magnet,

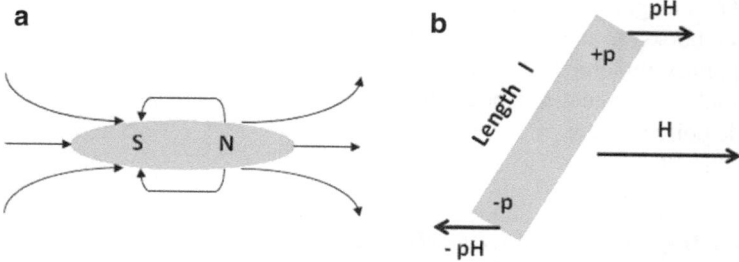

Fig. 1.1 (**a**) Magnetic field representation outside a magnet or magnetized material, and (**b**) couple of magnetic force pH acting on a bar of length l placed in a magnetic field **H**

the magnetic field of the magnet will magnetize the material. Consequently, the magnetic field is sometimes called a *magnetizing force* schematically represented by lines, also called *lines of force* shown in Fig. 1.1a. These lines of force or magnetic field lines are drawn such that lines outside the magnet radiate outward from the north pole, and a compass needle would be tangent to them. Magnetic field lines leave the north pole and enter the magnet again at the south pole, attracting a ferromagnet toward the magnet, even if the two are a certain distance apart.

Consider a bar magnet of length l with magnetic poles p and $-p$ at its ends, placed in a uniform magnetic field (Fig. 1.1b). In this case, the couple of magnetic forces gives rise to a torque L

$$L = -pl\mathbf{H}\sin\theta,\tag{1.3}$$

where θ is the angle between the direction of the applied magnetic field H and the direction of the magnetization M of the bar magnet. In this case, the product pl is the magnetization M of the bar. The work done by the torque gives rise to a potential energy U in the absence of frictional forces

$$U = -MH\cos\theta.\tag{1.4}$$

This equation is particularly important in the next chapter, when magnetic domains and the realignment of their magnetization toward an externally applied magnetic field are discussed. It can be seen that the potential energy has a minimum value when $\theta = 0$.

When the bar length l tends to zero value, and if the strengths of the magnetic poles p approach infinity simultaneously, the system is called a *magnetic dipole*. Alternatively, a magnetic dipole can also be defined by a circular electric current of infinite intensity spanning an area of zero dimensions. Regardless of how we look at it, the magnetic dipole is only a mathematical concept, merely useful for the definition of some magnetic quantities. For instance, the magnetic moment m of the magnetic dipole is

$$m = M\,dV,\tag{1.5}$$

where M is magnetization and dV is the unit volume. This equation is considered in earlier books as the definition for M. It is important to emphasize the fact, and hopefully remember later in this book, that if the magnetization is constant throughout the magnetized body, the latter is considered homogeneous from a magnetic point of view.

1.1.2 Magnetic Field Intensity and Magnetic Flux Density

Magnetism has come a long way since these early interpretations mentioned here. And yet, more complex concepts can sometimes be reduced to these basic ideas, so that an easier understanding is facilitated at that moment. As such, the purpose of this section is to name the two most basic quantities of magnetism, *magnetic field intensity* and *magnetic field density* in the way they were historically introduced. However, in order to describe the magnetic properties of materials, and have a quantitative measure of them, we need more than just two quantities. Therefore, for a more rigorous treatment, the reader is directed to the variety of magnetism or electromagnetics related books that deal with these concepts in depth. It suffices to say that the strength of the field of force discussed in the previous section is known as the *magnetic field strength*, *magnetic field intensity*, or *magnetizing force*, and is denoted, very often, by the symbol H. In the cgs system of units, the magnetic field intensity H is measured in *oersteds* (*Oe*), where 1 *Oe* is defined as the field strength at 1 cm of the *unit pole*. On the other hand, a magnetic field can also be produced by an electric current, and is therefore defined in terms of current as one *Ampere-turn per meter* (A/m), the measuring unit for magnetic field intensity in the MKS system. Magnetic field intensity is a *vector* denoted in bold letters by H, and therefore, has both direction and strength. The direction is given by that in which a north magnetic pole starts to move when subjected to a magnetic field, or that indicated by the north end of a compass needle.

The concentration of lines of force at any point is known as *magnetic flux*, whereas the flux per unit area is the *magnetic flux density* represented by the symbol B, or in vector form B. The units for magnetic flux density are *gauss* (*Gs*) in the cgs system of units, and *weber*/(*meter*)^2 (*Wb*/m^2) or *tesla* (*T*) in the MKS system, or the SI system of units developed later. The magnetic flux density is traditionally defined by the relation

$$B = H + 4\pi M. \tag{1.6}$$

The quantity B–$H = 4\pi M$ attains a saturation value. The factor 4π is the area of a sphere, and its presence in the equation is attributable to a unit field caused by a unit pole everywhere on the surface of a sphere of unit radius enclosing the pole. Alternative to (1.6), the magnetic flux density can also be defined from the movement of charge or current flow, and the laws of magnetostatics.

1.1.3 Magnetic Susceptibility and Magnetic Permeability

In order to determine whether a material is magnetic or not, we need at least one quantity that can describe the material's magnetic behavior under an applied field. The quantity, termed *magnetic susceptibility* χ, characterizes the magnetic response of a material through the relationship

$$M = \chi H_0, \tag{1.6}$$

where M is the magnetization, also known as magnetic moment per unit volume and H_0 is the applied magnetic field intensity. The subscript "0" is often added to H to highlight the fact that this field is an applied or external field. Magnetic susceptibility is usually a tensor and a function of both field H_0 and magnetization M. For a magnetically isotropic material, M is parallel to H_0, so that χ is reduced to a scalar quantity. The SI units of χ are *henry/meter* (H/m).

A $B-H$ hysteresis loop is obtained by plotting magnetic field density B against magnetic field intensity H. An $M-H$ curve is obtained in a similar way, where magnetization M replaces magnetic flux density B. In both cases, if the field intensity H is increased from zero to a high value and then decreased, the original curve is not retraced (Fig. 1.5). This means that the material acquires a magnetic history where the magnetic flux density B does not follow its original values on the initial curve. If the hysteresis loop is symmetrical about the origin and the material is magnetized to saturation, the value of field intensity H_c for which flux density is zero is known as *coercivity*, often used in magnetic recording to determine the usability of a material. On the other hand, the value of flux density B_r for which intensity H is zero, is called *retentivity*, while that of remanent magnetization is known as *remanence*, although quite often the two are interchanged, and B_r is called "remanence" by some authors. Either way, the remanent magnetic flux density, or remanent magnetization, when the external field has been brought back to zero, are important quantities for determining the quality of a magnetic material.

Another quantity that is determined from the $B-H$ hysteresis loop, and can give us some insight into the magnetic properties of a material, is its *magnetic permeability* μ, obtained through the ratio B/H, although, technically, it is a derivative on the $B-H$ hysteresis loop. The magnetic permeability of a material represents the relative increase in magnetic flux caused by the presence of the material itself, in an external field. The unit for the permeability μ is the same as for χ. Hence, it is possible to measure χ in units of the permeability of vacuum μ_0 which is a constant. In this case, the measured dimensionless quantity is called *relative susceptibility* and is denoted by χ_0.

$$\chi_0 = \frac{\chi}{\mu_0}. \tag{1.7}$$

The values for relative susceptibilities range from 10^{-5} (very weak)–10^6 (very strong magnetism) [2]. In some cases, the relative susceptibility is negative. Or, the relationship between M and H is not linear, so that χ_0 depends on H. The behavior of χ_0 leads to various types of magnetism.

1.2 Classification of Magnetic Materials

1.2.1 Diamagnetism

The most common way of classifying magnetic properties of materials is by their response to an applied magnetic field; therefore, both relative permeability and relative susceptibility can be used to differentiate between classes of materials. As such, it is said that materials that can be magnetized to a certain extent by a magnetic field are called *magnetic*. We start with *diamagnetism*, which is a weak form of magnetism, attributed mainly to the orbital motion of electrons that create a magnetic moment when viewed classically as a "current loop." A magnetic flux is induced in the diamagnetic material when an external magnetic field is applied to it. However, the induced flux counters the change in the external field so that diamagnetic materials show an antiparallel magnetization with respect to the direction of the applied magnetic field, opposing the latter, according to Lenz's law. Therefore, the magnetization of a diamagnetic material is proportional to the applied magnetic field, as seen in Fig. 1.2a, and they have a negative and very weak relative susceptibility of about 10^{-5}. Their permeabilities are somewhat less than one. Many metals and most nonmetals are diamagnetic. It is interesting to point out that, if only a few magnetic atoms exist in the material, their influence is enough to overshadow the diamagnetism, so that nonmagnetic atoms become spin–polarized by neighboring ferromagnetic atoms. This effect is explored nowadays, in practice, in certain nanoscale spin electronic devices.

1.2.2 Paramagnetism

For a certain class of materials known as *paramagnetic materials*, the relative permeability is only slightly greater than one. It is independent of magnetic field strength, and will decrease with increasing temperature, provided it is not temperature independent. Many metals fall into this class, such as platinum or palladium, and also salts of iron or of the rare earth metals, or elements like sodium, potassium, and oxygen. Ferromagnets become paramagnetic above the *Curie temperature* T_c. When an external magnetic field is applied to a paramagnetic material, a weak induced magnetization is produced parallel to the field. The induced magnetization that is proportional to the external field, nevertheless, stays

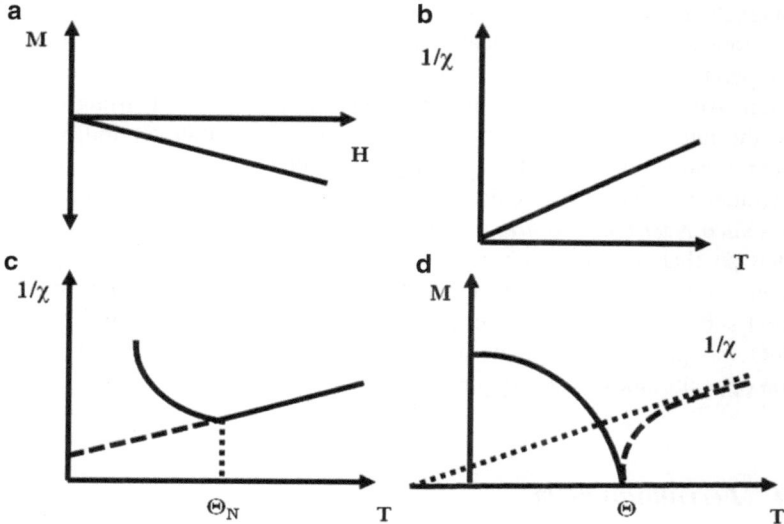

Fig. 1.2 (**a**) Linear relationship between magnetization and applied magnetic field (intensity) in a diamagnetic material, (**b**) Curie–Weiss law of paramagnetism, where the susceptibility is inversely proportional to absolute temperature, (**c**) variation of susceptibility with temperature for an antiferromagnetic material, and (**d**) Spontaneous magnetization decreases with temperature in ferrimagnetic materials because the arrangement of spins is disturbed with increases in temperature. Upon reaching a critical temperature, the spontaneous magnetization vanishes and the specimen exhibits paramagnetism where the susceptibility decreases with increasing temperature

positive, unlike in diamagnets. On the other hand, the susceptibility as a quantity is inversely proportional to absolute temperature T, and this inverse proportionality is also known as the *Curie–Weiss law* (Fig. 1.2b). For paramagnets, the relative susceptibility is a positive 10^{-3}–10^{-5}. *Paramagnetism* is in certain ways similar to ferromagnetism (see Sect. 1.2.5 further below), in that it is attributed to unpaired electron spins just as much. Nevertheless, due to a different electronic configuration these spins are free to change their direction, which is unlike the spins in ferromagnets. In paramagnets, spins assume random orientations at certain temperatures as a consequence of thermal agitation.

1.2.3   Antiferromagnetism

Antiferromagnetism is a type of magetism with an ordered arrangement of antiparallel aligned spins on different sublattices, such that the antiferromagnetic structure has no net spontaneous magnetization. Antiferromagnetic materials have small permeabilities and are, therefore, often classified as paramagnetic. Antiferromagnets exhibit a small positive relative susceptibility that varies with temperature similar to paramagnetism when higher temperatures are reached; however, this dependence has a unique shape below a critical temperature (Fig. 1.2c). This is because below this temperature, the electron spins are arranged antiparallel so that they cancel each other out. An externally applied magnetic field is faced with a strong opposition due to the interaction between these spins, resulting in a susceptibility decrease with temperature, in contrast to paramagnetic behavior. Therefore, even though antiferromagnetic order may exist at lower temperatures, it vanishes at and above the so-called *Néel temperature*, Θ_N (Fig. 1.2c), when the spins become randomly oriented so that the susceptibility decreases as the temperature is raised. The exchange interaction in antiferromagnets acts to anti-align neighboring spins on different sublattices. For instance, the antiferromagnetism of certain oxides such as MnO was proven to originate in a strong exchange interaction between metal ions on opposite sides of an oxygen ion.

1.2.4 Ferrimagnetism

Ferrites exhibit a kind of magnetism known as *ferrimagnetism* (Fig. 1.2d) that is in some ways similar to both ferromagnetism and antiferromagnetism. In ferrimagnetic materials, since ions are placed on two different types of lattice sites, such that spins on one site type are oppositely oriented to spins on the other lattice site type, they tend to be compared to antiferromagnetic materials. However, one of the resultant magnetizations on the two lattice sites is stronger than the other, so that the result is a total non zero, spontaneous magnetization. Since antiferromagnets are not strongly magnetic, and ferrimagnets have a spontaneous and non-negligible magnetization, ferrimagnetic materials are often compared to ferromagnets. And similar to ferromagnetic materials, an increase in temperature brings about a disturbance in the spin arrangement that culminates in completely random orientation of spins at the Curie temperature. At this temperature, the ferrimagnet loses its spontaneous magnetization and becomes paramagnetic (Fig. 1.2d). Ferromagnetic materials also have a Curie point above which they exhibit paramagnetic behavior [3].

Example 1.1. Comment on the consequences of adding different concentrations of Ge to Fe_3O_4 thin films

Answer: Magnetite, a ferrimagnetic material with extensive history and diverse technical applications, finds use today in advanced areas such as spintronics or spin valves due to its half–metallic character. Nevertheless, the tunnel magnetoresistance behavior on which these modern applications are based (discussed in later chapters) encounters some problems when Fe_3O_4 is used as an electrode grown epitaxially on a MgO single crystal. In this case, the interface between the magnetite electrode and tunnel barrier is complicated and difficult to control. To solve some of these problems, researchers have prepared the magnetite film by sputtering of a composite target with added Ge, in order to suppress the iron deficit which occurs during the sputtering process [4]. Ge is thermodynamically stable in a magnetite matrix, and it was confirmed that by adding ~5 at% Ge, a thin magnetite film is still obtained. Furthermore, these magnetite thin films are ferrimagnetic, even though Ge has been added. Nevertheless, higher concentrations (~14%) of Ge result in the magnetite films becoming paramagnetic. The advantage of adding Ge seems to be that it ensures formation of only the magnetite phase, whereas sputtering the magnetite target without Ge results in two phases, magnetite and hematite. The latter fact is detrimental to the electrode/tunnel barrier interface characteristics of spintronics applications.

1.2.5   Ferromagnetism

We should finally mention *ferromagnetism;* however, given that so many concepts in this book apply directly to ferromagnetic materials, they will be frequently discussed, and, therefore, no section alone can be dedicated to them. In essence, ferromagnetic materials have a permeability that depends on the field strength and on the previous magnetic history (see Hysteresis Loop mentioned above). They approach magnetic saturation as the field strength continues to increase, meaning that the material can only be magnetized to a finite limit. Ferromagnets contain spontaneously magnetized magnetic domains, where a magnetic domain is an entity with a total domain magnetization. This is a small, magnetized region containing many atoms with individual magnetic moments that align parallel to each other, against the forces of thermal agitation. The domain magnetization of one domain is usually oriented differently with respect to the total domain magnetization of neighboring domains. A historically early and rather crude representation of magnetic domains is shown in Fig. 1.3.

We could ask – why would magnetic domains be spontaneously magnetized? The formation of magnetic domains is due to minimization of the total energy, as will be discussed in the next chapter. This implies that, under every condition of the ferromagnet, the domain structure will strive to remain stable, and will, therefore, change until it finds this stability point if onditions change. We know from quantum mechanics [5,6] that the spontaneous domain magnetization is a result of unpaired electron spins from partially filled electronic shells. These spins align parallel to each other as a result of a strong *exchange interaction*. Since the arrangement

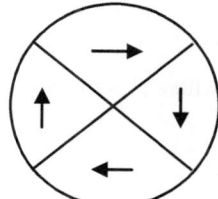

Fig. 1.3 Historical interpretation of domain structure in a ferromagnetic material. Note that magnetostriction is not taken into account

of spins depends on temperature, so does the spontaneous domain magnetization. When the total resultant magnetization for all magnetic domains is zero, the ferromagnetic material is said to be *demagnetized*. However, an applied magnetic field changes the total resultant magnetization from zero to a saturation value. When the magnetic field is decreased and thus, reversed in sign, the magnetization of a ferromagnetic material does not retrace its original path of values, with the material exhibiting so called *hysteresis*. A strong ferromagnet exhibits a relative susceptibility of the order of 10^6, which is a large value as compared to other types of materials. It explains why ferromagnets can be easily magnetized, while other kinds of magnetic materials fall short on their ability to respond to a magnetic field. The spontaneous magnetization of ferromagnets disappears above a certain temperature called *Curie temperature* T_c, when they become paramagnetic. Technically, ferromagnets are considered a subclass of paramagnetic materials; nevertheless, in time they have been placed in a class of their own. Historically, ferromagnetic materials were the only ones considered "magnetic"; however, this interpretation has changed recently, given that so many other types of materials respond to a certain extent to a magnetic field.

1.3 Origins of Magnetism

1.3.1 Angular Momentum and Its Role in Magnetism

To understand the origins of magnetism, we have to look at the magnetic moment of atoms which originates from electrons in partly filled electron shells. This atomic magnetic moment is determined by a fundamental property known as the *angular momentum* [5,7]. Each individual electron within an atom has an *angular momentum* L, associated with its orbital motion, and an intrinsic or *spin angular momentum S*. Hence, there are two sources of the atomic magnetic moment (a) currents associated with the orbital motion of the electrons and (b) the electron spin. In an n-electron atom, the individual atomic orbital angular momenta couple together to give a total orbital angular momentum L, and the individual atomic spin angular momenta couple together to give a total spin angular momentum S. As such, the orbital and spin angular momenta each have a magnetic moment expressed as [5–8]

$$\vec{\mu}_L = -\beta \vec{L},$$ (1.8)

$$\vec{\mu}_S = -2\beta \vec{S},$$

where β is the *Bohr magneton*. The total magnetic moment μ is then given by

$$\vec{\mu} = -\beta(\vec{L} + 2\vec{S}).$$ (1.9)

For an n-electron atom, the $2n$ angular momenta couple together to give a total angular momentum, J, whose exact properties depend on the details of the coupling parameters. It should be noted that L, S, and μ precess about the total angular momentum J.

The magnetic moment of atoms in magnetic materials is largely given by the spin, rather than the orbital motion of electrons. This specifically refers to materials containing iron–series transition metal atoms (e.g., Fe, Co, Ni, and YCo_5), or ferrimagnetic oxides (e.g., Fe_3O_4 and NiO). For these typical magnetic materials, the spin moment $\vec{\mu}_S$ is equal to the number of unpaired electron spins, while the orbital moment $\vec{\mu}_L$ is very small. The latter is usually of the order of 0.1β due to th orbital motion of electrons being "quenched" by the *crystal field*, a concept discussed in a subsequent section. The component of μ perpendicular to J averages to zero over a time significantly larger than the precession period. As such, when a magnetic field is applied, only the component of μ parallel to J is sensed. That parallel component is usually denoted by μ_J [5–8]. We should emphasize that the angular momentum state of an atom is characterized by eigenvalues of J^2, which is $J(J + 1)$. Thus, using the properties of angular momentum operators and the law of cosines, we have

$$\mu_J^2 = g^2 J(J + 1)\beta^2. \tag{1.10}$$

Choosing the z component of J, that is J_z with eigenvalues $m_j = J, J - 1, \ldots, -J$, the magnetic moment along z is given by

$$\mu_{J_z} = -g\beta J_z, \tag{1.11}$$

where g, the Landé g-factor or spectroscopic splitting factor, is[1]

$$g = 1 + \frac{J(J + 1) + S(S + 1) - L(L + 1)}{2J(J + 1)}. \tag{1.12}$$

Nevertheless, the Landé g-factor results from the calculation of the first-order perturbation of the energy of an atom, when a weak external magnetic field acts on the sample. Normally, the quantum states of electrons in atomic orbitals are degenerate in energy, and therefore, the degenerate states all share the same angular momentum. However, if the atom is placed in a weak magnetic field, the degeneracy is lifted [8]. Furthermore, this dimensionless g-factor relates the observed magnetic moment μ_{J_z} of an atom to the angular momentum quantum number m_j, and the fundamental quantum unit of magnetism, that is, the Bohr magneton.

1.3.2 Exchange Coupling and Exchange Bias

At this stage, we need to comment on the interaction of spins with each other since this phenomenon was already mentioned a few times. Heisenberg stated, in 1928,

---

[1]For a rigorous derivation of above results, please see any introduction to quantum mechanics [5]–[6], or more specialized books on electric and magnetic susceptibilities [7].

that the force which makes the spins line up is an exchange interaction of quantum mechanical nature, and this view is still being upheld these days. As such, two atoms with spins S_i and S_j have a potential energy U_{ij}

$$U_{ij} = -2J\,\mathbf{S_i} \cdot \mathbf{S_j},\qquad\qquad(1.13)$$

where J is the exchange integral. A positive value for J is obtained when the two spins are parallel and the energy is at a minimum. In contrast, an antiparallel alignment of spins gives a negative exchange integral J. The exchange interaction arises due to Pauli's exclusion principle which permits an orbit to be occupied by one "up" and one "down" spin, while two electrons with the same kind of spin cannot approach each other closely. This means that two "up" or two "down" spins cannot belong to the same electron, and therefore, the mean distance between two electrons with parallel spins is different from that between electrons with antiparallel spins. In quantum mechanical terms, the expectation value of the energy of two spins differs for two spins of the same orientation than for those of different orientation when their wave functions overlap.

Example 1.2. Comment on the "flow of spin," also known as *spin current*

Answer: Spin has been researched and considered the origin of magnetism for roughly 80 years, in contrast to *charge*, the other degree of freedom of an electron that has been studied for about 200 years. The "flow of charge" has been successfully utilized for a long time; however, many readers may not be that familiar with the "flow of spin" which has attracted much attention and exploitation only in relatively recent years. A charge current contains both "up" spin and "down" spin electrons, where "up" and "down" are with respect to the quantization axis. In contrast to a charge current, a spin current is given by the *difference* in flow between "up" and "down" spin electrons. In ferromagnetic and ferrimagnetic materials, the number of "up" spin and "down" spin electrons differs, making those electrons *spin polarized*. Thus, spin current and charge current flow together in these materials.

A nonconserved quantity, the flow of spin changes through the exchange of angular momentum. It can be generated by electric, magnetic, or optical signals, although its generation is not as straightforward as the one of charge current. On the other hand, spin currents are easily annihilated through spin relaxation when viewed over a length of time, meaning that spins flip in time intervals ranging from picoseconds to nanoseconds. Concurrently, a spin current can be lost through spin diffusion when considering the spatial aspect of the flow. The spatial range in which spin current flows in the material is known as the *spin diffusion length*, and is usually between 100 nm and 1 μm, depending on the material (e.g., metal and semiconductor), structure (e.g., nanostructure), and experimental conditions (e.g. temperature). This type of spin current that diffuses through a material is not accompanied by charge movement. Therefore, it is often referred to as *pure spin current*; nevertheless, electrons still move. Whether pure or carrying electric charge, spin current constitutes the foundation for many devices belonging to a new class of electronics, *spintronics*. More advanced chapters in this book discuss a few aspects of spin current generation with particular emphasis on the topic of maintaining its flow.

Two separate ferromagnetic layers can become "exchange coupled" when separated by a thin, nonmagnetic spacer layer. However, this process is not straightforward, as one might expect. Most interfaces of bilayers of ferromagnetic or ferrimagnets are either uncoupled or ferromagnetically coupled. Only under special circumstances do the bilayers become exchange coupled. The mechanism of exchange coupling is treated in terms of the Ruderman–Kittel–Kasuya–Yoshida (RKKY) interaction or the theory of spin dependent quantum well effects [9]. This means that when the two ferromagnetic layers are exchange coupled, an antiparallel alignment of spins occurs, of the type observed in antiferromagnetic materials. In this case, the exchange integral J in relationship (1.13) is negative and fairly large in value. For instance, the strongest interlayer exchange coupling constant of $J \approx -5\,\text{erg/cm}^2$ has been observed in a Co/Ru/Co system with a $\sim 0.5\,\text{nm}$ thick Ru layer [10]. Coupling strength, as well as its sign are both dependent on spacer thickness. Most reported studies have so far been for metallic systems, but this trend is changing. These days, magnetic oxides or combinations of magnetic oxides and metals enjoy a lot of attention. The interlayer exchange coupling constitutes the basis for controlling magnetization alignment in many modern magnetic applications, such as magnetic recording media, magnetic sensors, and magnetic random access memories. An interlayer exchange coupled system is referred to as a *synthetic antiferromagnet*, a topic that will be further discussed in later chapters.

There is a class of magnetic materials known as *half-metallic* compounds, named so because they only have one channel of spin; hence, the flow of spin they generate is either "up" or "down." Ideally, these materials show 100% *spin polarization* due to one spin channel being present at the Fermi level, with a band gap for the other spin channel. Half-metallic compounds are usually magnetic materials in either their ferromagnetic or ferrimagnetic states. These include CrO_2, the so-called *half*-Heusler NiMnSb, the *full*-Heusler Co_2MnSi, and the pervoskite Sr_2FeMoO_6. Some of these materials, or the class they belong to, will be discussed further in subsequent chapters. These materials have a total, positive magnetic moment and are good candidates for spintronics applications. Nevertheless, researchers have found another interesting class of materials that displays the half–metallic character but has no net magnetic moment, although they show full spin polarized charge transport. In this case, these antiferromagnetic systems would be relatively insensitive to applied magnetic fields, and, therefore, could possibly lead to a different type of spintronics devices.

For instance, a group [6] in Japan has successfully fabricated the half–metallic antiferromagnet $SrLaVMoO_6$ by modifying the half-metallic ferrimagnet Sr_2FeMoO_6. They assumed that Fe^{3+} is antiferromagnetically coupled to Mo^{5+}, and this fact was the starting point for their investigation. Stochiometric amounts of $SrCO_3$, La_2O_3, V_2O_5, and MoO_3 were mixed, calcined twice, ground, and sintered, all steps being performed under certain temperature conditions. The $SrLaVMoO_6$ specimen displayed a majority phase of an ordered perovskite with a small impurity phase. Nevertheless, the majority phase showed a negative paramagnetic Curie temperature, indicating that the exchange interaction of the magnetic moments is antiferromagnetic. A linearly increasing magnetization with temperature ruled out the existence of ferromagnetic or ferrimagnetic components in the sample. Also, the resistivity shows metallic temperature dependence, attributed to the large Mo $4d$ bandwidth. The V^{3+} ion is assumed to be antiferromagnetically coupled to Mo^{4+} and therefore responsible for the total net zero magnetization. X-ray photoelectron spectroscopy studies support this assumption.

All data thereby suggested that the obtained specimen [6] was a conducting antiferromagnet. However, the observed spin polarization of 50% was well below 100% of the ideal half-metallic compound, which may also be due to the actual measurement method used. Similar spin polarization measurements on Co-based Heusler compounds have previously indicated comparable polarization values. In any case, this study [6] indicates that a high spin polarization may co-exist with a zero net magnetic moment. This is quite unexpected in materials for which identical "up" spin and "down" spin band structures without exchange splitting are assumed, where spin polarization and net magnetization are zero. It shows that progressive knowledge of magnetics has resulted in some compounds being manufactured nowadays with both exchange and crystal field splitting, as well as non-identical, special band structures for "up" and "down" spin channels.

Exchange bias is related to exchange coupling, and occurs at the interface between a ferromagnet and an antiferromagnet. This type of coupling usually happens when the materials are field-cooled through the *Néel temperature* Θ_N of the antiferromagnetic material. The experimental signs are recognized in the $M-H$ hysteresis curves, when magnetizations shift along the field axis. The shift is assumed to be due to the exchange coupling at the interface between "soft" ferromagnet spins and "hard" antiferromagnet spins. Some researchers argue that uncompensated spins at the antiferromagnet surface also play a role in this effect, as these spins are "frozen in" and, therefore, not reversed when the applied magnetic field is acting. Uncompensated spins in antiferromagnets originate in surface imperfections such as intermixing, nonstoichiometry, and structural roughness, where the antiferromagnetic order is disturbed. For instance, reports [11] on NiO/Co, IrMn/Co, and $PtMn/Co_{90}Fe_{10}$ identified uncompensated Ni or Mn spins located at the antiferromagnetic–ferromagnetic interfaces. As such, it was determined through hysteresis loop measurements that only ~4% of the ferromagnetic monolayer is tightly pinned to the antiferromagnet, therefore not rotating in an external field. Furthermore, a quantitative correlation was found between the size of the pinned interfacial magnetization and the macroscopic magnetic exchange bias field. In these experiments, high sensitivity X-ray magnetic circular dichroism spectroscopy was used in a total electron yield detection because the 1/e probing depth makes this method responsive to interfacial surfaces [12]. The exchange bias phenomenon has applications in magnetic recording, as will be seen in later chapters.

1.3.3 Partition Function Z

A few words should also be said about the partition function which is particularly a concept of interest in multi-particle structures, as well as other areas of magnetics. Consider a simple paramagnet where the atoms do not interact with each other. In this case, the only contributions to the Hamiltonian $\hat{H}$ come from their interaction with the applied magnetic field $\boldsymbol{H}_0$. Because the atoms are identical, only the Hamiltonian for a single atom needs to be considered

$$\hat{H} = -\vec{\mu}_J \cdot \vec{H}_0. \tag{1.13}$$

Choosing $\boldsymbol{H}_0$ to be along the z-axis, we can write

$$\hat{H} = -\mu_{J_z} H_0. \tag{1.14}$$

The eigenvalues of $\hat{H}$ are then

$$E_m = -g\beta m_j H_0. \tag{1.15}$$

The partition function Z is then defined by

$$Z = \sum_n e^{-E_m/kT} = Tr(e^{-\hat{H}/kT}) \tag{1.16}$$

or, in our case,

$$Z_J = \sum_{m_j=-J}^{J} e^{g\beta m_j H_0/kT}. \tag{1.17}$$

As stated, the partition function Z is an important quantity when dealing with multi-particle structures, as it encompasses the statistical properties of the entire system. It depends on a number of factors, such as the system's temperature, the angular momentum quantum number, external magnetic field etc. Furthermore, it is a sum of all states based on their individual energies while determining how the probabilities are divided among the various states that compose the system.

Example 1.3. Give an example of the usefulness of the partition function

Answer: For instance, the magnetic $(g\beta J H_0)$ and thermal (kT) energies can be expressed in terms of the partition function. As such, denoting by x, the ratio of magnetic and thermal energies

$$x = \frac{g\beta J H_0}{kT}, \tag{E1.1}$$

the partition function becomes

$$Z_J(x) = \sum_{m_j=-J}^{J} e^{m_j x/J} = \frac{\sinh\left(\frac{2J+1}{2J}x\right)}{\sinh\left(\frac{1}{2J}x\right)}. \tag{E1.2}$$

This expression of the partition function allows calculation of the expectation value of the magnetic moment μ_{J_z}, a quantity observed experimentally.

Consider a sample of N atoms, where the magnetization is given by

$$M = N\left\langle \mu_{J_z}\right\rangle = Ng\beta\left\langle J_z\right\rangle \tag{E1.3}$$

and

$$\left\langle J_z\right\rangle = \frac{Tr(J_z e^{-\hat{H}/kT})}{Tr(e^{-\hat{H}/kT})}. \tag{E1.4}$$

In this case,

$$M = Ng\beta\frac{Tr[J_z e^{\hat{H}/kT}]}{Z_J(x)}. \tag{E1.5}$$

This expression can be reduced to the useful form

$$M = Ng\beta J B_J(x), \tag{E1.6}$$

where, B_J is called the *Brillouin function* [5–8]. This function describes the dependency of the magnetization on the applied magnetic field, temperature, and the total angular momentum quantum number, and hence it is a useful concept. It is used to derive important laws of magnetism, such as the Curie–Weiss law mentioned earlier.

1.3.4   Intrinsic and Induced Anisotropy

A magnetically anisotropic material is one whose magnetic properties depend on the direction in which they are measured. It is an interesting fact that a magnetic material can display various types of anisotropy, but nevertheless the anisotropy of most magnetic materials is of magnetocrystalline origin. This anisotropy is a magnetic property determined by crystal structure and is the only *intrinsic* anisotropy, while other types of magnetic anisotropy are *induced*. Magnetocrystalline anisotropy arises in a magnetic material because each atomic moment is acted on by a *crystal field*, proportional to the local magnetization of its environment. This means that if an atomic moment were to be removed from its environment, it would leave behind a magnetic field. This field is produced by the surrounding spins, and is a manifestation of the local symmetry of the crystal, and hence the name "crystal field." When an external field tries to reorient the spin of an electron, the orbit of that electron also tends to reorient, as spin and orbit are coupled. Nevertheless, the orbit of the electron is also strongly coupled to the crystal lattice, and, therefore, resists reorientation. The competition between the spin–orbit coupling and the crystal field gives rise to magnetocrystalline anisotropy, which manifests itself in the energy required to overcome the spin–orbit coupling.

Magnetocrystalline anisotropy is one of those material properties that affects the shape of the hysteresis curve, allowing us to make a general distinction between permanent magnets and magnetically softer materials. This is because in essence, magnetocrystalline anisotropy manifests itself as a variation of the magnetic properties of the material with a specific, crystallographic orientation, being largest along preferred crystallographic directions. For instance, an iron crystal has directions of the form $<100>$ as preferred crystallographic directions, along which magnetic moments from magnetic domains tend to align. Magnetic saturation can be achieved in iron, in these easy directions of magnetization with fields of only a few tens of Oe. In contrast, nickel crystals display easy directions of magnetizations along $<111>$ directions of form, and so do all the cubic ferrites with the exception of cobalt ferrites and their derivatives which exhibit $<100>$ easy axes. Magnetocrystalline anisotropy is, thus, a valuable characteristic of a magnetic material in as much as it allows a materials engineer to design a device who exploits this fact in a commercial application.

Permanent magnets, such as $SmCo_5$ or $Nd_2Fe_{14}B$, require a high magnetic anisotropy in order to maintain the domain magnetization in a chosen direction. This preferred magnetization orientation through anisotropy is permitted in permanent magnets by the electronic configuration that gives rise to a particular interaction between the crystal field and the spin–orbit coupling. Specifically, the magnitude of the magnetocrystalline anisotropy depends on the ratio of crystal field energy and spin–orbit coupling [13]. As such, the crystal field acts on the orbits of the inner shell d and f electrons, and given that spin–orbit coupling is a relativistic phenomenon, it is most pronounced for inner-shell electrons in heavy elements, such as rare-earth $4f$ electrons. This fact results in a rigid coupling between spin and orbital moment in heavy elements [14]. On the other hand, in magnetic materials such as Fe, Ni, and Co, the magnetocrystalline anisotropy is due to $3d$ electron spins, in contrast to the magnetocrystalline anisotropy for rare earth elements that originates in the $4f$ shells. In fact, the strong magnetocrystalline anisotropy in permanent magnets is given by the comparatively small electrostatic interaction of the unquenched $4f$ charge clouds with the crystal field. The absence of quenching means that typical single ion anisotropies (rare earth ions) are much larger than $3d$ anisotropies. This strong magnetocrystalline anisotropy is exploited in advanced permanent magnets, where it leads to very high coercivities, for instance, 4.4 T in $Sm_3Fe_{17}N_3$-based magnets. Therefore, a large number of magnetic applications are based on rare earth metal alloys.

Unlike magnetocrystalline anisotropy, induced anisotropies are due to some external material treatment which has more or less directional properties. For instance, most ferromagnetic materials are known to exhibit a uniaxial anisotropy after heat treatment in a magnetic field. This is quite often known as the *magnetic annealing effect*. It was discovered in the first half of the twentieth century, and has since then been used in many laboratory studies or engineering applications. Similarly, mechanical processing such cold rolling or cold working can also induce anisotropy, termed in the former case as *roll magnetic anisotropy*. This type of anisotropy is also purposely exploited in industrial products. For instance, 3.25% silicon steel employed as a core material for power transformers, had started to be investigated and used decades ago, for the crystal orientation and the silicon content of the steel. A (110) [001] texture can be obtained by a sequence of cold rolling and annealing operations so that the [001] easy direction is parallel to the rolling direction, while a magnetically harder direction [110] is present in the transverse direction. As such, electrical equipment is designed so that the magnetic flux is carried mostly in the rolling direction. Further research has been performed in an attempt to increase the Si content beyond the 3.25%; however, many barriers had to be surmounted (see problem question P1.10 at the end of the chapter). Nevertheless, it was observed that for obtaining optimum results for the improved silicon steels with higher Si content, the magnetic annealing field has to be strong enough to saturate the sample in the desired direction, while the temperature and time conditions must be sufficient to accomplish short range Si diffusion. Within these limits, most combinations of annealing field and temperature will yield the desired results [15]. It was apparent from these studies that magnetic properties perpendicular to the rolling direction are more sensitive to magnetic annealing than those parallel to the rolling direction, irrespective of Si content.

Example 1.4. What are the effects of a transverse annealing field on an amorphous, Co-rich magnetic alloy?

Answer: Magnetic properties of amorphous, Co-rich magnetic alloys depend strongly on the induced transverse magnetic anisotropy which is obtained after magnetic annealing in a transverse field. A group of researchers [16] investigated a few magnetic properties such as magnetic remanence, coercivity, and domain structure of the magnetostrictive compound $Co_{71.5}Fe_{2.5}Mn_2MoSi_9B_{14}$ after a two-step annealing process. This process involved first annealing the samples at 570 K without a magnetic field, and then cooling them to room temperature in a range of transverse fields spanning 8–160 kA/m. The higher annealing fields resulted in a lower magnetic remanence, as well as lower coercivity values, in comparison to smaller annealing fields. At the same time, in those specimens with an existing longitudinal magnetic anisotropy, a reorientation of this anisotropy axis was noticed toward the transverse direction after annealing in higher magnetic fields, a fact which was expected. High annealing magnetic fields also cause a strong magnetic ordering in the domain structure, thus explaining the lower remanence and coercivity values. It was noted that the cooling rate contributed to the magnetic annealing process, altering magnetic properties. The higher the cooling rate, the longer the time required for magnetic ordering.

1.3.5   Magnetostriction

Aside from exchange coupling, spontaneous magnetization, or magnetocrystalline anisotropy, other intrinsic magnetic properties such as magnetostriction also have origins in atomic scale magnetism [17]. Although they manifest themselves on length scales of a few angstroms, their consequences already reach bulk values at $\sim$1 nm and are, therefore, exploited in today's nanotechnology applications. Magnetostriction was first discovered in 1842, by Joule who noticed a change in the length of an iron rod that was magnetized in a weak field. As such, under the influence of the magnetic field, the shape of the iron rod changed due to a magnetic property that was coined *magnetostriction*, and is, nowadays, often denoted by λ. This change in shape is schematically shown in Fig. 1.4; however the change is exaggerated, as this type of deformation is very small, only of the order of 10^{-5}–10^{-6}, or even smaller in weakly magnetic materials. It was observed that when the specimen elongates under an applied magnetic field, its volume remains constant, implying that a *transverse* magnetostriction exists. Indeed, the transverse magnetostriction is about half the value of the longitudinal magnetostriction, and of opposite sign.

The existence of magnetostriction is attributed to the spin–orbit coupling of valence electrons in ferromagnets, and as such, when spins change direction to align with domain magnetization, the orbits change shape to conserve angular momentum, because electron orbits are coupled to spins. Since electron orbits are also coupled to the crystal lattice, the lattice inside a magnetic domain deforms spontaneously in the direction of domain magnetization. For instance, a single iron crystal that is magnetized to saturation in the [100] direction will increase in that direction due to magnetostriction. In this case, the strain due to magnetostriction will continue to increase with the applied magnetic field until it reaches a saturation value. This value of saturation magnetostriction can be positive, negative, or in some alloys, zero. Additionally, magnetostriction will saturate along a specific crystallographic axis. For example, we have $\lambda_{100} = 19.5 \times 10^{-6}$, and $\lambda_{111} = -18.8 \times 10^{-6}$ in a single cubic crystal, where λ_{100} and λ_{111} are the saturation values of the longitudinal magnetostriction in the directions [100] and [111], respectively [18]. Quite often, an "isotropic saturation magnetostriction" denoted

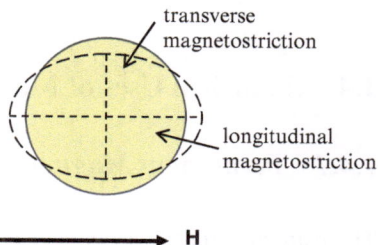

Fig. 1.4 Elongation of a ferromagnetic object in the direction of an applied magnetic field

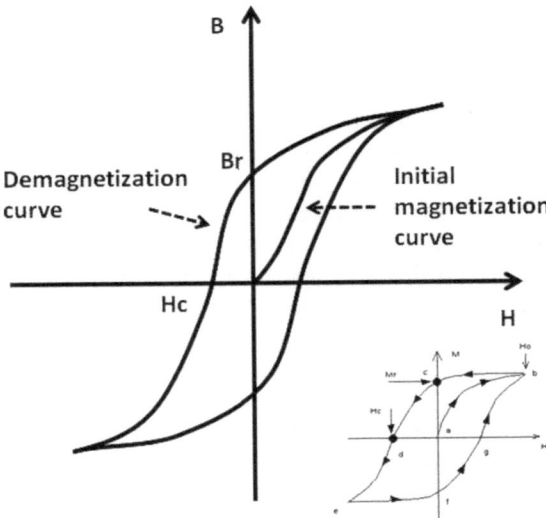

Fig. 1.5 Hysteresis loop that includes the initial magnetization curve. Also shown are coercivity H_c and remanent magnetic flux density B_r. A hysteresis loop can be obtained by plotting magnetic field density B vs. applied magnetic field intensity H, or as shown in the inset, by plotting magnetization M vs. applied magnetic field intensity H

λ_s and equal to $\lambda_s = \lambda_{100} = \lambda_{111} = -7 \times 10^{-6}$ is assumed, although it was not found to be representative of experimental results [19].

Many magnetic materials of practical importance undergo additional magnetic treatment, such as magnetic annealing, in order to enhance their magnetic properties. Silicon steels are a group of such materials that have traditionally been improved through further processing, and their magnetostriction has also been altered through mechanical and magnetic treatments. It was, thus, observed that the magnetostriction of these materials is positive but rather small in the rolling direction without magnetic annealing, and that upon magnetic annealing, the magnetostriction became even smaller and negative. Alloys with reduced Si content had lower magnetostriction values without magnetic annealing but their magnetostriction was closer to zero upon annealing [15].

1.4   Historical Uses of Magnetic Materials

1.4.1   Permanent Magnets

There are materials that can be magnetized which remain magnetized after the magnetizing field has been removed. These materials store magnetic energy permanently without getting drained through repeated use, thus being employable in a variety of instruments. Humanity has long been fascinated with this interesting class of materials known as *permanent magnets*, as they are able to produce constant external magnetic fields available at any time. Nevertheless, any discussion about permanent magnets has to first answer the question as to what distinguishes this class of materials from other magnetic compounds. The clue lies in the second quadrant, between remanence and coercivity, in the so-called *demagnetization curve*. The demagnetization curve is part of the *hysteresis loop* (Fig. 1.5), which is obtained by plotting magnetic field density B, or magnetization M of the material being magnetized versus the applied magnetic field H. The hysteresis loop reveals the magnetic history of a ferromagnetic material, and is, therefore, very useful in assessing a magnetic material. From the hysteresis loop, we can determine the two quantities, retentivity B_r (or remanent magnetization M_r) and coercivity H_c, as well as the product $B_{max}H_{max}$ of the maximum field values. These characteristics are commonly used by engineers in evaluating the quality of permanent magnetic materials [20].

For instance, the product $B_{max}H_{max}$ is related to the minimum volume of a magnet that is required to produce a given field in a given gap in the magnetic circuit. Similarly, engineers are interested in the magnetic flux density B_r, because it indicates how much magnetic flux or magnetization has been retained after the externally applied field is brought to zero on the demagnetization curve of the hysteresis loop, after magnetic saturation has been attained. The capability of the material of retaining a magnetization is an indication of the ability of the material to be permanently magnetized, useful for magnetic memory applications. On the other hand, the value of the magnetic field intensity H_c, necessary to bring the magnetization of the material to zero, is an indication of how magnetically *hard* or magnetically *soft* the material is. It measures the resistance of the ferromagnetic material to becoming demagnetized. Soft magnetic materials are usually employed as transformer cores, or used in motors and generators, with some example materials discussed in the next section. Magnetically soft materials need to have a high permeability, low coercivity, and a small hysteresis loss. Conversely, permanent magnets have to be magnetically hard in order to display a high coercivity and a high remanence. Nevertheless, as it may already be known to the reader, certain magnetic properties are not sufficient on their own for a material to be considered a permanent magnet. Geometry counts as well, because a permanent magnet needs "free" magnetic poles in order to provide an external field. The magnetic flux circuit cannot close within the material; it has to be able to leak outside the material by offering a gap in the circuit. Furthermore, the energy stored in the field in the air gap is directly proportional to BH, which is also called the *energy product*. As such, permanent magnets always have open circuit geometry in order to be operational.

Historically, the most familiar magnetic material may have been the lodestone, followed by high carbon steel magnets of the type used in compass needles. Nevertheless, one of the best magnetic steels employed for a while in magnetic applications was a type of steel made with 30–40% Co, as well as W and Cr. When magnetic recording started to flourish, it required its own magnetic materials with particular magnetic characteristics. Consequently, Co–Pt–Fe alloys gained ground, given their excellent remanence value and low coercivity, in spite of being very expensive. New developments in rare earth metal alloys such as $SmCo_5$ and others followed soon, giving better magnets at lower costs, with rare-earth magnets still enjoying many uses in today's modern magnetic applications. Nevertheless, aside from magnetic recording, permanent magnets have traditionally enjoyed employment in loudspeakers, electric meters, magnetic particle separation, microwave devices, nuclear magnetic resonance, linear accelerators, and a variety of lifting or holding devices, to name just a few. The group of materials that were most used as permanent magnets were the Al–Ni–Co–Fe alloys (i.e., alnicos) with additions of Ti. Adding Ti gave these materials an exceptionally high coercivity, a very desirable quality in many applications. On the other hand, there were other permanent magnets that may have competed with the alnicos because of their magnetic properties. In spite of displaying lower flux densities and higher coercivities than the alnicos, they had their own share of the market because some applications could function solely on this class of materials. These permanent magnets that were considerably used in the past were the Ba or Sr hexaferrites, also known as "ceramic magnets" [1].

Many times in history when a magnetic field was required in a particular environment or space, permanent magnets were employed due to their convenience as energy storage devices and quite often, due to lack of a better alternative. These days, many industrial applications requiring permanent magnets such as ion pumps or focusing devices no longer rely on a particular magnetic material or magnetic compound, but rather make use of electromagnets or a superconducting magnetic circuit. Unfortunately, these constructions necessitate a large expenditure of electric power and also generate a considerable amount of heat. Although these latter human inventions are still related to magnetism in some way, they are not part of the class of magnetic materials that are based on the magnetic phenomena and principles we are discussing in this book. As such, superconducting magnets are not a topic we intend to cover here, due to the highly specialized branch of engineering that they represent. Separate treatments of this subject and its underlying physics are required, and the reader is thus encouraged to seek further information elsewhere. Nevertheless, a few words can be said about the significance of superconducting magnets, and why it is worth for the readers to pursue further investigation on their own.

A group of researchers [21] have recently demonstrated a cryogen-free 23 T superconducting magnet where the insert coil was a 7.5 T $YBa_2Cu_3O_7$ conductor. This achievement is particularly timely due to troublesome experimental barriers encountered in today's magnetics experiments, where really high magnetic fields are required for very precise measurements. High magnetic field experimental research contributes significantly to the development of fundamental research in many areas of magnetics. Yet, it is difficult to maintain a high static field over a long period of time, especially while performing high precision measurements. Many problems arise when trying to generate high static fields of 20 T or more, especially since huge electrical powers in the range of 20–30 MW are necessary to power a water-cooled resistive magnet of a hybrid magnet system. Many institutions cannot afford such costs. Hence, superconducting magnets can help with these applications, and, nowadays, their magnetic field capabilities have reached impressive high values. The mentioned researchers [22], managed to demonstrate functionality of a unique cryogen-free superconducting magnet that can be maintained in a room temperature bore without using liquid helium or nitrogen. By incorporating an $YBa_2Cu_3O_7$ (Y123) coated conductor tape, they were able to improve on an earlier version of their design and obtain the generation of a 23 T magnetic field. The coil has a total inductance of 244 H, and is capable of storing a magnetic energy of 8.7 MJ, thus showing quite remarkable characteristics. There is also a low Curie temperature background magnet composed of four Nb_3Sn coils and one NbTi coil. The magnetic field generated by either coil is directly related to the coil current density. The whole assembly has demonstrated superior use in as far as it generates a maximum 23.4 T field and, thus exceeds magnetic field generation of similar, but liquid helium cooled, superconducting magnets.

1.4.2   Silicon–Iron Alloys. Wound Cores

The electrical industry has benefited significantly from alloys of silicon and iron, with specific applications depending on the silicon content. As such, low percentage alloys (1.5–3.5% Si) have enjoyed employment in motors, generators, and relays, while higher percentage alloys (3–5% Si) have been used for high-efficiency motors and power transformers [23]. Therefore, we need to say a few words about the magnetic properties of this important group of materials. Silicon–iron alloys, also known as silicon steels, are usually melted in an open hearth, an electric arc, or an induction furnace. A rigid control of furnace conditions is performed, especially during refining when oxygen is introduced. Because magnetic properties of silicon steel are so important in a variety of industrial products, it is necessary to produce a low loss material. A large number of researchers have thoroughly studied these alloys over many years. It was discovered as early as the beginning of the twentieth century that, adding more silicon to the already known silicon–iron alloys made them magnetically softer, inasmuch as their coercivity was only half of that of iron, a standard material used till then in transformer cores. Surprisingly, it was soon observed that their permeability was also increased, while hysteresis losses and eddy currents were diminished.

Today we know that silicon–iron alloys possess a high magnetic permeability, and display low hysteresis losses in the cold rolling direction. This fact can be attributed to the directionality of the domain structure which shows antiparallel domains along the rolling direction, rendering particular magnetic properties of these type of steels along this direction. Therefore, any factor that destroys or improves this antiparallel domain structure directly affects magnetic properties in this direction. As a matter of fact, only grain-oriented silicon steels particularly of (110) [001] orientation are of interest for many applications; therefore, many studies and production procedures (e.g., specially designed annealing furnaces) refer exclusively to these type of textured materials. Furthermore, domain refining methods and the mechanisms associated with them have been at the forefront of many improvement techniques. As such, it has been found [24] that there is an interaction between the tensile stresses created during rolling and the newly formed sub domains, the latter consisting of transverse domains resulted from magnetic "free" poles and internal stresses.

A large number of electrical steels that are being produced are used for wound cores. Core losses in silicon steels depend on saturation values of magnetic flux density B achievable in these materials. It has been assumed for a while that larger 180° domain wall spacing due to larger grains leads to increase in anomalous eddy currents and, hence, higher losses. Thus, by refining magnetic domain structure, the domain wall spacing is reduced by counteracting the increase in magnetostatic energy. That, in turn, reduces anomalous eddy currents, allowing a larger saturation flux density, which will then decrease losses. Nevertheless, domain refining needs to be heat resistant, as wound cores are subjected to stress relief annealing. Furthermore, a secondary recrystallization technique is required during material fabrication to ensure a highly oriented crystallographic structure, so that a large number of magnetic domains have the same magnetization orientation.

Another way of reducing eddy current and hysteresis losses is by reducing the thickness of the silicon steel sheet down to a minimum value. Some studies report a 40 μm optimum thickness for use of the silicon steel at 50 Hz frequency, whereas higher frequencies require even smaller thicknesses of ∼10 μm ([25] and references therein). To further investigate sheets of the thinner kind, two types of silicon steel sheets were compared, where the starting material had been fabricated in two different ways [24]. The first sheet was obtained by cold rolling a 300 μm thick conventionally fabricated silicon steel sheet down to a 5 μm thickness. During this process, the sheet was annealed at 1,100°C for 1 h in vacuum while at a thickness of 10 μm to decrease work hardening. The second type of sheet was also obtained by cold rolling, this time down to 8 μm, however, a hot rolled 2 mm thick silicon steel sheet was chosen as a starting material. Similarly, the second sheet was annealed at 1,250°C for 10 h in vacuum at a thickness of 300 μm to restore a highly necessary crystallographic texture. Results indicated that the conventionally fabricated silicon steel sheet of 5 μm thickness had a saturation magnetic flux density of 1.96 T and a coercivity of 6.6 A/m, both illustrating good magnetic properties. Surprisingly, the initially hot rolled steel sheet that was further processed by cold rolling and annealing also showed a similar saturation magnetic flux density of 1.95 T, indicating that the final thickness of the silicon steel sheet plays a more important role than the fabrication details of obtaining it.

Conventional transformers have historically used grain oriented silicon steel sheets. However, some fabrication studies have demonstrated the process of obtaining silicon steel wires instead of sheets, with the lowest possible magnetic losses [26]. A silicon–iron wire of 7.2 mm diameter was drawn down to 0.4 mm in several stages. A few heat treatment steps were necessary in order to relieve stress and recrystallize the deformed material. A magnetic anisotropy direction along the wire is highly desired, as well as a large saturation magnetization. However, a coldly deformed wire acquires a strong <110> texture parallel to the wire axis, obviously not a direction of easy magnetization. It was noticed that in silicon steel sheets, a small grain size leads to the disappearance of the dominant (001) <110> component during processing, a reduction in the (111) <110> component, and the appearance of a strong (111) <112> texture component. On the other hand, a large grain size >100 μm in the original material, leads to a strong (011) <100> crystallographic component, suitable for magnetic applications. This latter result obtainable in silicon steel sheets was desired for drawn silicon steel wires. As such, through a sequence of deformation and annealing processes, decarburization and recrystallization, it was possible for researchers [25] to induce a grain growth at crucial stages of fabrication. Unfortunately, the crystallographic texture of the silicon steel wires did not quite have the desired (011) <100> crystallographic component, but showed an intermediate magnetization axis with a somewhat better permeability compared to that of small grain sized traditional silicon steel sheets. Hence, they felt encouraged to continue their experiments, this time by applying magnetic annealing treatments. It is hoped that magnetic annealing will lead to an improvement in magnetic properties by increasing the grain size, and maximizing <100> crystallographic directions along the wire axis. An improvement in magnetic losses is also highly desired.

1.4.3   Nickel-Iron Alloys. Permalloys

Magnetic recording and, in particular, recording heads are topics that have fascinated several generations. A magnetic recording head should preferably have a high magnetic moment, low coercivity, zero magnetostriction, a large electrical resistance, no internal stresses, and a high corrosion resistance. However, none of these properties are achievable in one single material, and manufacturers have to make compromises. Ni–Fe alloys have enjoyed long time use in magnetic recording, in spite of new materials having entered the recording market since. It was noticed early on that, unfortunately, Ni–Fe alloys have limited applicability in their "as is" state. This is because they are weakly magnetoresistive at room temperature, requiring low temperatures and fields in excess of 20 kOe. Magnetoresistance and its applications are subjects discussed in detail in a more advanced chapter in this book but nevertheless, it suffices to say that room temperature functionality and sensitivity to low magnetic fields are highly desirable attributes. As such, Ni–Fe alloys display a weak magnetoresistance at ambient temperatures. This fact is detrimental to magnetic recording applications that have to function at room temperature while responding to small fields.

   Fortunately, the invention of electroplating baths, as well as improved fabrication processes for magnetic recording heads brought progress to IBM in the earlier days of magnetic recording. As such, magnetic heads used a plated inductive permalloy head [23] which was the only selected material for this purpose. "Traditional" permalloy has 20% Fe and 80% Ni, being a material with a high magnetic permeability, low coercivity, and an almost zero magnetostriction. Permalloy has been historically used in transformer laminations, where it is wrapped around the insulated copper conductors of telecommunication cables, reducing signal distortion by improving inductive compensation of cable capacitive reactance. But the low magnetostriction of permalloy is particularly significant for the earlier magnetic recording heads, where variable stresses in thin films would have otherwise caused a large variation in magnetic properties. The permalloy head had the advantage that, it was able to meet the strict requirements for both reading and writing processes, especially given its soft magnetic properties [20]. Permalloy heads became quickly established as they were easily plated into films, had negligible magnetostriction, and good corrosion resistance, displaying a saturation flux density [27] of 1 T.

A certain nickel–iron alloy with a specific elemental content also gained ground in magnetic recording in the previous century, particularly as areal densities of disk drives increased. $Ni_{45}Fe_{55}$ also known as 45 Permalloy is a magnetically soft alloy, not as good as "traditional" permalloy, but with an increased magnetic moment. This larger magnetic moment attributed to the higher Fe content allowed writing in higher coercivity media, which was a significant improvement at that time. Ni–Fe alloys became heavily researched and improved upon, encouraged by the fact that the magnetic moment for binary NiFe alloys was found [28] to increase monotonically with Fe content up to about 65% Fe composition. The better corrosion properties of this alloy were an added bonus when compared to Co-based high moment alloys. A disadvantage was, however, that efforts to optimize the material in order to improve the writeability process were noticed to deteriorate readability. Inductive writing was used, and such a process necessitates specific magnetic properties which differ from those of reading. The inductive writing with these heads required a larger magnetic anisotropy, a smaller coercivity as well as negligibly low magnetostriction. After some serious improvements, $Ni_{45}Fe_{55}$ or 45 Permalloy became the new standard in the thin film head industry [29]. In fact, the read/write challenge of NiFe heads was overcome when the newly emerged magnetoresistive heads made their appearance. These heads were both read and write heads, as a read sensor [30] was used in conjunction with an inductive write head [31], leaving a small physical separation between the two. Such separation of processes allowed tremendous optimization of writing and reading in an individual manner. Consequently, the soft magnetic property requirements of the inductive writer material became less stringent. However, in order to achieve even higher recording densities, larger magnetic moments were needed.

The magnetic recording industry turned for a while to soft magnetic ternary alloys, such as CoFeCu with high Co content in excess of 75%. They displayed a low magnetostriction, reduced internal stresses, and saturation flux densities as high as 2 T. IBM developed these materials for write heads [32], in an effort to replace "traditional" permalloy and $Ni_{45}Fe_{55}$. It was noticed that the addition of Cu reduced Barkhausen noise while it also refined grain sizes, and, as such, improved domain configuration. These latter topics will be discussed in subsequent chapters, and, therefore, we will not go into further detail here. Unfortunately, the fabrication process of these CoFeCu ternary alloys was not straightforward, as it was challenging to deposit the alloy and obtain a uniform composition. This was because a bath was used which contained a reduced concentration of a diffusion-controlled element such as Cu. Also, the weak corrosion resistance made CoFeCu ternary alloys impractical in magnetic recording. Because these alloys did not display a visible advantage over $Ni_{45}Fe_{55}$ in terms of head performance, they were not implemented in magnetic recording. Partly because of these reasons, IBM decided to consider other soft magnetic ternary alloys such as CoNiFe instead of CoFeCu [33]. It was noticed that CoNiFe films displayed a magnetic flux density higher than 2 T when electrodeposited galvanostatically from a sulfate/chloride bath similar to the plating bath used for the fabrication of $Ni_{45}Fe_{55}$. Consequently, more than 250 alloy compositions were obtained by changing the Ni^{2+}, Fe^{2+}, and Co^{2+} concentrations and varying the current density for a given bath composition. However, this CoNiFe plating system was not consistently reproducible in the high moment region of the ternary diagram. More reliable data was obtained for samples with compositions around $Co_{44}Ni_{27}Fe_{29}$ displaying a slightly lower magnetic flux density ($\sim$2 T), a coercivity of 1.2 Oe, and a low internal stress (115 MPa). It was important that most samples underwent thermal annealing treatments without damage to their structure [34].

In the end, IBM managed to bring significant improvements to magnetic recording heads from permalloy to CoFe alloys, by developing innovative methods of electrodepositing soft magnetic alloys that displayed magnetic flux densities between 1 and 2.4 T. This re-evaluation process began at some point, when it was decided to have another look at NiFe binary alloys with higher iron content [28], in spite of challenges encountered earlier. It had been noticed that magnetic properties of bulk alloys differ from those that have been electrodeposited, especially if these materials have been annealed at high temperatures. Therefore, electrodeposition of NiFe with compositions ranging from $Ni_{35}Fe_{65}$ to $Ni_{15}Fe_{85}$ was performed according to patented recipes. These processes included using sulfate/chloride baths containing boric acid, saccharin, and a surfactant, while electrodeposition occurred in magnetic fields of $\sim$800 Oe, at temperatures varying between 15°C and 30°C [35]. A thermal annealing process was also performed at 225°C for 8 h [36]. Following this fabrication process, the films were noticed to display higher values of internal stress and a larger magnetostriction than CoFeNi. However, in spite of these shortcomings, the films were more corrosion resistant and easier to manufacture. In time and with further investigation, IBM obtained $Ni_{32}Fe_{68}$ and $Ni_{20}Fe_{80}$ alloys with saturation flux densities of 2 T and 2.2 T, respectively [37]. Furthermore, they noticed that plating a nickel–iron alloy with 50–70 wt% Fe in a galvanostatic bath with a proprietary additive leads to magnetic flux densities of 2.4 T [38]. These significantly improved plated alloys exhibited a large positive magnetostriction (+45 ppm), and high internal stresses of up to [22] 850 MPa.

1.4.4   Compounds with Rare Earth Elements

Of historically great interest, and even more so these days, compounds of a rare-earth metal and a transition metal such as Mn, Fe, Co, or Ni have been extensively studied all over the world and employed as practical magnets. These materials display large anisotropy constants, by far larger than other magnetic materials, and are of great interest in magnetic recording as well as in our modern conveniences such as cell phones. Nevertheless, rare earth magnets may have also been known to the reader from other applications requiring strong magnets, such as nondestructive testing or audio speakers. As such, the neodymium alloy $Nd_2Fe_{14}B$ may be the most familiar permanent magnet used extensively in audio devices, and, for a while, in most computer hard drives. The strong magnetic brushes used in magnetic flux leakage techniques are also made of neodymium alloys. Although these compounds have the highest magnetic field strength B, another rare earth compound, $SmCo_5$, has outperformed them in some applications, due to better resistance to oxidation and higher Curie temperature. Thus, $SmCo_5$ has created a niche in applications where high field strength is needed at high operating temperatures. Unfortunately, $SmCo_5$ magnets are brittle and prone to cracking and fracture, limiting their use.

Rare earth elements are not particularly scarce, but the terminology refers to the difficulty and high cost of separating them from each other. Ores usually contain more than one element, and these elements are chemically very similar. Demand fixes the price, as well as world economic interests; thus, any new technological development requiring magnetic materials can lead to these materials making news headlines. It should also be noted that several alloys in this rare earth compound category are not usually thought of as permanent magnets, because they are often not single magnets but rather conglomerate structures of a large number of micro or nanoscale magnets. Magnetic recording is a good example of an area of application that rests on these materials, and where individual entities have complex magnetic structures that fall into more than one classification group.

Rare earth compounds contain lanthanide elements which are ferromagnetic metals, but with an individual Curie temperature per element that is below room temperature. Fortunately, as stated earlier, these elements form alloys with transition metals such as Fe, Ni, and Co, and the resulting alloy can reach Curie temperatures well above room temperature. A significant advantage of rare earth compounds is that their crystalline structures have a very high magnetic anisotropy and, therefore, easy to magnetize in a particular direction, at the expense of being magnetized in any other direction. The atoms of rare earth elements have high magnetic moments due to the incomplete filling of the $4f$-shell, which can contain as many as seven unpaired electrons with aligned spins. $4f$-Electrons in these orbitals are strongly localized and can thereby, easily retain their magnetic moments. This fact stands in contrast to magnetic moments in other orbitals which are often lost because of a strong overlap with their neighbors, thus leading to zero net spin. The combination of high atomic magnetic moments with a high magnetic anisotropy results in high magnetic strength, leading to a variety of applications, some discussed later in this book.

To appreciate the vast world of magnetic materials, one has to look at the many subfields of magnetics and the multitude of phenomena that govern them. Audio, video, digital applications, transformers, brakes, cordless devices, and even toys, all benefit from the fascinating materials that respond to an external magnetic field, with some materials able to retain a magnetic flux density of their own. From disks, drums, and recording tapes not usually thought of as permanent magnets, to powerful brushes or magnetoresistive heads, each category involves its own type of magnetic materials, thus bringing its own contribution to the diversity of magnetic effects exploited for the use of humankind. The purpose of this book is therefore, to offer the reader a glimpse into this world, and, in the process, facilitate gaining some knowledge of magnetics and how this scientific/engineering branch evolved over the years. Some questions are ultimately answered: Why do regions within materials become magnetized in one direction while the others can barely be magnetized, or how can magnetism benefit recording and playback of digital information, to how can we ignore the charge of an electron and use its spin instead, in order to develop new types of electronic devices? In the end, it may thus seem less curious that of all the magnetic materials available to humans, only particular ones were essentially chosen, the reasons for these choices becoming hopefully clearer.

Tests, Exercises, and Further Study

P1.1 Identify the different kinds of magnetism from the schematic representations in Fig. P1.1.

P1.2 From a historical perspective, why was the concept of spin introduced, and how was it experimentally verified?

P1.3 A large number of materials are considered magnetic and are commercially employed in devices. However, only three pure metals exhibit ferromagnetism at room temperature. Many magnetic alloys contain at least one of these ferromagnetic metals. What are these three metals, and where does their ferromagnetism originate?

P1.4 In Fig. 1.2c, the magnetic field is applied parallel to the spin axis. Explain what happens to the spins when the magnetic field is applied perpendicular to the spin axis.

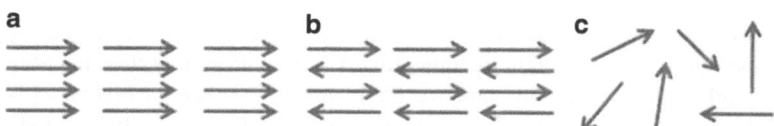

Fig. P1.1 Different kinds of magnetism

P1.5 What are ferrites? Give a few examples.

P1.6 What are perovskite–type oxides and what kind of magnetism, if any, do they display?

P1.7 How are spontaneous magnetizations distributed in a ferromagnet?

P1.8 What constituted the first direct observation of ferromagnetic domain structure and what interpretation was it given?

P1.9 Can magnetostriction influence domain structure?

P1.10 Comment on the influence of the Si content on the magnetic properties of silicon steel. What are the barriers encountered in obtaining higher Si content in commercial silicon steels? *See for instance* [15].

P1.11 What is the observed domain structure in a silicon steel with a (110) [001] crystallographic texture obtained by a sequence of cold rolling and annealing operations?

P1.12 Nanocrystalline ribbon alloys with thickness below $10\,\mu$m are sought after in electronic devices due to their soft magnetic properties displayed at high frequencies. Particularly, these materials have a high effective permeability and a high magnetic saturation around $1\,$MHz. They need to have flat B–H curves but also minimum core losses. Because magnetic annealing can change their magnetic properties, this effect on these types of alloys warrants further study. What is the effect of annealing in a magnetic field on the magnetic properties of ultrathin Fe-based nanocrystalline alloys? *See for instance* [39].

P1.13 Hot rolling of rare earth permanent magnets is a type of mechanical processing that has the advantage of not having to handle magnetic powders. In addition to fabrication steps that involve hot rolling, rare earth permanent magnets also undergo magnetic annealing in order to alter their magnetic properties. What are the effects of magnetic annealing on hot rolled rare earth permanent magnets? *See for instance* [40].

P1.14 Amorphous Co-based thin films are known for their good high frequency characteristics when integrated in miniaturized devices, such as transformers or inductors. Their soft magnetic properties can be controlled with high precision, particularly when used in a high frequency magnetic field. As such, Co–Nb–Zr thin films display an in-plane magnetic anisotropy. The in-plane magnetic anisotropy is an important attribute of these films because they are used at high magnetic fields where the dominant magnetization reversal process occurs through magnetization rotation. The high frequency magnetic field is applied parallel to the hard axis. The permeability along the hard axis depends strongly on the in–plane anisotropy. Controlling this anisotropy and decreasing coercivity are necessary for minimizing hysteresis losses. Studies show that the in-plane magnetic anisotropy can be controlled by magnetic annealing. Comment on how the annealing field angle influences the in plane magnetic anisotropy. *See for instance* [34].

P1.15 $Nd_2Fe_{14}Si_3$ is a permanent magnet compound, a Si solid solution in the binary intermetallic Nd_2Fe_{17}. Specifically, Nd_2Fe_{17} crystals have a spontaneous magnetic moment of $39\mu_B$ per formula unit and a Curie temperature of 348 K. The c-axis is the hard magnetization axis, while [120] is an easy magnetization axis, estimated to exceed 20 T. These materials belong to a wide class of rare earth compounds with a high content of 3d transition metal which combines the localized magnetism of the rare earth sublattice with itinerant magnetism of the transition metal sublattice. Comment on the magnetic properties of this compound. *See for instance* [41].

P1.16 Ferrimagnetic oxides are important magnetic materials due to their use in transformer cores, microwave antennas, and some magnetic recording devices. Their magnetic properties can be modified nowadays in nanosized structures of the same material as the bulk compound. Nanotechnology is known for the ability to obtain properties in the nanomaterials not seen in their bulk counterpart. Comment on the *reasons* behind the tailored structural, magnetic, and electrical properties of nanosized $Mn_{1-x}Zn_xFe_2O_4$ used in high frequency applications. *See for instance* [42].

P1.17 Comment on the thickness dependence of the Curie temperature of ferromagnetic thin films. *See for instance* [43].

P1.18 Magnetostrictive materials that contain transition metals, such as iron, are known for their large magnetostriction values observable at low saturation fields. Furthermore, these materials have high mechanical strength, a good ductility, and are cost effective. Comment on the magnetocrystalline anisotropy of Fe–Ga alloys, which is responsible for the large magnetostriction. *See for instance* [44].

P1.19 Carbon nanotubes can be filled with Fe and ferrite nanoparticles. Comment on the properties of these novel compounds and their applications. *See for instance* [45].

P1.20 Ferromagnetic shape memory alloys enjoy widespread use as magnetic actuators due to the strain that can be induced in these materials by a magnetic field. Ni–Mn–X ($X =$ In, Sn, Sb) Heusler alloy systems are particularly interesting as ferromagnetic shape memory alloys because of the transformation they undergo when a magnetic field is applied, from a ferromagnetic to a nonmagnetic martensitic phase. A very good shape memory effect is observed in the martensitic phase, once the transformation has occurred. Comment on the magnetic properties of this martensitic phase. *See for instance* [46].

Hints and Partial Answers to Problems in Chap. 1

The answers to these problems are, in many instances, just a hint as to where further information is to be found. They contain the main idea, but the reader needs to find the references that are sometimes recommended for each exercise and complete the answer through further bibliographic reading and study. Reference numbering follows the one used in that particular chapter.

P1.1 From Fig. P1.1: (a) Ferromagnetism; (b) Antiferromagnetism; (c) Paramagnetism.

P1.2 The concept of spin was introduced to explain the multiplet structure of atomic spectra. It was initially seen as a spinning motion of the electron around its own axis with angular momentum $\pm\hbar/2$. The magnetic moment due to spin was experimentally verified by the Zeeman effect, which revealed a spectral line split into two or three lines when atoms were placed in a magnetic field. The spin angular momentum changes by the unit $\hbar/2$, and the unit of spin magnetic moment is a Bohr magneton.

P1.3 These metals are iron, cobalt, and nickel. Each of these elements has an unfilled $3d$ electronic shell which constitutes the origin of their ferromagnetism. *See* [1–3] *for further details.*

P1.4 When the magnetic field is applied perpendicular to the spin axis, magnetization of the antiferromagnet takes place by the rotation of each spin from the direction of the spin axis. As a consequence, the susceptibility becomes independent of temperature. This is valid in case of a single antiferromagnetic crystal, since in a polycrystal, the susceptibility is the average between the two cases (i.e., variation of susceptibility with parallel and perpendicularly applied field). Nevertheless, above the transition temperature, the susceptibility decreases with an increase in temperature, independently of the direction of the applied magnetic field.

P1.5 Ferrites are a group of iron oxides with the general formula $MO{\cdot}Fe_2O_3$, where M is a divalent metal ion such as Mn^{2+}, Fe^{2+}, Co^{2+}, Ni^{2+}, Cu^{2+}, Zn^{2+}, Mg^{2+}, or Cd^{2+}. The best known ferrite is magnetite Fe_3O_4, also known as $FeO{\cdot}Fe_2O_3$. The type of divalent ion M, determines how strong the spontaneous magnetization of the ferrite is. Mixed ferrites are also possible, obtained by mixing two or more kinds of M^{2+} ions. Ferrites have the so-called "spinel" structure.

P1.6 The type of mineral known as perovskite has usually the composition $CaTiO_3$ and is non-magnetic. However by replacing Ti with Fe, a ferrimagnetic perovskite–type oxide is obtained of the general formula $MFeO_3$ where M is a large metal ion such as La^{3+}, Ca^{2+}, Ba^{2+}, or Sr^{2+}. These oxides have a cubic crystal structure and are a type of magnetic material used in many modern devices, as will be discussed in later chapters. *See published articles on magnetic pervoskites for further details.*

P1.7 Atomic magnetic moments in ferromagnetic materials order in such a way as to give rise to magnetic domains. Within a domain, atomic magnetic moments are aligned, resulting in a spontaneous magnetic domain magnetization. The spontaneous domain magnetization of one domain is usually oriented differently with respect to the spontaneous domain magnetization of neighboring domains. The spontaneous domain magnetization of all domains gives rise to a total magnetization of the ferromagnet, which depends on the temperature. When the total resultant magnetization for all magnetic domains is zero, the ferromagnetic material is said to be demagnetized.

However, an applied magnetic field changes the total resultant magnetization. *See* [1–3] *for further details.*

P1.8   Ferromagnetic domain structure was observed directly in 1931 by Bitter who applied a drop of a ferromagnetic colloidal suspension to the polished surface of a ferromagnetic specimen. The suspensions contained numerous fine ferromagnetic particles that rearranged so that they formed domains on the surface of the specimen. The micrographs were interpreted by Bitter to reveal inhomogeneities in ferromagnetic substances, and were, unfortunately not seen as images of domain structure, as the latter were thought to be of smaller size than the observed structures. It was not until 1949 that Williams, Bozorth, and Shockley finally observed well defined domain structures on Si–Fe crystals. *See* [2] *for further details.*

P1.9   Magnetostriction can influence domain structure. As such, in a ferromagnetic material with a positive magnetostriction, a magnetic domain will have a tendency to expand parallel and contract perpendicular to the direction of domain magnetization. Given that the magnetostrictive strain has a low value of about 10^{-5} it is not strong enough to break the crystal at the domain boundaries. However, magnetoelastic energy is stored in a crystal, in part, due to magnetostriction.

P1.10 Si–Fe alloys exhibit decreasing magnetization saturation, magnetic anisotropy, and magnetostriction with increasing Si content. At 3.25% Si, an optimum electrical resistivity is obtained that limits eddy current losses. Nevertheless, at higher Si concentrations, further reduction in eddy current losses is obtained, concurrent with a lower anisotropy and magnetostriction that help in reducing noise problems when magnetic flux deviations take place, such as at transformer core corners or induced by strains due to core assemblies. Unfortunately, silicon steels with Si content higher than 3.25% are very brittle and, therefore, almost impossible to cold roll. Researchers have tried to develop methods of introducing higher Si concentrations in these steels, but with a lower impact on their brittleness. Methods such as adding Si by diffusion in a gaseous atmosphere have shown some success and an improved performance over the starting material, especially after further annealing the sample in a magnetic field.

P1.11 By using magneto-optical Kerr observations, it was determined that a predominantly antiparallel domain structure is formed, aligned with the rolling direction which is also the magnetic easy axis. When these domains were exposed to a cyclic and longitudinal magnetic field, magnetization reversals took place by motion of antiparallel domain walls which have a higher mobility. It was concluded that the directionality of the domain structure remains basically unchanged by the magnetization process. *See indicated reference and P1.10 for further details.*

P1.12 The magnetic properties of ultrathin $Fe_{78}Al_4B_{12}Nb_5Cu$ and $Fe_{83}B_9Nb_7Cu$ ribbons were investigated after annealing in a transverse magnetic field of 1.5 kOe. The magnetic field was continuously applied during all heating and cooling stages. Effective permeability and core loss were measured with an impedance analyzer, and a $B-H$ analyzer, respectively. It was noticed that the alloy with added Al is more sensitive to magnetic annealing, displaying improved properties, such as lower core losses and higher effective permeability. This is indicative of significantly different domain structures between the two types of alloys. *See indicated reference for further details.*

P1.13 Rare earth permanent magnets of the type $Pr_{15}Fe_{78}B_{5.5}Cu_{1.5}$ (at%), and the related compound $Pr_{15}Fe_{79}B_{5.5}Ga_{0.5}$ (at%) were annealed in a two step process. In the second compound, Cu was replaced with Ga. Prior to annealing; the compounds were fabricated from ingots by hot rolling at 950°C until a reduction ratio of 76% was obtained. Note that no magnetic field was applied during annealing; however, a magnetic field was used to magnetize and demagnetize the magnets after the two step annealing process. Magnetic properties of the two-step annealed magnets were compared to those of samples that only underwent a conventional single-step annealing process. The two step annealing process involved annealing at an elevated temperature of $1,000°C$, prior to the next annealing process carried out at 475°C, a temperature which is near the eutectic temperature of the rare earth rich phase. It was determined that the two step annealing process resulted in improved magnetic properties of the permanent magnets, such as coercivity, remanence, and squareness of $B-H$ curve. This stands in contrast to the conventional single step annealing process that results only in an improved coercivity, but does little for other magnetic properties. Microstructural investigations revealed that a two step annealing process leads to a better definition of the network structure of the two phases, the matrix phase and the rare earth phase surrounding the former. Thus, the disorder in the matrix phase brought about by hot rolling is responsible for the reduced crystal alignment, and, therefore, the poor magnetic properties observed prior to annealing. *See indicated reference for further details.*

P1.14 Sputtered, amorphous Co–Nb–Zr thin films were investigated for their in-plane magnetic anisotropy. Researchers observed that local anisotropy dispersions form in the plane of these films that degrade their magnetic properties, especially at high frequencies where the material is functional. It was found that the in-plane magnetic anisotropy can be changed by magnetic annealing, and that the change depends on the angle of the applied magnetic field with respect to the surface of the film. As such, the magnetic field is applied obliquely, at an angle α with the plane of the film (Fig. P1.14). After magnetic annealing at this angle, the magnetic easy axis of the film is no longer in plane, but along the direction of the annealing field. By suppressing local anisotropy dispersions that occur in the plane of the film, a degradation of the frequency characteristics of magnetic permeability are avoided, thus rendering the material highly functional at MHz frequencies. *See indicated reference for further details.*

Fig. P1.14 Annealing in an
oblique magnetic field

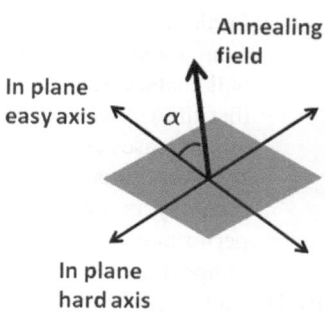

Annealing
field

In plane
easy axis α

In plane
hard axis

P1.15 The Nd_2Fe_{17} compound is one of the widest groups of rare earth intermetallic base units used for permanent magnets and magnetostrictors, given
its magnetic properties. For instance, $Nd_2Fe_{14}Si_3$ has a large magnetic
anisotropy and magnetostriction originating from the rare earth sublattice.
On the other hand, the transition metal sublattice renders the compound a
high spontaneous magnetic moment and a low Curie temperature. These
particular magnetic properties can be traced back to the atomic structure
of the compound, specifically distances between Fe atoms in the sublattice,
which are shorter than those in typical bcc Fe crystals. The short distances
between Fe atoms give rise to negative exchange interactions which lower
the Curie temperature. However, any Fe atom substitution with magnetic
or nonmagnetic atoms of a different kind (e.g., Si) destroys the negative
exchange interaction balance, resulting in an increase in Curie temperature.
This is the principle behind obtaining permanent magnet materials, by
manipulating the Fe, rare earth, and nonmagnetic atom content, and their
arrangement on sublattices. *See indicated reference for further details.*

P1.16 From the study in the indicated reference, it seems that Zn substitution for
Mn has a major impact on the dielectric properties of the mixed ferrite. In
particular, the size of the samples and method of preparation influence the
properties obtained. As such, Zn is found to occupy the octahedral sites
of nanosized manganese zinc ferrites prepared by co-precipitation methods,
pulsed laser deposition, or high energy ball milling. This alters cation
distributions, reduces lattice parameters, modifying exchange interactions,
and thereby the magnetization values, while increasing Curie temperature.
None of these effects seem to be observed in the bulk counterpart of these
materials, not even in micron-sized compounds.

P1.17 Ferromagnetic materials are described by the Heisenberg model (see (1.13)),
which predicts the basic characteristics of a ferromagnetic phase transition.
Ferromagnetic thin films can, thus, be analyzed using this model. In the
indicated reference, a study was undertaken to predict the Curie temperature
of such ferromagnetic thin films. A few differences can be noted between
thin films and bulk materials. A thin film is anisotropic by nature, and
hence, its magnetic properties are expected to be different from the bulk.

Furthermore, since interaction partners can be from the same atomic layer or from a neighboring layer, this gives different results for thin films than for bulk materials, as thin films create a "surplus" monolayer. Calculations show that the Curie temperature is lower for a reduced number of monolayers, but increases as the number of monolayers increases. The Curie temperature for thin films approaches that of bulk materials when the number of monolayers approaches infinity. This explains why so many experiments performed nowadays on nanoscale ferromagnetic materials occur at low temperatures.

P1.18 The magnetocrystalline anisotropy of Fe–Ga alloys is believed to be due to their electronic structure, in particular, the $3d$ electrons of Fe. The indicated reference shows that the electron cloud is not isotropic, leading to anisotropy due to the exchange interaction between different atoms. The authors calculated an orbit projected density of states for Fe atoms which led to the conclusion that the B2-like structure had the highest magnetic anisotropy. Results are in good agreement with previous theoretical studies and experimental values of the magnetic anisotropy constant, which assumed that the anisotropy of Fe–Ga is due to the electron cloud of the Fe atoms.

P1.19 Fe and ferrite filled carbon nanotubes form magnetic composites that are used as microwave absorbers tuned by matching the dielectric loss and the magnetic loss by changing the ratio of ferrite. It is known that hexagonal ferrites have a high saturation magnetization and high complex permeability values in a wide frequency range. Furthermore, carbon nanotubes enjoy an increased popularity because of their low density and high complex permittivity values for microwave absorption and electromagnetic interference shielding. Fe has a higher coercivity than ferrite, and adding Fe to these composites seems to make a difference in the microwave absorption quality. The microwave absorbing properties of these compounds have been found to be superior to those absorbers containing only carbon nanotubes or ferrite. This is due to a better electromagnetic matching of the dielectric loss of the carbon nanotubes with the magnetic loss of the ferrite and additional Fe filling. *See indicated reference for further details.*

P1.20 High magnetic field measurements were carried out in $Ni_{50}Mn_{50-x}In_x$ alloys for different x concentrations in order to study the shape memory effect. The strong spontaneous magnetization was found to decrease with In concentration, suggesting that Mn atoms substitute at In sites, coupling ferromagnetically to the magnetic moments at existing Mn sites. This determines the observed magnetic properties which differ for different In concentrations, with no ferromagnetic phase observed for $x = 14$ and below. *See indicated reference for further details.*

References

1. B.D. Cullity, *Introduction to Magnetic Materials*, 2nd edn. (Addison-Wesley, New York, 1972)
2. S. Chikazumi, *Physics of Magnetism* (Wiley, New York, 1964)
3. R. Becker, W. Döring *Ferromagnetismus* (Springer, Berlin, 1939)
4. S. Abe, S. Ohnuma, Appl. Phys. Express **1**, 111304 (2008)
5. D.J. Griffiths, *Introduction to Quantum Mechanics* (Prentice Hall, New Jersey, 1995)
6. H. Gotoh, Y. Takeda, H. Asano, J. Zhong, A. Rajanikanth, K. Hono, Appl. Phys. Express **2**, 013001 (2009)
7. W. Greiner, *Quantum Mechanics, An Introduction*, 2nd edn. (Springer, Berlin, 1994)
8. Van Vleck, *Theory of Electric and Magnetic Susceptibilities* (Oxford University Press, Oxford, 1965)
9. *Ultrathin Magnetic Structures II*, ed. by J. Bland, B. Heinrich (Springer, Berlin, 1994)
10. S.S.P. Parkin, Phys. Rev. Lett. **67**, 3598 (1991)
11. H. Ohldag, A. Scholl, F. Nolting, E. Arenholz, S. Maat, A.T. Young, M. Carey, J. Stöhr, Phys. Rev. Lett. **91**(1), 017203 (2003)
12. W.J. Antel, F. Perjeru, G.R. Harp, Phys. Rev. Lett. **83**, 1439 (1999)
13. A. Aharoni, *Introduction to the Theory of Ferromagnetism* (Oxford University Press, Oxford, 1996)
14. J.M.D. Coey (ed.), *Rare-Earth Iron Permanent Magnets* (Oxford University Press, Oxford, 1996)
15. G.L. Houze Jr., S.L. Ames, W.R. Bitler, IEEE Trans. Magn. **6**(3), 708 (1970)
16. R. Kolano, N. Wojcik, W.W. Gawior, M. Kuzminski, IEEE Trans. Magn. **30**(2), 1033 (1994)
17. J.S. Smart, *Effective Field Theories of Magnetism* (Saunders, Philadelphia, 1966)
18. C. Kittel, J.K. Galt, Solid State Phys. **3**, 437 (1956)
19. M.J. Sablik, G.L. Burkhardt, H. Kwun, D.C. Jiles, J. Appl. Phys. **63**(8), 3930 (1988)
20. P.C. Andricacos, L.T. Romankiw, Magnetically Soft Materials in Data Storage: Their Properties and Electrochemistry, in *Advances in Electrochemical Science and Engineering*, vol. 3, ed. by H. Gerischer, C. Tobias (VCH, New York, 1994)
21. K. Watanabe, S. Awaji, G. Nishijima, S. Hanai, M. Ono, Appl. Phys. Express **2**, 113001 (2009)
22. H. Xu, J. Heidmann, Y. Hsu, Electrochem. Soc. Proc. **PV2002–27**, 307 (2003)
23. L.T. Romankiw, D.A. Thompson, Thin film inductive transducer, US Patent 4,295,173 October 13 (1981)
24. N. Takahashi, Y. Ushigami, M. Yabumoto, Y. Suga, H. Kobayashi, T. Nakayama, T. Nozawa, IEEE Trans. Magn. **22**(5), 490 (1986)
25. M. Nakano, K. Ishiyama, K.I. Arai, IEEE Trans. Magn. **31**(6), 3886 (1995)
26. T. Yonamine, M. Fukuhara, N.A. Castro, F.J.G. Landgraf, F.P. Missell, IEEE Trans. Magn. **44**(11), 3954 (2008)
27. R. Bozorth, *Ferromagnetism* (Van Nostrand, Princeton, NJ, 1951)
28. N. Robertson, H.L. Hu, C. Tsang, IEEE Trans. Magn. **33**(5), 2818 (1997)
29. L.T. Romankiw, Electrochem. Soc. Proc. **PV90–8**, 39 (1990)
30. D.A. Thompson, Proc. AIP Conf. Magn. Magn. Mater. **24**, 528 (1975)
31. L.T. Romankiw, US Patent 3,908,194, Sept. 23 (1975)
32. J-W Chang, P.C. Andricacos, B. Petek, P.L. Trouilloud, L.T. Romankiw, Electrochem. Soc. Proc. **PV98–20**, 488 (1999)
33. E.I. Cooper, C. Bonhôte, J. Heidmann, Y. Hsu, P. Kern, J.W. Lam, M. Ramasubramanian, N. Robertson, L.T. Romankiw, H. Xu, IBM J. Res. Dev. **49**(1), 103 (2005)
34. H. Tomita, T. Sato, T. Mizoguchi, IEEE Trans. Magn. **30**(3), 1336 (1994)
35. J.V. Powers, L.T. Romankiw, US Patent 3,652,442, March 28 (1972)
36. T. Osaka, M. Takai, K. Hayashi, Y. Sogawa, Y. Ohashi, Y. Yasue, M. Saito, K. Yamada, IEEE Trans. Magn. **34**, 1432 (1994)
37. M. Ramasubramanian, J. Lam, A. Hixson-Goldsmith, A. Medina, T. Dinan, N. Robertson, T. Harris, S. Yuan, Electrochem. Soc. Proc. **PV2002–27**, 298 (2003)

38. C. Bonhôte, H. Xu, E.I. Cooper, L.T. Romankiw, Electrochem. Soc. Proc. **PV2002–27**, 319 (2003)
39. J.Y. Park, S.J. Suh, K.Y. Kim, T.H. Noh, IEEE Trans. Magn. **33**(5), 3799 (1997)
40. T. Yuri, T. Ohki, IEEE Trans. Magn. **29**(6), 2752 (1993)
41. A.V. Andreev, S. Yoshii, M.D. Kuz'min, FR deBoer, K. Kindo, M. Hagiwara, J. Phys. Condens. Matter **21**, 146005 (2009)
42. E.V. Gopalan, K.A. Malini, D.S. Kumar, Y. Yoshida, I.A. Al-Omari, S. Saravanan, M.R. Anantharaman, J.Phys. Condens. Matter **21**, 146006 (2009)
43. R. Rausch, W. Nolting, J. Phys. Condens. Matter **21**, 376002 (2009)
44. L. Zheng, C-B Jiang, J-X Shang, H-B Xu, Chin. Phys. B **18**(4), 1647 (2009)
45. X. Gui, W. Ye, J. Wei, K. Wang, R. Lv, H. Zhu, F. Kang, J. Gu, D. Wu, J. Phys. D Appl. Phys. **42**, 075002 (2009)
46. RY Umetsu, Y. Kusakari, T. Kanomata, K. Suga, Y. Sawai, K. Kindo, K. Oikawa, R. Kainuma, K. Ishida, J. Phys. D Appl. Phys. **42**, 075003 (2009)

Chapter 2
Micromagnetism and the Magnetization Process

Abstract Many magnetic properties depend strongly on the micromagnetic structure of a material and for this reason different materials display their own quite unique behavior under an applied magnetic field. In ferromagnets, magnetic domain configurations change with an applied magnetic field or stress through the displacement of domain walls. This makes micromagnetic characteristics of ferromagnets with their magnetic domains and walls responsible for extrinsic magnetic properties such as remanence and coercivity. They are also the cause for the hysteresis events observed in ferromagnetic materials. Due to the variety of magnetic properties and their manifestations in different phenomena, separate branches of condensed matter physics have emerged over time. Consequently, to understand diverse and complex developments in the broad field of magnetics, some fundamental concepts need to be discussed before moving on to more advanced subjects.

Because so many technological applications are based on a distinctive, yet controllable response of materials to a magnetic field, it is necessary to include a chapter in this book describing a few basic principles and observations attributed to the microstructure of magnetic materials, specifically to ferromagnets. For instance, what are magnetic domains and what happens when they move? What factors determine wall movement and how can that be exploited in applications? Why does spontaneous division into magnetic domains occur in the first place? By answering these questions, we will be able to understand why clusters are single domain magnets, and why their large surface-to-volume ratio leads to strong diameter dependence of intrinsic properties such as anisotropy and magnetization. We will see why particle size determines domain configurations and the mechanism of magnetization reversal within magnetic domains. Or, why magnetic thin films are related to micromagnetic structures such as domains and domain walls.

Upon grasping some basic knowledge of magnetics, it will become clearer why single domain, also known as "monodomain magnets" tend to be superparamagnetic, particularly at high temperatures, an aspect that is significant in magnetic recording. We will find out why nanostructured thin films with intermediate or high coercivities are used in permanent magnets or traditional magnetic

C.-G. Stefanita, *Magnetism*, DOI 10.1007/978-3-642-22977-0_2,
© Springer-Verlag Berlin Heidelberg 2012

recording. We will become more familiar with an interesting phenomenon known as magnetic Barkhausen noise which is based on magnetic microstructures, in particular, domain wall pinning by obstacles. Some technological applications of the thin film type exploit magnetic domains, or thickness-dependent domain wall and coercive phenomena. Many magnetic nondestructive evaluation techniques exploit other specific micromagnetic properties to detect flaws and strains on the surface of engineering components. On the other hand, traditional magnetic tapes for information storage still struggle with energetic losses created in the material because of their microstructure, and need to minimize these losses.

In search of more performant storage media, magnetic recording has turned back to some micromagnetic concepts such as magnetic domains, albeit physically scaled down. It is important to note at this point that phenomena of interest in contemporary magnetic applications (including magnetic storage) have reached nowadays submicron scales, requiring dimensional shrinking to nanometer size or below. In order to make the shift to smaller and smaller scales in magnetics, the reader needs to undergo this process of stepwise understanding, where did we start in magnetics and what intermediate stages are involved. Micromagnetics is therefore a necessary transition in this progression.

2.1 Magnetic Domains in Ferromagnets

2.1.1 The Process of Division into Magnetic Domains

In ferromagnetic materials, individual atomic magnetic moments have a tendency to align parallel to each other, keeping at a lower value a type of energy known as the *exchange energy*. The origin of the *exchange energy* E_{exchange} can be found in spin–spin interactions that constitute the basis for ferromagnetism [1]. When the spins of unpaired electrons are parallel, the exchange energy is minimum. However, the parallel alignment of spins and thereby atomic magnetic moments increases another kind of energy, the *magnetostatic energy* $E_{\text{magnetostatic}}$. Large external magnetic fields are created, a process shown schematically in Fig. 2.1a. Therefore, to decrease the magnetostatic energy, a division into magnetic domains with antiparallel magnetizations occurs while also magnetic walls start to form between these domains (Fig. 2.1b). In this configuration, the exchange energy is somewhat increased; however, the magnetostatic energy is lowered. Several magnetic domains are thereby formed within the material so that each domain will contain individual magnetic moments. These moments add up to a total magnetization in each magnetic domain.

Further division into magnetic domains tends to occur; nevertheless, this brings the other energies out of balance. The crystal symmetry of the ferromagnet gives rise to a *magnetocrystalline (anisotropic) energy* $E_{\text{magnetocrystalline}}$ that seeks a minimum value, as well. The minimum for this energy is attained when the magnetization of a magnetic domain is aligned with a preferred crystallographic direction, such as

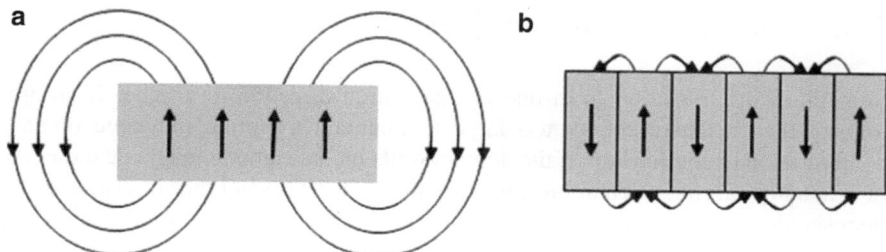

Fig. 2.1 Formation of magnetic domains occurs through a multistep process due to a minimization competition between five different energies. Shown in this figure are (**a**) alignment of individual atomic moments, resulting in an increase in magnetostatic energy (*larger flux lines*). Consequently, a large external magnetic field is created; and (**b**) Occurrence of division into magnetic domains with antiparallel magnetizations decreases magnetostatic energy and reduces flux lines, allowing closure through neighboring domains. Note the formation of magnetic walls between the magnetic domains

< 100 > in iron. These directions are also known as *axes of easy magnetization*. All these divisions into magnetic domains and magnetic moment alignments create a crystal lattice strain, related to the direction of regional magnetization through the *magnetoelastic energy* $E_{\mathrm{magneotelastic}}$. This energy reaches a minimum when the lattice is deformed such that the magnetic domain is elongated or contracted in the direction of its magnetization.

The magnetic wall which just formed between domains with antiparallel magnetizations introduces an energy of its own, energy associated with the wall itself. This is the fifth energy in our energy balance, attributed to the fact that magnetic walls have a certain energy per unit area of surface, and unit thickness of wall. It arises because of those atomic moments that are not parallel to each other, or to an easy axis. The wall energy E_{wall} increases the exchange energy which is highest in the wall vicinity. This exchange energy also known as an exchange force acts only over one or two atomic distances [2], and arises from the Pauli exclusion principle as a quantum mechanical effect based on the degree of wavefunction overlap. The five basic energies involved in the formation of magnetic domains are thus:

$$E = E_{\mathrm{exchange}} + E_{\mathrm{magnetostatic}} + E_{\mathrm{magnetocrystalline}} + E_{\mathrm{magnetoelastic}} + E_{\mathrm{wall}}. \qquad (2.1)$$

Increases and decreases in these five energies have consequences for the equilibrium of the crystalline lattice in the material such that not all energies can be minimum at the same time. Formation of a certain magnetic configuration is the outcome of the *sum* of the five basic energies being minimized, although the energies themselves may not be at their minimum. As a consequence of this minimization competition between five different energies, ferromagnetic materials divide spontaneously into magnetic domains.

2.1.2 Formation of Domain Walls

Sometimes the transition from one magnetization direction to another is abrupt, making the exchange energy too large to maintain a certain magnetic domain configuration in equilibrium. If the domain walls include atomic magnetic moments of gradually varying orientation (Fig. 2.2), this ensures a smoother transition to an opposite domain magnetization direction. Such a transition decreases the exchange energy, especially for a predetermined width of the transition layer.

Example 2.1. Give an estimate of the size of a magnetic domain

Answer: Consider, for instance, perpendicular magnetic domains in granular $La_{0.6}Sr_{0.4}MnO_3$ perovskite films with thickness $< 40\,nm$. For these structures, the domain width δ is given by [3]

$$\delta = \pi \sqrt{\frac{JS^2}{Ka}}, \qquad (E2.1)$$

where J is the exchange coupling constant, S is the spin quantum number, K is the magnetic anisotropy constant, and a is the lattice constant. In this case, $J = 3.0 \times 10^{-22}\,J$, $K = 2.0 \times 10^4\,J/m^3$, and $S = 3/2$ giving $\delta \approx 30\,nm$ [3]. Magnetic domain size can vary in compounds of the same type depending on the granularity and strain of the substrate on which these films are grown. For instance, the substrate could create a tensile strain which results in an in-plane magnetization for the domains that form in the immediate vicinity of the substrate. On the other hand, the magnetization of top grains (further away from the substrate) is oriented in a perpendicular direction. The formation of in-plane magnetized domains near grain boundaries would result in the circulation of magnetic flux, thus suppressing magnetostatic energy. See [3] and also P2.10).

The magnetocrystalline energy needs to maintain a minimum; therefore, it tends to keep atomic magnetic moments aligned along one of the easy directions of the crystal axes. The net magnetization thus follows a certain crystallographic axis, and is said to give rise to an easy axis of magnetization along it. The ferromagnet can be magnetized along that crystallographic direction without much difficulty. At least, this is the case in crystallographically textured ferromagnets. If the ferromagnetic

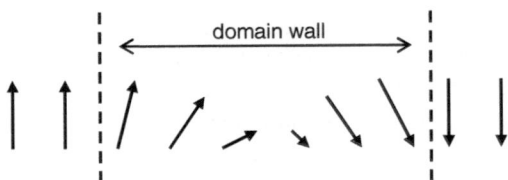

Fig. 2.2 A smoother transition to opposite domain magnetization is obtained when magnetic domain walls contain atomic magnetic moments of gradually varying orientation

materials consist of grains of random crystallographic orientation, an easy axis of magnetization is still possible, nevertheless it will be determined mostly by material processing which is know to alter magnetic domain configurations and thereby their magnetization orientation.

The transition layer known as a domain wall is usually of two types, although other wall configurations are possible, depending on the crystallographic structure of the material, as well as some processing factors. In cubic structures such as *fcc* or *bcc*, the commonly encountered types of walls are *Bloch walls*[4] and *Néel walls* [5]. In Bloch walls, atomic magnetic moments rotate outside the plane of the magnetic moments. On the other hand, Néel walls are known for their atomic moments remaining in plane while the rotation occurs. Since domain magnetizations tend to align with preferred crystallographic axes, domain walls separating domains of different orientations can be classified as 180°, 90° (iron) or 109°, 71° (nickel), depending on the angles these crystallographic axes make in a specific lattice. Some walls of different orientation occur in closure domains which are created when the material divides into magnetic domains that allow more of the magnetic flux to stay within the material. Such a configuration minimizes magnetostatic energy. Figure 2.3 shows schematically a few types of domain walls.

Magnetostatic fields and domain configurations can be imaged through colloidal suspensions termed [6] *ferrofluids*, although for quite some time, many research laboratories have been using more sophisticated techniques such as magnetic force microscopy (MFM). Nevertheless, if MFM is not readily available, ferrofluids such as Fe_3O_4 or $BaFe_{12}O_{19}$ offer a viable alternative, at least in some cases. Ferrofluids are known to be stable substances, typically formed of ~ 10 nm particles which are immersed in hydrocarbons or some other organic liquids, as water-based ferrofluids are more difficult to produce. The reader may be familiar with ferrofluids, as sometimes they are used as liquids in bearings.

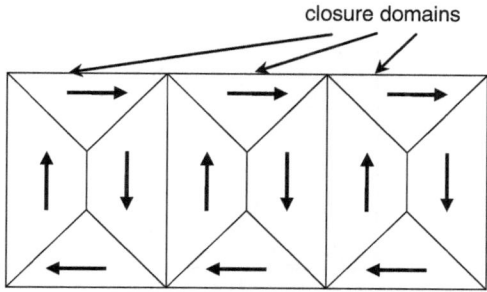

Fig. 2.3 Formation of 90° closure domains in iron. The closure domains are perpendicular to the 180° domains, illustrated here with vertical domain magnetizations

2.2   The Process of Magnetization in Ferromagnets

2.2.1   Initial Magnetization Curve

Upon the application of an external magnetic field to an initially demagnetized bulk ferromagnet, the net magnetization increases, while the rate of increase at a given magnetic field value is determined by a number of factors. Among these, we include the external field orientation with respect to each individual domain orientation, the magnitude of spontaneous magnetization, the defect structure of the material, as well as the anisotropy and geometry of that particular specimen being investigated. To start our analysis, let us have a look at Fig. 2.4, which is a schematic representation of the initial leg of a *hysteresis loop*, a closed magnetization curve typical of a bulk polycrystalline ferromagnet.

The initial region designated "1," right at the beginning of this curve is called the region of *reversible domain wall motion*. In this portion of the curve, a very small magnitude of an applied field intensity increases slightly the value of the net magnetic field density. However, upon removal of the applied field, the net value of the magnetic field density returns macroscopically to zero in this region. The recovery of a net zero magnetization is an approximation on a macroscopic scale because there is no way that the recovery can be made on a microscopic scale as well. On the other hand, in region "2" of the magnetization curve, also termed the region of *irreversible domain wall motion*, the net magnetization increases at its highest rate so that domain walls acquire new positions, and their initial place can no longer be restored. In this region, changes are so drastic that a polycrystalline ferromagnet starts to experience domain magnetization rotations.

With further increases in magnetic field intensity, the magnetization curve enters region "3" which begins even before the end of the region of irreversible domain wall motion, so that regions "2" and "3" overlap. Region "3" is therefore known as region of *irreversible domain wall motion and domain magnetization rotation*.

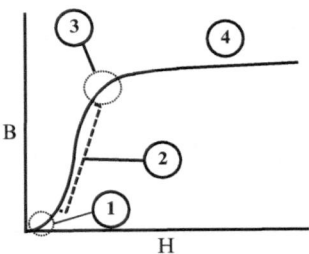

Fig. 2.4 Different regions can be distinguished on the initial leg of a magnetization curve that is typical of a bulk polycrystalline ferromagnet. They represent the region of (1) reversible magnetic domain wall motion, (2) irreversible domain wall motion, (3) reversible domain wall motion and domain magnetization rotation, and (4) only domain magnetization rotation

Nevertheless, as the applied field intensity continues to increase, the net magnetization increases solely due to *domain magnetization rotation*, entering region "4." Finally, the specimen's magnetization approaches and reaches *technical saturation*. Remember this initial description of the different regions on a magnetization curve, as the detailed effects that determine the shape of the magnetization curve will be discussed in the next section.

2.2.2 Magnetic Domain Wall Motion

The potential energy density within a magnetic domain can be described by

$$U = -\vec{M}_s \cdot \vec{B} \tag{2.2}$$

and is given in units of Joule/Volume = Newton × meter/meter^3 = Newton/meter^2. We see from this relationship that a magnetic field applied to a magnetic domain produces a uniform pressure throughout the domain. In this case, if the angle between magnetization and applied magnetic field density is different in two neighboring domains, there will be a net pressure on the wall dividing them. The net pressure (or stress) is the force causing the magnetic domain wall to move as illustrated in Fig. 2.5. The magnetic domain wall encounters barriers in its motion that need to be overcome in order for the wall to continue its motion. There are a number of mechanisms that resist domain wall motion, and these will be discussed in more detail in the following sections.

Figure 2.6 shows a schematic of domain wall motion with potential energy barriers that the wall encounters during its motion. Specifically, when a magnetic field is applied to a bulk ferromagnet, a large number of magnetic domains begin to move while facing different types of barriers.

The potential energy barriers depicted in Fig. 2.6 can be thought of as a series of barriers averaged over the entire ferromagnetic object, while the position of the wall can be considered to represent the net magnetization of the object. In this case, the domain wall motion-resisting force produced by the potential barriers is

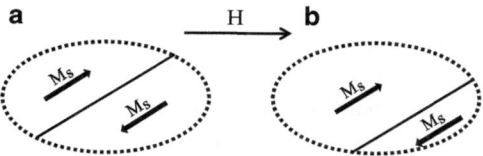

Fig. 2.5 Volume expansion of an energetically favored domain due to domain wall motion. The domain wall motion is a result of the net force on the wall. Schematic shows two domains of which only one expands at the expense of the other one: (**a**) domains in the absence of the force, and (**b**) domains expand/contract upon application of a field which leads to the exerted force

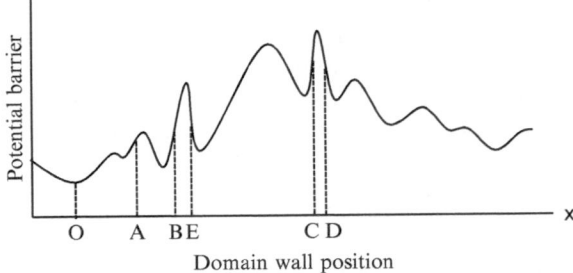

Fig. 2.6 Schematic view of potential barriers created by lattice defects present in a crystalline material. The force resisting the magnetic domain wall motion is proportional to the slope at a given point on the potential barrier curve

$$F(x) = -\frac{dV(x)}{dx}, \tag{2.3}$$

where $V(x)$ is the potential produced by the presence of lattice defects. Expression (2.3) for the domain wall motion–resisting force is actually the slope of the potential barrier at a given location. An externally applied magnetic field-induced force of (2.2) must exceed that of (2.2) in magnitude in order to cause the magnetic domain wall to jump over a barrier and resume its motion.

Let us consider at this moment the domain wall positions indicated in Fig. 2.6 in relation to the magnetization curve of Fig. 2.4. Initially, in the totally demagnetized state, the domain wall is located at position labeled O. The application of a small external field moves the wall to the right of the curve, a condition we choose. It then climbs up the potential barrier near the position labeled O, but at some distance from it. If the external magnetic field is removed at this point, the domain wall will return to position O. This is the region indicated as "1" in Fig. 2.4. Of course, returning to the initial demagnetized state is only meant in a macroscopic sense, i.e., a state that represents all the individual domain wall positions averaged over the entire sample, not a state valid for individual domain walls. In case a sufficiently strong external field is applied to the specimen, the domain wall will jump over to position A. With a slight increase in the applied field, it jumps to position B, and then to position C.

The distance between a given domain wall position and O is proportional to the net magnetization of the sample. Hence, a sudden increase can occur in the net magnetization due to a relatively small increase in the applied field. The domain wall movement between B and C emphasizes the fact that *the critical factor is the slope and not the height of the potential barrier*. Removing the applied magnetic field at this stage does not bring the domain wall back to O. The process of net magnetization increase at this stage corresponds to the "irreversible domain wall motion" region indicated in Fig. 2.4. The resistance against the domain wall motion is frequently called "pinning," and a location such as C is called a "pinning site."

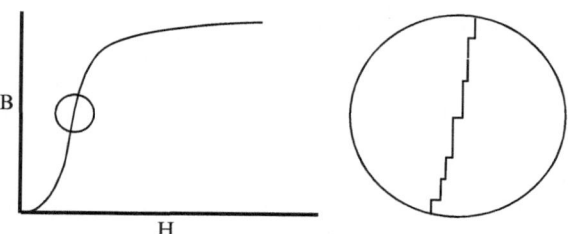

Fig. 2.7 An irregular step-like pattern is seen in a magnified magnetization curve within the region of irreversible domain wall motion. These discontinuities are due to sudden jumps of domain walls over potential barriers

Once domain walls are driven to a location that is slightly left to B, an increase in the applied field is needed to move the walls past B. There is hardly any increase in magnetization until the wall reaches B; therefore, the slope of the B–H curve is flat. Once the wall passes B, it abruptly moves to a location in the vicinity of C without any further increase in the applied field. The slope of the B–H curve is in this case practically infinite. Region 2 in Fig. 2.5 is the cumulative effect of a large number of such abrupt irreversible domain wall motions, creating step-like discontinuities in region 2 of the B–H curve, as shown in detail in Fig. 2.7. These microscopic discontinuities were first discovered by Barkhausen [7], and the next section elaborates on this phenomenon.

2.2.3 Magnetic Barkhausen Noise

We have seen in previous sections that the magnetization process is not smooth, but rather consists of random sequences of irreversible domain wall movements. This phenomenon looks and sounds like noise when detected by different experimental equipment, such as an oscilloscope or a loudspeaker. It was first described by Barkhausen in 1919, and hence its name *magnetic Barkhausen noise* (MBN), and is a phenomenon that is investigated statistically. MBN is observable in region "2" (Fig. 2.4) of the magnetization curve of a ferromagnetic material, which is also the steep part of the curve in Fig. 2.7. Quite often in experiments, MBN is detected through random voltages usually picked up by a coil during the magnetization of the material [8], and processed further through adequate instrumentation. It is termed "magnetic" to distinguish it from *acoustic Barkhausen noise*, the latter being based on magnetoacoustic emission [9, 10]. The Barkhausen noise discovery was the first experimental, and at the same time indirect verification of the magnetic domain structure of ferromagnets. The fact that this noise exists is proof that the magnetization process occurs through small discontinuous flux changes, while it was later established that it corresponds to changes in domain volume.

A non-magnetized ferromagnetic material contains a large number of magnetic domains with randomly orientated magnetizations that add up to a zero bulk net magnetization [11]. When an external magnetic field is applied, the magnetizations of the domains tend to align with the field. Domains whose magnetizations are aligned most closely with the external field increase in volume at the expense of domains less favorably oriented. When this happens, domain walls move to new locations toward adjacent domains, and the specimen becomes magnetized. In order for domain walls to move, the local coercive field needs to be overcome by the component of the magnetic field along the domain wall. This local coercive field is often associated with microscopic pinning barriers. Upon removal of the external magnetic field, domains do not necessarily revert back to their original configuration, as previously discussed. If domain walls have encountered pinning sites while moving, some energy was expended in order to overcome the pinning. If the wall has successfully passed the pinning site, it will not return to its original position once the field is no longer acting. MBN is therefore associated with this irreversible "jump" of domain walls over local obstacles acting as pinning sites, such as grain boundaries, dislocations, inhomogeneities, or other imperfections. All lattice irregularities are likely to cause delays in domain wall movement, leading to uneven and discontinuous changes in magnetization [12].

Following the discovery by Barkhausen, researchers set out in the 1920s and 1930s of the twentieth century to measure magnetization curves for single crystals of iron, and later other magnetic materials. It was observed that MBN varies with crystallographic orientation, and that magnetization takes place easier along some directions. In the decades that followed, MBN was actively studied all over the world, and data on this phenomenon started to accumulate. As such, it became known that microstructural changes, prior mechanical processing, grain size, stress state, impurities etc. have an unquestionable influence on the magnetic Barkhausen noise. For this reason, Barkhausen noise developed in time as a nondestructive method of evaluating ferromagnetic materials, and MBN techniques started to diversify and became more reliable for testing. The fascination with this phenomenon continues these days, as MBN has earned a place in the nondestructive testing industry, helping other magnetic evaluation methods to gain a better understanding of defects in ferromagnetic structures. The next chapter elaborates a bit more on this subject, when other aspects of nondestructive testing are also discussed.

2.2.4   Magnetization Rotation

Magnetic domain wall motion ceases to exist when a wall reaches the surface, or strong pinning sites such as grain boundaries, or an array of inclusions. Upon the application of a stronger field, the magnetization vector in a domain will then orient away from the easy axis of magnetization in a grain, and against the anisotropy energy (Fig. 2.8). In a polycrystalline sample, the magnetization in certain grains is oriented closer to the direction of the applied field than in other grains. As a result,

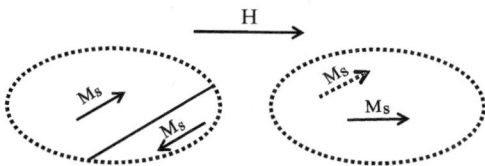

Fig. 2.8 Domain magnetization rotation against the anisotropy energy, upon reaching the end of the region of magnetic domain wall motion

the usual transition region "domain wall motion-magnetization rotation" extends further over a wider region of the applied field strength. The region in question corresponds to that indicated as (3) in Fig. 2.4. Beyond this region, the increase in the net magnetization is solely due to domain magnetization rotation, i.e., the region indicated as (4). The domain magnetization vector reorientation accompanies the domain along its motion, as well as the strain field localized next to the domain wall region which also travels together with the domain.

2.3 Mechanisms of Resistance to Domain Wall Motion

2.3.1 Lattice Strain Fields

Any kind of lattice defects such as inclusions, precipitates, dislocations, or grain boundaries can offer resistance to the motion of magnetic domain walls. The magnitude of local lattice strain fields can be severe in the vicinity of a lattice defect, and the elastic energy associated with this strain field can produce a substantial influence on domain wall motion. As a magnetic domain wall approaches an inclusion, the lattice structure within the domain wall is forced to conform to the deformed lattices near the defect. Therefore, the interaction between a magnetic domain wall and a defect is repulsive.

An example of a lattice strain field in the vicinity of a coherent inclusion which is embedded in a ferromagnetic matrix is given in Fig. 2.9a. The lattice parameters of the inclusion in this example are smaller than those of the matrix. Despite the deformation, crystal planes in the matrix run through the inclusion without discontinuity, and the relation between the matrix and the inclusion is "coherent." Lattice strain fields are also produced by grain boundaries, dislocations and various types of point defects. Figure 2.9b depicts the bending of a domain wall by the lattice strain field-induced resistance in the vicinity of the inclusion.

To estimate the magnitude of the domain wall resistance by an inclusion, we will first investigate the local lattice strains in the vicinity of the inclusion. In order to discuss a realistic situation, the experimental results of Tirry and Schryvers [13] are

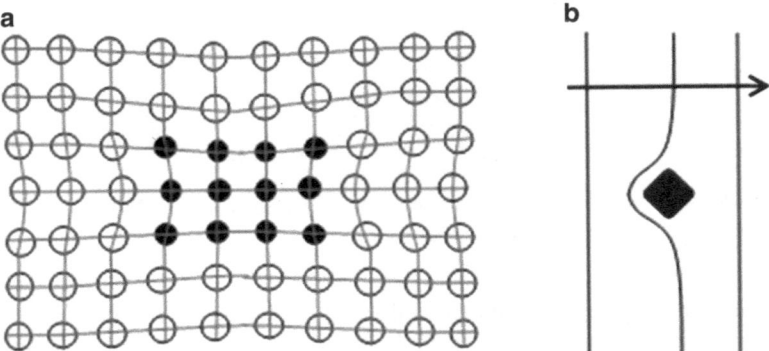

Fig. 2.9 (**a**) Schematic illustration of a lattice strain field in the vicinity of an inclusion with lattice parameters smaller than those of the matrix. (**b**) Bending of a domain wall by a hypothetical inclusion as the wall moves from left to right

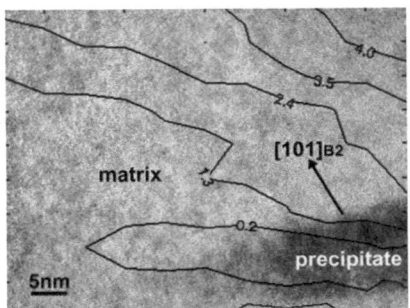

Fig. 2.10 Experimentally determined lattice strain field in a NiTi matrix due to a Ni_4Ti_3 precipitate that forms a coherent interface. The precipitate, indicated in darker gray is in a shape that is close to an ellipse. The strain values with respect to the interplanar spacing of unstrained matrix are obtained by subtracting 2.01%. Reprinted with permission from [13]

considered. In their work, a Ni4Ti3 precipitate[1] was grown in a NiTi matrix. The resulting strain field distributions in the vicinity of the precipitates were measured quantitatively using a high-resolution transmission electron microscope (HRTEM). The matrix has a cubic B_2 structure which consists of two simple cubic primitive cells, one occupied by Ni, and the other by Ti. Two simple cubic cells penetrate each other such that a Ni atom is located at the center of simple cubic unit cell of Ti and vice versa. The results are shown in Figs. 2.10 and 2.11, with the lines in Fig. 2.10 representing particular strain values.

---

[1] A precipitate is formed by a thermodynamic process of fast cooling, and it can be regarded as a particular form of an inclusion which is a general term.

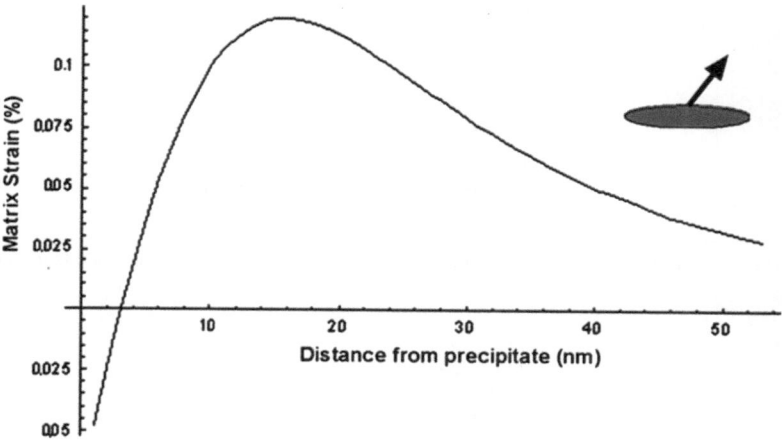

Fig. 2.11 The computed strain field due to Ni_3Ti_4 in the NiTi matrix along the [101] direction. Compressive strain is present in the vicinity of the precipitate since its planar spacing is smaller than that of the matrix. Reprinted with permission from [13]

These strain values are the percentage difference between the (101) interplanar spacing in the matrix and that of a corresponding plane in the precipitate, the latter being the reference. The actual strain values in the matrix are obtained by subtracting 2.01%. These values are referenced to the (101) planar spacing in the unstrained matrix. Thus, the percentage difference is between the (101) interplanar spacing of the unstrained matrix and the spacing of the corresponding plane of the precipitate. This means that the area in the matrix close to the precipitate is under compression.

The mechanical aspects involving the strain field near an inclusion are a very complex issue, and the details are beyond the scope of this book. However, we can use a simplified model in order to obtain some physical insight. First, we assume that the precipitate is spherical and its diameter is much larger than the magnetic domain wall thickness. Second, we assume that the strain distribution is an isosceles triangle as shown in Fig. 2.12. A simplified strain field in the vicinity of the precipitate is displayed in Fig. 2.13. See the following example for details of this simplified model and its results.

Example 2.2. Show that the strain amplitude is

$$s_{max}\left(\frac{d-r+2d_1}{d_1}\right) \text{ at } (x', y') \quad \text{when } r \geq d + d_1 \qquad (E2.2.1)$$

and

$$s_{max}\left(\frac{r-d}{d_1}\right) \text{ at } (x', y') \text{ when } \quad r \leq d + d_1, \qquad (E2.2.2)$$

where $r = \sqrt{y'^2 + z'^2}$.

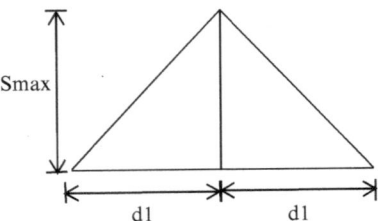

Fig. 2.12 A simplified distribution of strain in the matrix due to the presence of a precipitate

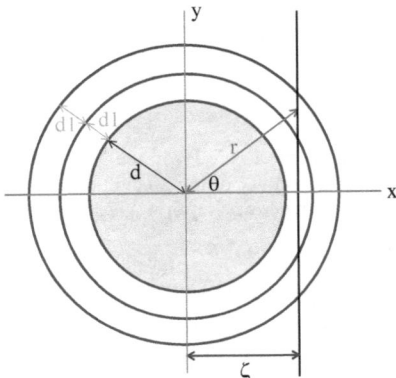

Fig. 2.13 A model for strain distribution around a spherical precipitate. The plane at ζ is the domain wall with a negligible thickness compared to the diameter of the precipitate which is denoted by d. The total range of strain is twice of d_1 and the distribution is radially symmetric about the circle with the radius of $d + d_1$

Answer: The energy density of a strain field is $\frac{1}{2}\sigma\varepsilon$, where σ is stress and ε is strain. The purpose here is to find out the location along the y-axis where the strain field energy is maximum as a domain wall travels against and over the precipitate (Fig. E2.2.1).

For a given value of y in the region $d + 2d_1 \geq r > d + d_1$, the strain energy within a domain wall of very small thickness is proportional to the quantity

$$S_{max}^2 2\pi\delta \int_0^z \max \left[\frac{d + 2d_1 - (y^2 + z^2)^{\frac{1}{2}}}{d_1} \right]^2 z dz, \qquad (E2.2.3)$$

where $z_{max} = \sqrt{y^2 + (d + 2d_1)^{\frac{1}{2}}}$ and δ is the domain wall thickness.
In the region $d > r \geq d + d_1$ at a given value of y, the strain energy in the domain wall will be proportional to the following two integrals,

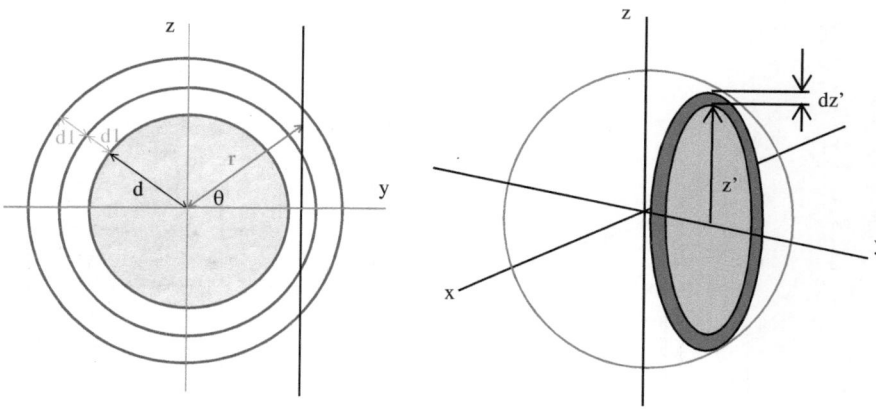

Fig. E2.2.1 Model for a strain distribution around a spherical precipitate. The diameter of the precipitate is denoted by d

$$S_{\max}^2 2\pi\delta \int_{z_{\text{int}}}^{z_{\max}} \left[\frac{d + 2d_1 - (y^2 + z^2)^{\frac{1}{2}}}{d_1}\right]^2 z dz + S_{\max}^2 2\pi\delta \int_0^{z_{\text{int}}} \left[\frac{(y^2 + z^2)^{\frac{1}{2-d}}}{d_1}\right]^2 z dz,$$

(E2.2.4)

where $z_{\max} = \sqrt{y^2 + (d + 2d_1)^{\frac{1}{2}}}$ and $z_{\text{int}} = \sqrt{y^2 + (d + d_1)^{\frac{1}{2}}}$.

Finally, for $r < d$, we have the strain energy proportional to

$$S_{\max}^2 d\pi\delta \int_{z_{\text{int}}}^{z_{\max}} \left[\frac{d + 2d_1 - (y^2 + z^2)^{\frac{1}{2}}}{d_1}\right]^2 z dz + S_{\max}^2 2\pi\delta \int_{z_{\min}}^{z_{\text{int}}} \left[\frac{(y^2 + z^2)^{\frac{1}{2-d}}}{d_1}\right]^2 z dz,$$

(E2.2.5)

where $z_{\min} = \sqrt{y^2 + d^2}$. Results of the calculations of this strain energy density vs. distance from precipitate are shown in Fig. E2.2.2.

2.3.2 Nonmagnetic Inclusions

Another possible mechanism for the occurrence of resistance to magnetic domain wall motion has its origin in the reduction of magnetic energy. As mentioned earlier, a domain wall possesses its own energy; therefore, the reduction of the domain wall energy is accomplished by reducing the domain wall area as shown in Fig. 2.14 for a nonmagnetic spherical inclusion. Free poles become attached to the inclusion, since the latter is totally enclosed within the body of the domain. A change in the distribution of the free poles occurs, and thereby a reduction in the magnetostatic energy takes place when the domain wall intersects the inclusion. Furthermore, the surface area of the domain wall is reduced, causing also a reduction in the surface

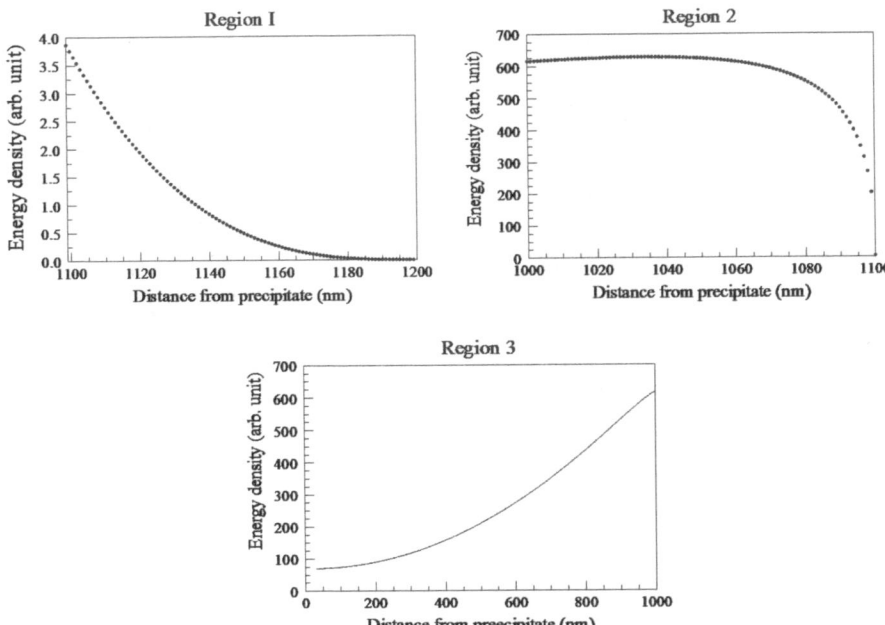

Fig. E2.2.2 Computed energy density of a strain field vs. distance from precipitate

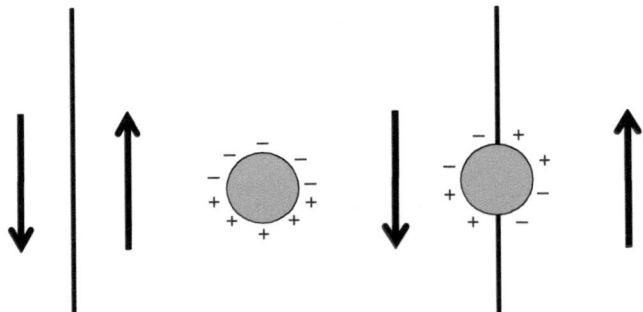

Fig. 2.14 Reduction of domain wall area in the presence of a nonmagnetic spherical inclusion. This process reduces domain wall energy

energy of the domain wall. Because the domain wall energy is reduced when the wall intersects an inclusion, the wall becomes pinned to the inclusion. Consequently, the magnetic field necessary for the movement of the domain wall is higher than that required in the absence of the inclusion.

Nickel is one of the most investigated ferromagnets, and as such, it is not surprising that detailed studies about the effects of nonmagnetic inclusions on domain wall motion have been undertaken on this material. For instance, a group of researchers [14] reported results on the effects of two kinds of nonmagnetic inclusions, alumina particles and voids, since structure-sensitive magnetic properties are influenced by these types of nonmagnetic particles. Voids are a particularly interesting type of nonmagnetic particle given that many magnetic materials exhibit some porosity which needs to be evaluated nondestructively. Alumina is a commonly used dispersion for improving high temperature mechanical properties of nickel. The researchers worked in this study with nickel compacts that either had voids or varying amounts of alumina present, in order to determine their influence on various magnetic parameters. Both voids and alumina particles were expected to affect the movement of domain walls during magnetization, by providing resistance to their motion, thereby influencing the hysteresis properties of the material.

Coercivity is usually one of the measured parameters because some theories assert that the largest effects are observed when the inclusions have a diameter about equal to the thickness of the domain walls, which is often the case. However, other opinions state that only particles about an order of magnitude larger result in significant pinning for domain wall motion. In any case, the presence of alumina particles in nickel introduces compressive strains in the vicinity of the particles, similar to the ones discussed in the previous section. Furthermore, these strain fields introduced by the alumina particles cause hindrances to dislocation movement, a fact which does not happen in the presence of voids. Thus, alumina particles and voids do not have the same effects on domain wall movement, impeding with their motion through different mechanisms. A distinction between the effects of particles and voids with respect to magnetic properties would be useful in certain magnetic nondestructive testing techniques used for creep-damaged materials where void nucleation is a critical stage in the initiation of failure, while fatigue damage due to particles acts as another factor responsible for failure of these materials.

The referenced study [14] found that the addition of alumina particles inhibited the grain growth during the sintering of the nickel compact, in addition to the clustering and agglomeration of the particles caused by electrostatic discharge during mixing. Nevertheless, the main factor responsible for the observed increase in coercivity was the presence of second-phase particles which acted as strong domain wall pinning sites, thereby causing an irreversible motion of domain walls. Thus, increased amounts of alumina present resulted in an almost linear increase in the coercivity. A similar effect on coercivity was noticed with an increase on the amount of voids which also acted as pinning sites for domain wall motion. It seems that for the same volume fraction of inclusions, alumina was more effective in increasing the coercivity than voids. Conversely, the initial permeability was reduced due to the increasing amounts of pinning sites which cause the domain walls to bend less when the initial magnetic field is applied. The initial permeability is one of those magnetic parameters that responds immediately to an applied field since it is measured at the very beginning of the magnetization cycle. Remanence, on the other hand, is a magnetic property measurable during the demagnetization process, thus any changes due to the presence of inclusions are observable later on the hysteresis curve. As such, inclusions cause demagnetizing fields which in turn decrease remanence. However, the larger number of pinning sites causes a simultaneous increase in remanence. The effect of the two factors is not the same, as the demagnetizing fields play only a minor role at low fields, whereas the increased number of pinning sites affects domain wall motion, the dominant magnetization mechanism around the coercive point. During demagnetization, domain wall motion is not as pronounced around the remanence point where demagnetizing fields become more important.

Example 2.3. Comment on the magnetic domain structure and the influence this has on the magnetic properties of iron meteorite. See for instance the study reported in [15]

Answer: Iron meteorite may not be a material of choice for magnetic device applications, nevertheless it has its place in the study of the solar system. As such, planetary scientists may have an interest in the Widmanstätten structure that was formed by the long range diffusion of Fe and Ni in an asteroid's core over a period of 4.6 billion years [16]. This remarkable formation process has resulted in unusual magnetic properties not observable in synthetic Fe–Ni alloys [15]. It should be noted that the Widmanstätten structure is regarded as Fe–Ni alloy segregated α (bcc Fe–Ni) and γ (fcc Fe–Ni) lamellae on a micrometer scale where the Ni concentration in the γ lamellae increases rapidly toward the interface [16]. The Widmanstätten structure is assumed to be responsible for the large magnetic anisotropy and strong coercivity displayed in iron meteorites which are a special type of Fe–Ni alloy. In this unique alloy, a tetrataenite phase is formed, described as a chemically ordered Fe–Ni alloy with $L1_0$-type superstructure. This phase was found to have an unbelievably large coercive force, and also a magnetic anisotropy energy that is one order of magnitude larger than that of Fe, permalloy, or pure Ni. As such, iron metoerite is a hard ferromagnet with a strong anisotropy, in spite of the fact that common Fe–Ni alloys are soft ferromagnetic materials. A transversal striped domain structure was observed in the α lamellae of the iron meteorite material [15], in contrast to the wide rectangular domain structure with a sharp domain wall, normally displayed by bcc-Fe. Over the α/γ interface, a fine elongated domain structure was observed parallel to the $[110]_{bcc}$ direction, with a parallel or antiparallel magnetization along that direction. The magnetizations on both sides of the interface aligned opposite to each other, and orthogonal to the domain wall, such that a large amount of magnetostatic energy would be required to demagnetize the material. Normally, in a typical $180°$ domain structure, the magnetization aligns parallel to the wall in order to reduce the magnetostatic energy. This was not the case for the iron meteorite which displayed a completely different domain configuration as compared to the common Fe–Ni alloy. Unfortunately, in order to exploit the magnetic properties offered by iron meteorite due to its unique magnetic domain configuration, researchers would have to be capable of reproducing the formation conditions of the iron meteorite material in the asteroid's core, a task not easily accomplished.

2.3.3 Artificial Pinning and Depinning

Field-driven and current-driven artificial pinning and depinning have become topics of interest for many research groups around the world, and a large number of publications have thus appeared. Among the many studies being currently undertaken, the ones of most significance for our discussion are those dealing with magnetic systems of nanoscale dimensions and the controlled depinning of domain walls contained within them. Magnetic domain walls have recently attracted renewed attention in view of contemporary high density magnetic storage applications where controlled wall movement is highly desired. Particularly, the manipulation of domain walls shows great potential in logic and magnetic storage devices which require artificial pinning sites in dense arrays of ferromagnetic systems. Controlled movement of domain walls along nanowires is a subject that will be discussed in more detail in Chap. 8. The pinning of domain walls is a necessary step in controlling the position of the domain wall along the nanowire. Unfortunately, magnetic domain configurations can become unstable at small separation distances where domain walls can depin without an externally applied field. For instance, transverse magnetic domains in different configurations have been reported [3] to interact by attracting each other when present in adjacent nanowires, due to the existence of a nonzero spontaneous depinning field. Nevertheless, the mechanisms of such spontaneous depinning processes are not clear, and they warrant further elucidation if external control over pinning and depinning of domain walls is wanted, especially in closely packed magnetic systems.

Consider, for instance, nanostrips of a magnetic material such as permalloy, with well-defined notches in carefully chosen places along the strip (Fig. 2.15). These are sites for artificial depinning of domain walls. Concurrently, there is a magnetostatic interaction between domain walls of neighboring permalloy nanostrips which depends on the separation between the strips. For this reason, we would like to know the effect this interaction has on the artificial depinning, especially since nanostrips are usually closely packed. A recent report [17] contains details of such an investigation of the magnetostatic interaction between domain walls in neighboring ferromagnetic nanowires, and its effect on artificial domain wall pinning in preselected notches.

The permalloy nanowires and the domain walls within them were simulated micromagnetically in order to predict the influence of the magnetostatic interaction on the depinning field and depinning mechanism. Wires of 20 nm thickness were considered, of the simplest non-trivial minimum energy configuration. This configuration consisted of a magnetic structure with a transverse domain wall pinned in the middle of the notch, similar to Fig. 2.15. Given the orientation of the domain wall and the adjacent magnetic domains, four subconfigurations were taken into account: head-to-head and tail-to-tail with antiparallel chiralities (Fig. 2.15), head-to-head and tail-to-tail with parallel chiralities, head-to-head and head-to-head with antiparallel chiralities, and head-to-head and head-to-head with parallel chiralities. The magnetostatic interaction was assumed to occur due to volume and surface

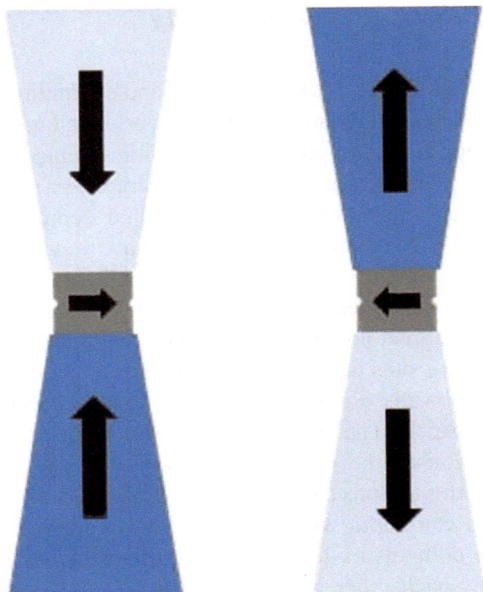

Fig. 2.15 Schematic of two neighboring magnetic nanostrips with double symmetrical notches at specific sites along the strip. Depicted are head-to-head and tail-to-tail magnetizations in two domains within the same nanostrip, and a pinned domain wall in the middle of the notch, similar to the experiment described in [17]

charges, and these two sources were considered in the calculation. Nevertheless, given the two sources are quite different, they result in different effects. As such, surface charges resulted in a dipolar field connected with the transverse domain wall, whereas volume charges result only in a monopolar field [17]. The dipolar field acts on the transverse domain wall in the middle of the notch as a homogeneously magnetized, isolated particle with a dipole field polarity given by the domain wall orientation which favors a parallel alignment of in-plane magnetized transverse walls. On the other hand, the volume charges and the associated monopolar field act on the domain wall depending on its type rather than its orientation. For this reason, two domain walls of the same type have volume charges of equal sign, thereby repelling each other. Consequently, the magnetostatic interaction between transverse domain walls in neighboring nanowires is a result of the complex interplay of charges of different sign and strength, originating in two different sources.

The researchers [17] found that an applied field along the axis of the nanowires resulted in head-to-head and tail-to-tail domain walls moving in opposite directions, where the head-to-head and tail-to-tail with antiparallel chiralities, and head-to-head and tail-to-tail with parallel chiralities domain wall subconfigurations show higher depinning fields due to the monopolar field attraction between domain walls of different type. The depinning was calculated to occur at the same field value for both nanowires, thus in a single step. Conversely, the remaining two domain wall subconfigurations displayed a two-step depinning process where depinning in one wire occurred at a field value lower than the field expected in an isolated wire. Similarly, the other domain wall was depinned at a field typical for a single wire, indicating it was not influenced by the neighboring wire. Unfortunately, these subconfigurations were not found to be stable enough for small distances between the wires, as the magnetostatic field was sufficiently strong to depin the domain wall from the notch. Overall, the depinning mechanism was found to be determined by the domain wall type and not by the relative chirality, and as such it was due to volume charges. Furthermore, the magnetostatic interaction on account of the volume charges reduces the energy barrier for pairs of equal domain wall type, and increases it in the case of different wall types. As such, the pinning potential of a domain wall at a notch is highly influenced by the magnetostatic field of a domain wall in an adjacent wire, where the magnetostatic interaction can be attractive or repulsive, depending on domain wall configuration. This altered depinning mechanism is expected to affect the stability of devices operating on the basis of domain wall propagation, especially when a domain wall in a neighboring wire can be depinned even in the absence of an applied magnetic field. (*Note:* See also Problem P2.14 for another study on domain wall depinning in permalloy nanowires which emphasizes lowering the threshold in current-driven artificial depinning [18]).

Tests, Exercises, and Further Study

P2.1   What are magnetic domain walls? What are wall thickness and energy density estimates, and what are they saying about the structure of the wall?

P2.2   Which magnetic properties are structure-sensitive, and which ones are not? To which magnetic property changes is MBN responsive?

P2.3   What distinguishes acoustic Barkhausen noise from magnetic Barkhausen noise?

P2.4   How do grain boundaries affect magnetic domains and their walls?

P2.5   Microstructural features such as grain size, dislocations, and precipitates influence the magnetic as well as mechanical properties of ferromagnetic materials. The effect of dislocations on MBN is discussed in the next chapter. Grain size has only a secondary bearing on the bulk magnetic properties of ferromagnets. Comment on the influence of precipitates on MBN as they are responsible for most of the variations in magnetic properties as seen through MBN behavior.

P2.6   Comment on the effects of an increase or decrease in temperature in the region of a domain wall.

P2.7   Acoustic Barkhausen noise is believed by many research groups to be based on processes of nucleation and annihilation of residual and closure domains occurring at the "knee" of the magnetization curve. Unlike magnetic domains responsible for MBN, these domains are non-180°. Give an example of a study of acoustic Barkhausen behavior around a complex type of defect, such as the fracture site of a steel sample. *See for instance* [19].

P2.8 Spintronics is a new branch of electronics that relies on spin rather than charge of the electron for their devices to function. Some later chapters in this book discuss a few basic topics of interest related to this category of devices. Magnetic domain wall motion in ferromagnetic wires is an intensely studied topic in contemporary magnetism because of its important potential applications to spintronics, including information storage and logic circuits. Report on some results regarding current-driven domain wall motion, as opposed to magnetic field driven wall motion. *See, for instance, [20] where current-driven domain wall motion is put to the test in a structured Co/Ni wire with perpendicular magnetic anisotropy.*

P2.9 When a domain wall is pinned by defects, its position at the pinning center becomes metastable in an external magnetic field, so that if the barrier height is low enough, the wall tunnels out of the local energy minimum. These quantum phenomena in mesoscopic systems have led to new nanodevice applications that control domain wall motion, and are based on the possibility of modulating the wall pinning potential. As such, it has been proposed to asymmetrically modulate the exchange coupling along a ferromagnetic/nonmagnetic/ferromagnetic multilayer nanowire with quantum interference in order to achieve asymmetric propagation of a domain wall. It should be noted that quantum interference occurs because the exchange coupling between ferromagnetic films separated by a nonmagnetic metal layer does not decrease monotonically with increasing spacer layer thickness, but exhibits an oscillatory dependence on thickness of ferromagnetic/antiferromagnetic succession and coupling. This oscillatory dependence is due to exchange interactions brought about by the spin-dependent confinement of electrons which leads to quantum interference as a function of spacer layer thickness. Comment on some reports of experimental investigations of asymmetric propagation of a domain wall in single multilayered wires whose spacer layer has a thickness gradient resulting in asymmetrical coupling from side to side. *See, for instance, [21].* Some of these concepts (or rather their applications) are discussed in more advanced chapters in this book. This problem serves as an example of an early introduction of the reader to applications of topics such as ferromagnetic/antiferromagnetic coupling, tunneling through a barrier, and confinement of electrons.

P2.10 Mn-based oxides with a perovskite structure are of interest in colossal magnetoresistance (CMR) applications (e.g., magnetic random access memories, tunneling magnetoresistance heads), and are therefore studied by many research groups. Their performance relies on the dynamics of magnetic domains; consequently, domain dimensions and structure in these materials are considerably investigated with various imaging tools. Comment on some of these imaging techniques and what they reveal about magnetic domain configurations in perovskite materials. *See, for instance [22], for details and some interesting figures.*

P2.11 For those spintronic devices that operate on domain wall motion, it is important to know the fundamental principles governing the dynamics of domain wall motion. Report on some results of studies of experimental observations regarding domain wall dynamics while staying within the conceptual domain wall motion framework of this chapter. There is no need to elaborate on more advanced topics in spintronics, or go into details about the magnetism of the effects on which these techniques are based. The goal of this exercise is to point out to the reader that domain wall motion is a complex event that is stochastic in nature. *See, for instance,* [23] *and references therein; however, other related publications can be considered.*

P2.12 How can you measure actual lattice deformations in the matrix around the Ni_4Ti_3 precipitates discussed in this chapter, and what technique do you use to determine the relative differences in interplanar spacings? Give a few details. *See* [8].

P2.13 The magnetic material $Sm(Co,Fe,Cu,Zr)z$ has shown permanent magnet capabilities, especially the precipitation-hardened compound with low Cu content. Comment on the magnetic properties of this material, on reasons why these are particularly observable at room temperature. *See, for instance,* [24].

P2.14 See also P2.11. Some spintronic devices are based on the spin torque effect in order to obtain a spin polarized current. Some stages of operation in these devices require the motion of domain walls which is strongly dependent on at least a couple of parameters of the spin torque effect. These days, many research laboratories are dedicated to investigating electric current-driven domain wall movement rather than magnetic field induced. The wall movement occurs above a threshold current value, and it is necessary that the value of this current is lowered. As such, many groups study the dependence of the necessary threshold current for domain wall motion and conditions of depinning from a pinning potential. A recent study performed on Gd-doped permalloy wires shows that an increased Gd concentration results in a marked reduction in threshold currents due to an enhanced non-adiabatic spin-transfer torque. Comment on these results as they relate to domain wall motion topics discussed in the present chapter, without giving details about how this is applied to the functionality of spintronic devices. The whole point of this exercise is to show the reader that pinning and domain wall motion are very much subjects of interest in today's advanced magnetic applications. *See* [18] *for further details.*

P2.15 Ultrathin films and quasi-two-dimensional ferromagnetic materials display fascinating magnetic properties not found in their bulk counterparts. Comment briefly on a recent experimental finding known as a *spin reorientation transition (SRT)* in thin ferromagnetic films. How can such a transition be induced and what are its consequences? *See, for instance,* [25].

P2.16 Comment briefly about differences in magnetic properties between the two structural phases of Fe, fcc, and bcc.

P2.17 What kind of domain walls are most likely to contribute to MBN and why?

P2.18 What is the typical energetic distance over which pinning barriers for domain wall motion act?

P2.19 A field applied to a ferromagnetic material induces a magnetization within it
by movement of domain walls that exist within domain structures. Where is
MBN most active and why?

P2.20 Name three factors that are attributed to modifying magnetic properties of
steel materials when a stress is applied below the yield limit. *See for instance*
[26–30].

Hints and Partial Answers to Problems in Chap. 2

*The answers to these problems are in many instances just a hint as to where further
information is to be found. They contain the main idea, but the reader needs to
find the references that are sometimes recommended for each exercise and complete
the answer through further bibliographic reading and study. Reference numbering
follows the one used in that particular chapter.*

P2.1 A domain wall is an interface between regions in a ferromagnetic material
where the domain magnetization has different orientations. As such, within
the wall, magnetic moments change directions. This change is not abrupt.
Just like the magnetic domains themselves, the domain walls are a consequence of an interplay of energies trying to reach a minimum value. Domain
walls are several hundred $\mathring{A}$ thick (e.g., $300\mathring{A}$ in iron) and have energy
densities of a few hundred nJ/cm^2, which is quite large for such a small
entity. The energy density within a wall is so large because the wall is very
thin. It is about five times the crystal anisotropy energy density of a $< 111 >$
saturated iron crystal.

P2.2 Structure-sensitive magnetic properties are coercivity, susceptibility, hysteresis loss, and remanence. Structure-insensitive magnetic properties are
saturation magnetization, saturation magnetostriction, magnetic moment per
atom, Curie point and some types of magnetic anisotropy. MBN is sensitive
to changes in structure-sensitive magnetic properties.

P2.3 Acoustic Barkhausen noise is caused by changes in magnetostrictive strain
that come about with the irreversible motion of non-180° domain walls. On
the other hand, MBN is caused by abrupt changes in magnetization with an
applied magnetic field due to the irreversible motion of 180° domain walls.

P2.4 Grain boundaries affect magnetic domains when the latter form within a
grain with a different orientation than in the neighboring grain. This is
because the two neighboring grains can have a different crystallographic
orientation. Magnetic domain walls give rise to MBN when an applied
magnetic field causes domain walls to become pinned at the grain boundary.
In this case, the center of the domain wall bows out, and as the field is
increased, the wall is unpinned and moves irreversibly within the grain. Some
theories assert that closure domains are generated at grain boundaries when
the crystallographic orientation changes across the grain boundary. As the
magnetization changes, closure domains act as nucleation sites for both 180°

and non-180° domain walls, affecting both MBN and acoustic Barkhausen noise.

P2.5 In a ferromagnet such as steel, the most important compositional factor is the amount of carbon and its distribution, present mainly as Fe_3C precipitates. However, these pinning sites such as Fe_3C precipitates impede the motion of domain walls, decreasing permeability or susceptibility, and at the same time, increasing coercivity and hysteresis loss. Thus, magnetic properties such as susceptibility, coercivity, and hysteresis loss are largely dependent on the number of pinning sites present within the ferromagnet. Increased carbon content causes higher coercivity and hysteresis loss by increasing the amount of precipitates and reducing the grain size, thus increasing the number of pinning sites which affects MBN.

P2.6 When the temperature increases or decreases in a ferromagnetic material, it leads to a change in the magnetic energy conditions in the region of a domain wall. The change in energetic conditions can be attributed to the fact that the magnetic properties of the ferromagnet depend on temperature. These magnetic properties include saturation magnetization, crystal anisotropy constant, and saturation magnetostriction. Also, the thermal expansion of the polycrystal leads to changes in anisotropy and the distribution of stresses, while magnetically active defects change their local distribution. When the Curie temperature is reached, entering the transition from the ferromagnetic state to the paramagnetic state, the domain wall structure is annihilated.

P2.7 *From the indicated reference:* A fracture site displays a complex distribution of stresses that can be characterized magnetically through either MBN or acoustic Barkhausen emissions. In this case, the researchers compare both types of characterization. Results of the acoustic Barkhausen measurements suggest that domain nucleation occurs through an enlargement of residual domains in the vicinity of the fracture, due to the pinning of domain walls. A fracture region is work-hardened and contains numerous defects, such as cracks and voids. These provide additional sites for the formation of closure domains which are usually confined by non-180° walls. Previous studies show that defects produce a large demagnetizing field, enhancing domain nucleation. At maximum magnetic field, the confined residual domains are not completely annihilated due to the strong demagnetizing field surrounding the fracture site. Due to incomplete closure domain annihilation at saturation, a reduction in acoustic Barkhausen emissions is noticed. Results of this study show that the signals due to acoustic Barkhausen emissions are sensitive to stress. A good agreement with MBN data is found.

P2.8 *See indicated reference.* The authors demonstrate that single current pulses can control precisely the position of the domain wall in a structured Co/Ni wire with perpendicular anisotropy. The wall position is defined by notches in the wire, while the wall moves from notch to notch with a velocity of 40 m/s. Observations are done by MFM. The perpendicular magnetic anisotropy is useful as far as lowering the threshold current density is concerned. This threshold current is needed for the wall to overcome the pinning field.

P2.9 *See indicated reference.* A three layered wire of $Fe_{19}Ni_{81}/Au/Fe_{19}Ni_{81}$ was grown on thermally oxidized silicon substrates using electron beam (e-beam) lithography, ultrahigh vacuum e-beam evaporation to create a wedge-shaped 0–12 nm Au spacer layer, as well as 50 nm and 10 nm thick ferromagnetic layers. The magnetoresistance of the wire depends on the relative orientation of the magnetic moments of the ferromagnetic layers, and under certain circumstances, a giant magnetoresistance (GMR) effect can be observed. The latter will be discussed more in Chap. 6. It suffices to say for now that the magnitude of the GMR is directly related to the exchange coupling, and that the domain wall position in the wire can be determined by the change in electrical resistance. This change was measured with a standard four-point dc technique at room temperature. The switching field for the parallel and antiparallel alignment of the magnetic moments was noticed to exhibit an oscillatory dependence that was attributed to quantum interference. Results indicated that no defects prevented domain wall propagation toward the opposite edge; however an unusual magnetization switching process was observed in the wire with the thickness gradient. The domain wall nucleated from the blunt edge and moved rapidly to the other edge indicating that the asymmetric wedge structure plays a crucial role in domain wall displacement. The domain wall also seems to stop at a position where the antiferromagnetic coupling is stable, and the energy barrier is between the local minimum and maximum. Results show unambiguously an asymmetric propagation of a single domain wall that stands in contrast to the motion that happens when overcoming pinning from a defect.

P2.10 If scanning mode atomic force microscopy (AFM) can resolve individual grains a few tens of nanometers in size, MFM can image magnetic domains by detecting their magnetization. Domain magnetization can be engineered to be in-plane or out-of-plane, depending on the technological application of that particular material. MFM can pick up whether the grains are single domain, where the domain boundaries are, and how the magnetic field is distributed across magnetic domain boundaries. In the indicated reference paper, researchers' report on low temperature (78 K) MFM probing of magnetic domain structure in granular $La_{0.6}Sr_{0.4}MnO_3$ perovskite films with thickness < 40 nm and a resolved grain structure of $\sim$50 nm. Images revealed that magnetic domains were perpendicular with either in-plane or out-of-plane antiparallel magnetization, and that their magnetic field distribution gradually inverted across magnetic domain boundaries, the transition spanning $\sim$250 nm. Results also showed that a single magnetic domain structure was formed in these grains, and that intergrain exchange coupling is weaker than intragrain coupling which is responsible for pinning magnetic domain walls. Other studies used photoelectron emission microscopy (PEEM) and observed in-plane domains varying in size between 1 and 5 μm.

P2.11 See indicated reference. Some techniques for experimental observation of domain wall dynamics include (a) measuring the average domain wall velocity through optical Kerr microscopy (see Chap. 4), (b) observing domain wall motion through measurement of the anisotropic magnetoresistance effect (see Chap. 6), (c) detecting domain wall motion electrically by using the anomalous Hall effect, or (d) examining domain wall motion through time and space-resolved detection of a domain wall by using the tunneling magnetoresistance (TMR) effect (see Chaps. 6–8). All these techniques study submicron ferromagnetic structures such as nanowires, while they demonstrate the complex and stochastic nature of domain wall motion. Most conventional techniques for domain wall motion observation are averaging techniques and give limited information about the dynamics of the motion.

P2.12 The actual lattice deformations in the matrix around the Ni_4Ti_3 precipitates are measured using an HRTEM, while the relative differences in interplanar spacings are determined by fast Fourier techniques applied to the HRTEM images obtained along two simple crystallographic zone directions of the matrix. *See [8] for further details.*

P2.13 Microstructure, the orientation of the basic crystallographic cell which is rhombohedral, as well as temperature are known to influence magnetic properties of Sm(Co,Fe,Cu,Zr)z. This compound is of particular interest to permanent magnet manufacturers. The group of researchers mentioned in the reference have undertaken a micromagnetic finite element simulation in order to investigate the magnetic properties of the compound, including the coercivity mechanism which is so important in a permanent magnet. The coercivity mechanism is controlled by domain wall pinning which is determined by the difference in domain wall energy density between the two phases. By studying demagnetization processes, they found that the pinning effect weakens gradually as the thickness of the cell boundary decreases. On the other hand, pinning strengthens gradually as the cell size decreases. The coercivity mechanism was seen to be due to differences in magnetocrystalline anisotropy between the cell phase and the cell boundary phase, and is influenced by cell alignment and temperature. Temperature increases result in a transformation at 650 K of the demagnetization mechanism from domain pinning to magnetization reversal. The Cu content plays an important role in the coercivity mechanism and its temperature dependence. See indicated reference for further details.

P2.14 A domain wall nucleation was induced during fabrication of the permalloy wires by saturating the wires with a longitudinal magnetic field along the axis of the wire, followed by a much lower reverse field. This process resulted in a domain wall being injected into the wire from an elliptical nucleation pad, and trapped at the pinning potential. It should be mentioned to the reader that injecting domain walls is a process used these days in experimental magnetic storage media. The researchers in the presently referenced paper investigated the structure of the pinned domain wall by means of X-ray magnetic circular dichroism photo-emission electron microscopy (PEEM), an advanced imaging technique which requires X-rays (and in this case a synchrotron). Pinning and depinning of the domain wall was detected by measuring the resistance of the wires. The threshold depinning current was measured as a function of longitudinal magnetic field. The permalloy wires displayed reduced pinning threshold currents for domain wall motion when two parameters that contribute to a non-adiabatic spin transfer torque were enhanced. Staying within the topics frame of the present chapter, we can say that one method for enhancing one of the parameters was by adding an impurity concentration such as Gd. It was found that only certain Gd concentrations have a significant effect on pinning potentials for domain wall motion. *See indicated reference for further details.*

P2.15 An *SRT* in thin ferromagnetic films can be induced by temperature, an applied magnetic field or by the film thickness itself. SRT is believed to be due to the formation of a stripe phase which contains ferromagnetic elongated domains of alternating direction with respect to the magnetization component perpendicular to the film. Stripe phase formation was attributed to a net magnetization perpendicular to the film being larger than the in-plane field, which is also smaller than its saturation value H_c. This results in an unstable collinear arrangement of spins which rearrange from in-plane to perpendicular to the film. A reverse rotation from perpendicular to in-plane can also occur, depending on the strength of the field component in the perpendicular vs. to in-plane direction. The origin of SRT is in the subtle interplay between the short-range exchange, anisotropy and long-range dipolar interactions between spins. *See indicated reference for further details and a mathematical treatment of the problem.*

P2.16 The difference in magnetic properties is attributed to structural changes between fcc and bcc Fe. The two phases of iron have different in-plane atomic spacings, i.e., $a_{fcc} = 2.527\,\text{Å}$ and $a_{bcc} = 2.866\,\text{Å}$. For instance, during growth of Fe crystals, the film stress experienced in the two phases is tensile in fcc, and compressive in bcc, thus magnetic domain configurations experience differences while forming. Also, with growth of the respective fcc and bcc crystals, misfit dislocations occur in distinct ways in the two lattices with different atomic distances. Periodic misfit distortions combine to create networks that develop in unique ways in fcc as compared to bcc crystals.

P2.17 The largest contribution to MBN is given by 180° domain walls because they have higher energy than other domain walls (e.g., 90° walls). Their motion occurs at lower flux densities, hence are easier to move as the specimen is being magnetized, and the domain moment rotation is over 180°.

P2.18 Pinning barriers within polycrystalline materials act within the energy landscape of microscopic grains as magnetic domain structures does not extend beyond the boundaries of a grain.

P2.19 The number and amplitude of Barkhausen jumps are at their highest values in the central region of the magnetization curve. This is also the region where they are most stress sensitive. The increased activity and stress dependence is due to the fact that 180° domain wall motion is the dominant magnetization mechanism in that particular part of the magnetization curve.

P2.20 Application of stress below yield is expected to modify magnetic properties of steel due to: reorientation of 180° domain magnetization in the direction of applied tensile stress, increase in 180° domain wall population, and modification of pinning energies with stress. *See indicated references for further details, as well as the next chapter.*

References

1. W. Heisenberg, Z. Physik **49**, 619 (1928)
2. W. Heisenberg, Z. Physik **49**, 619 (1928)
3. T.J. Hayward, M.T. Bryan, P.W. Fry, P.M. Fundi, M.R.J. Gibbs, M.-Y. Im, P. Fischer, D.A. Allwood, Appl. Phys. Lett. **96**, 052502 (2010)
4. F. Bloch, Z. Physik **57**, 545 (1929)
5. L. Néel, Ann. Geophys. **5**, 99 (1949)
6. W.-L. Luo, S.R. Nagel, T.F. Rosenbaum, R.E. Rosenzweig, Phys. Rev. Lett. **67**, 2721 (1991)
7. H. Barkhausen, Phys. Zeits. **20**, 401 (1919) cited in E. P. T. Tyndal, Phys. Rev. **24**, 439 (1929)
8. S. Chikazumi, *Physics of Magnetism* (Wiley, New York, 1964)
9. A.E. Lord, *Acoustic Emission*, in Physics and Acoustics, vol **9**, ed. by W.P. Mason, R.N. Thurston (Academic, New York, 1975)
10. M. Namkung, S.G Allison, J.S. Heyman, IEEE Trans. Ultrasonics Ferroelectrics Freq. Control **33**(1), 108 (1986)
11. D.C. Jiles, *Introduction to Magnetism and Magnetic Materials* (Chapman and Hall, New York, 1991)
12. J.C. McClure Jr., K. Schröder, CRC Crit. Rev. Solid State Sci. **6**, 45 (1976).
13. W. Tirry, D. Schryvers, Acta Mater. **53**,1041 (2005)
14. A. Ramesh, M.R. Govindaraju, D.C. Jiles, S.B. Biner, IEEE Trans. Magn. **32**(5), 4836 (1996)
15. M. Kotsugi, C. Mitsumata, H. Maruyama, T. Wakita, T. Taniuchi, K. Ono, M. Suzuki, N. Kawamura, N. Ishimatsu, M. Oshima, Y. Watanabe, M. Taniguchi, Appl. Phys. Express **3**, 013001 (2010)
16. C-Y Yang, D.B. Williams, J.I. Goldstein, Geochim. Cosmochim. Acta **61**, 2943 (1997)
17. F. Garcia-Sanchez, A. Kakay, R. Hertel, P. Asselin, Appl. Phys. Express **4**: 033001 (2011)
18. S. Lepadatu, J. Claydon, D. Ciudad, A. Naylor, C. Kinane, S. Langridge, S. Dhesi, C. Marrows, Appl. Phys. Express **3**, 083002 (2010)
19. D.H.L. Ng, C.C. Yu, A.S.K. Li, C.C.H. Lo, IEEE Trans. Magn. **31**(6), 3394 (1995)
20. T. Koyama, G. Yamada, H. Tanigawa, S. Kasai, N. Ohshima, S. Fukami, N. Ishiwata, Y. Nakatani, T. Ono, Appl. Phys. Express **1**, 101303 (2008)
21. A. Yamaguchi, T. Kishimoto, H. Miyajima, Appl. Phys. Express **3**, 093004 (2010)
22. S. Muranaka, H. Sugaya, T. Yamasaki, T. Fukumura, M. Kawasaki, T. Hasegawa, Appl. Phys. Express **2**, 063002 (2009)
23. K. Kondou, N. Ohshima, S. Kasai, Y. Nakatani, T. Ono, Appl. Phys. Express **1**, 061302 (2008)
24. R-J Chen, H-W Zhang, B-G Shen, A-R Yan, L-D Chen, Chin. Phys. B **18**(6), 2582 (2009)
25. A. V. Syromyatnikov, J. Phys. Condens. Matter **21**, 216009 (2009)
26. T.W. Krause, L. Clapham, A. Pattantyus, D.L. Atherton, J. Appl. Phys. **79**(8), 4242 (1996)
27. R.L. Pasely, Mater. Eval. **28**, 157 (1970)
28. L. Clapham, C. Heald, T. Krause, D.L. Atherton, J. Appl. Phys. **86**, 1574 (1999)
29. S. Tiito, Acta Polytech Scand. **119**, 1 (1977)
30. R. Rautioaho, P. Karjalainen, M. Moilanen, J. Magn. Magn. **68**, 314 (1987)

Chapter 3
Magnetic Nondestructive Testing Techniques

Abstract Much attention is dedicated to noninvasive magnetic evaluation of material body defects, stress configurations, or other types of environmental and age-induced flaws that may harm engineering components or entire industrial systems. With many groups trying to develop the next generation of portable and efficient testing tools, the application and re-evaluation of basic magnetic principles that make these techniques workable cannot be overlooked, especially given the race to outperform some dated characterization equipment. Thus, magnetics comes once more to the aid of many technical branches and their traditional nondestructive testing (NDT) needs, from nuclear, aerospace, or geophysical applications to modern biomedical sensors or transducers. It is hoped that a fresh new look given to some old established processes and experiments will reveal the path for future magnetic NDT developments.

Many young readers may not know, but magnetic particle inks and powders have been commonly used in the past and are still employed these days to detect a magnetic leakage field indicative of a surface crack or defect. The dry powder is most suitable for use on very rough surfaces, or for detecting flaws right underneath the surface. The best powders consist of particles of (what used to be a) small size, with a low coercivity and a high magnetic permeability. Additionally, the magnetic particles need to provide a good color contrast against the inspected surface. The overall principle of flaw detection using this method is based on the fact that the force of attraction exerted on the particles is proportional to the square of the magnetic permeability of the particle, but also the magnetic leakage field of the defect. The particles thus group in certain areas with higher flux leakage, forming an observable pattern on the surface of the body that is investigated. On the other hand, magnetic inks are suspensions of finely divided magnetic oxides in a liquid. The magnetic oxides in the liquid must also have a low coercivity to avoid coagulation once they find themselves in a magnetic field. Their particle size is about 0.5μ m, quite large these days, as compared to contemporary applications requiring only a few nanometers or below. Nevertheless, these magnetic inks with high permeability can provide visible indications of areas where magnetic flux is leaking, thus bringing a useful, quick, and cost-effective contribution to magnetic NDT.

C.-G. Stefanita, *Magnetism*, DOI 10.1007/978-3-642-22977-0_3,
© Springer-Verlag Berlin Heidelberg 2012

As diverse as the magnetic nondestructive industry has become over the decades, our attention is focused here particularly on techniques based on micromagnetics topics discussed in previous chapters, thus building on an already laid foundation, with the goal of reaching the nanorealm of applications discussed later in this book. As such, MBN will be revisited as a tool that gives information on the interaction between domain walls and stress configurations, or reveals details about compositional microstructure. Some complementary nondestructive testing techniques with more established roots in the nondestructive evaluation industry will also be discussed, such as remote-field eddy-current probing or magnetic flux leakage (MFL).

3.1 Barkhausen Noise as a Magnetic Evaluation Tool

3.1.1 Experimental Conditions for MBN

The circumstances under which magnetic Barkhausen noise is observed have been briefly mentioned in Chap. 2, including the strong connection between the Barkhausen effect and magnetic hysteresis. The magnetization curve is divided into characteristic regions, each with its own specific domain patterns and their changes. As such, the minimum amplitude of the alternating magnetic field that needs to be applied to the specimen should be close to the coercive field of the material that is being investigated. This is a necessary measure to ensure that magnetic domain walls become sufficiently "mobile" to "break away" from pinning points.

Magnetic Barkhausen noise may not be exactly the ideal instrument for non-destructive testing in industrial settings, given the setup, experimental design, and different signal processing methods this technique requires. Nevertheless, the insight it provides into the stress state of a ferromagnetic material, correlated with dislocation configurations, grain size, interstitial atoms, and/or microstructural changes makes this method of investigation well worth the effort of everyone embarking on this experimental journey. In spite of the variety of setups and parameter calculation used for MBN measurements in various laboratories, the Barkhausen noise experiments can usually be classified as being of two types. If the detection coil is placed on the surface of the specimen, the Barkhausen emissions are termed *surface Barkhausen noise*, whereas a coil wrapped around the specimen detects *encircling Barkhausen noise* [1, 2] (Fig., 3.1). Based on skin-depth calculations and experimental observations [3], the estimated depth for minimum penetration of the magnetizing field is roughly 1 mm, whereas the depth from which the MBN signal originates is ∼30 μm.

A typical MBN instrumentation consists of an excitation coil connected to a sweep current generator. The coil produces an alternating magnetic field which magnetizes an area on the sample surface and just beneath it. While the applied magnetic field is swept from lower to higher values and back along a magnetization curve, Barkhausen emissions are picked up by another coil with a large number

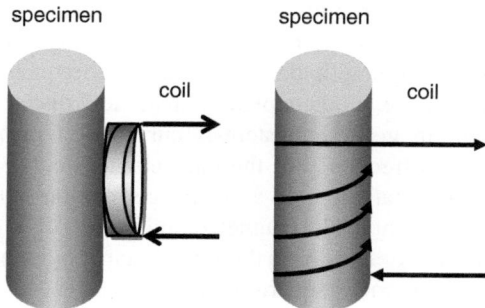

Fig. 3.1 Surface (*left*) vs. encircling (*right*) Barkhausen noise detection

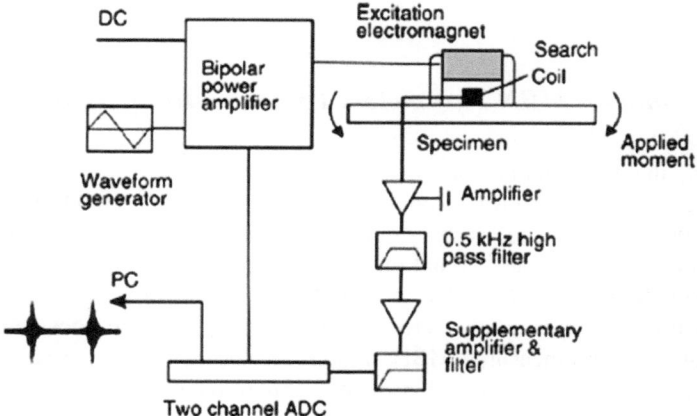

Fig. 3.2 *Schematic representation of a typical MBN measurement apparatus.* Reprinted from [4] (Copyright 2004) with permission from Elsevier

of turns of insulated copper wire wound around a ferrite cylinder, as shown in Fig. 3.1. The signal is then amplified and filtered, digitized, and further processed by software to extract a variety of MBN parameters discussed in the next section. An example setup is shown schematically [4] in Fig. 3.2. It should be noted that each laboratory sets its own experimental conditions so that extracted MBN data depend strongly on these unique environments in which they were obtained. As such, it is extremely difficult to compare raw data obtained from other laboratories, and for this reason, only qualitative evaluations are usually made between various research groups. Nevertheless, within the limits set by experimental differences, general trends in MBN signal should agree regardless in what setting they were recorded, especially since a large number of research groups have undertaken so many studies that certain patterns in Barkhausen emissions are observed again and again with a high degree of replication.

Micromagnetic Barkhausen emissions are quite dependent on excitation and detection frequencies. The input parameters that are usually known and controlled in an MBN experiment are amplitude, waveform, and frequency of the magnetic excitation field. For instance, a dc coupled to an ac field distort the magnetization curve of a specimen. In general, hysteresis curves of ferromagnetic materials depend on the excitation frequency of the magnetizing field. Also, we have seen in the previous chapter that there is a strong correlation between Barkhausen emissions and a specific point on the magnetization curve. A body of MBN research indicates that parameters extracted from the signal such as pulse height distributions, frequency spectra, and root mean square (rms) voltages depend strongly on dc bias fields, as well as excitation and detection frequencies. Simultaneous changes in one or more of generating or detecting parameters in an experimental MBN setup can therefore have drastic consequences on the interpretation of results.

3.1.2 Common MBN Parameters Extracted from Signals

As stated, the usual input parameters in an MBN experiment are mainly the amplitude, frequency, and waveform of the magnetic excitation field. It should also be noted that the amplitude is determined by the amount of current applied to the induction coil that is either wound on the specimen or around the magnetic core of a sensor. Sometimes, the alternating field is superimposed on a dc bias field which in turn becomes another crucial input parameter to be considered in the interpretation of the measured MBN signals. On the other hand, essential output parameters that affect the detection of these micromagnetic MBN emissions are the choices of filter types, upper and lower cutoff frequencies, and amplification. In most experimental setups, the output of the filter is connected to a full wave rectifier. Rectified MBN signals are then analyzed on a multichannel pulse height analyzer board. Nevertheless, variations in the experimentally used equipment are encountered in different research groups.

An analysis of the MBN signal has to be done with great care, as any measurement is highly sensitive to the rate at which the sample is being magnetized, as well as the sampling rate of the measurement system. It has been noticed, for instance, that the number of micromagnetic emissions at any given amplitude decreases when the dc offset field is increased. On the other hand, the use of the bias field reduces the number of higher amplitude Barkhausen events as the dc offset field is increased. This increase seems to be much more than the reduction in the number of low amplitude emissions. Some researchers commonly use rms voltage measurements for an analysis of MBN emissions. This is due to their simplicity and advantages because the only requirement is a digital voltmeter capable of measuring rms voltages at the desired frequency. However, others believe that rms voltages inherently contain less information than pulse height distributions or frequency spectra.

It is supposed that MBN is primarily due to 180° domain wall motion [5] because 90° domain walls have stress fields associated with them, and therefore cannot easily move. Magnetizations in 90° domain walls lie at right angles on either side of the wall, causing lattice spacings to be slightly larger in the direction of magnetization. Due to a resulting strain impeding with 90° domain wall motion, they are believed to be less competitive than 180° domain walls that have a higher velocity [6]. In most cases, two MBN contributions can be distinguished from which data can be extracted and analyzed. As such, a contribution to the signal comes from magnetic domains that are responsible for an easy axis, contribution denoted [7] α. These domain walls are the most responsive to the magnetic anisotropy in the material and are likely to move when a magnetic field is applied. The other contribution is from isotropically oriented domains whose magnetization orientation is away from the easy axis, and these domains and their wall movement are represented by β. The two contributions α and β can be incorporated into a so-called "MBN$_{energy}$," a quantity obtained from the measured MBN signal by observing the pattern on a graph. As such, α and β enter the mathematical expression of the "MBN$_{energy}$" as fitting parameters

$$MBN_{energy} = \alpha \cos^2(\theta - \varphi) + \beta, \tag{3.1}$$

where θ is the angle at which a magnetic field is applied, and φ is the direction of easy magnetization. The parameter α is obtained by subtracting β from the maximum "MBN$_{energy}$," due to β being a minimum in "MBN$_{energy}$," given that most domain walls contribute to the easy axis of magnetization. The "MBN$_{energy}$" can be attributed to the change in magnetic flux under the influence of the magnetic field component acting parallel to a particular domain wall. As such, a domain wall will move if its coercivity is overcome by the applied field component parallel to it. A large body of MBN work based on this parameter exists, work carried out mostly by the Applied Magnetics Group at Queen's University in Kingston, Ontario; therefore, interested readers are encouraged to consult the group's published literature (e.g., [7–14]).

To obtain further information from the "MBN$_{energy}$," the calculated quantity needs to be plotted on a polar (angular) graph. To this end, MBN measurements need to be taken at regular angular intervals, forming an angular or polar scan. The angle of measurement is with respect to the sweep field direction of the MBN sensor. Equation (3.1) is then fitted to the measured MBN data, while the resulting calculated "MBN$_{energy}$" is plotted on a polar graph. MBN is quite capable of detecting magnetic anisotropy in a ferromagnetic material, without regard to its origin. The shape of the graph indicates the direction of magnetic anisotropy because larger "MBN$_{energy}$" values are obtained in that particular direction. Figure 3.3 shows a typical polar graph for the "MBN$_{energy}$" when magnetic anisotropy is present. Note that the origin of the magnetic anisotropy is not important for obtaining the MBN data, whether the anisotropy was induced externally or is an intrinsic property of the material. In the absence of magnetic anisotropy, the "MBN$_{energy}$" plot is a perfect circle. If magnetic anisotropy is present such as, for instance, due to an

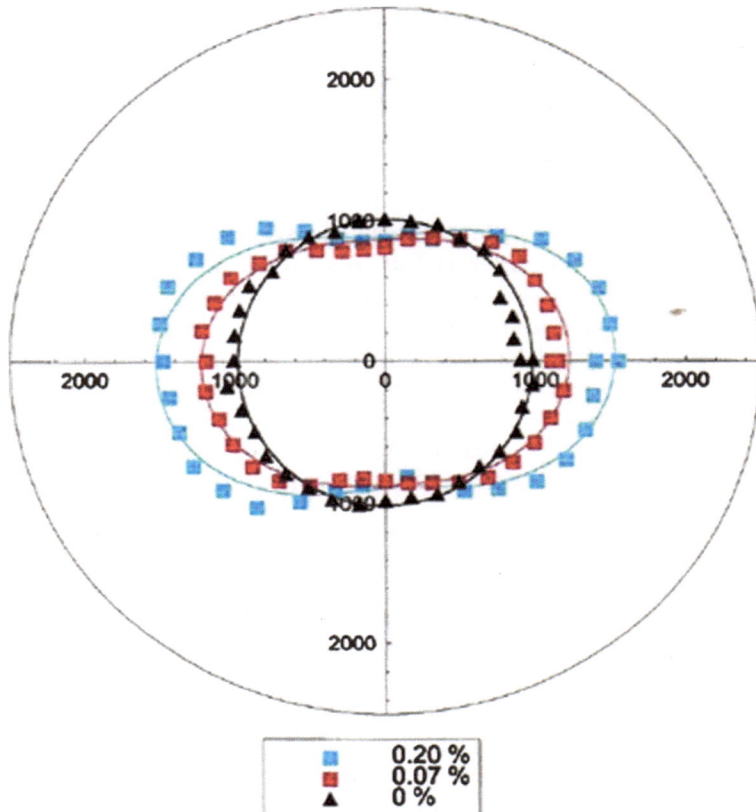

Fig. 3.3 Typical "MBN$_{energy}$" polar plot revealing a direction of easy magnetization (or magnetic anisotropy). A uniaxial tensile load is applied with the percent of uniaxial deformation indicated on the graph. At 0% deformation, the circle is not perfect, indicating that a magnetic anisotropy is present due to trapped residual stresses resulting from prior processing of the material. The plot is elongated in the direction of magnetic anisotropy when a uniaxial tensile stress is applied. Note that the initial magnetic anisotropy of the material is in a direction different from the one of the applied load [C.-G. Stefanita, L. Clapham, D.L. Atherton.]

applied uniaxial tensile load, the circle is elongated in the direction of the magnetic anistropy which is also the direction of the uniaxial tensile load, transforming the plot from a circle into an ellipse. If, for some reason, residual stresses are trapped in the material due to, for instance, prior processing, the "MBN$_{energy}$" plotted circle will be slightly elongated or otherwise distorted, depending on the direction in which these internal stresses act. Figure 3.3 is a typical example of an "MBN$_{energy}$" plot of a sample with trapped residual stresses.

A parameter termed *pulse height distribution* has also been extensively used to characterize the MBN signal. This parameter is obtained from the measured signal which consists of a collection of voltage pulses or "events" of varying amplitude. The pulses carry information about the magnetic state of the investigated material; however, only voltage pulses above and below a certain threshold are considered to assure the integrity of the analysis. The onset of an event is defined by a positive slope between two consecutive measurements crossing the positive voltage threshold. On the other hand, a similar positive slope crossing the negative voltage threshold indicates the end of it. The occurrence of events of different amplitude is represented in a graph termed a "pulse height distribution" where the absolute values of the event amplitudes are examined by height. An example of such a graph showing a pulse height distribution is given in Fig. 3.4. Actually, the quantity "MBN$_{energy}$" mentioned earlier is in fact obtained by calculating for each event the area between the time axis and the squared voltage pulse, and summing over all measured events. The significance of changes observed in pulse height distributions will be discussed in the next section.

A variety of MBN research groups have come up with their own parameters that they extract from the measured signals, such as amplitude, integral of the Barkhausen noise envelope, or results of pulse count analysis. For instance, the group of Zerovnik et al. [15] uses the squared rms voltage signal V_{rms}^2 as the most common Barkhausen noise characteristic for their investigations. As such, they successfully extracted information about the microstructure of hardened surface

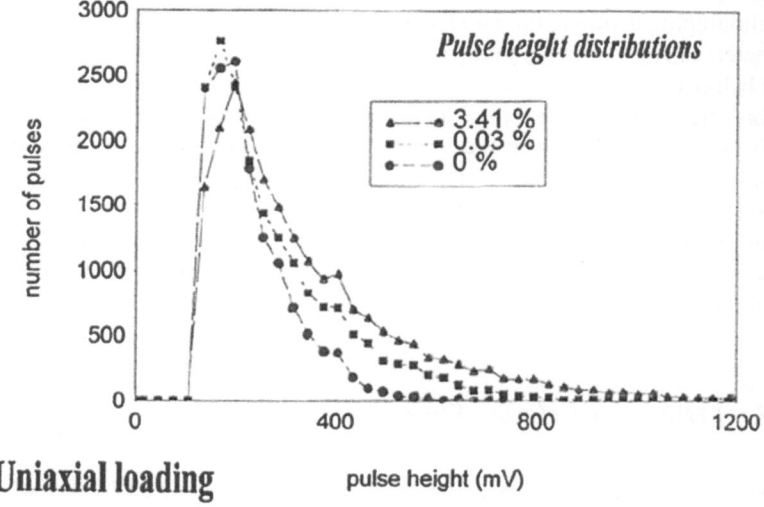

Fig. 3.4 An example of pulse height distributions extracted from MBN data. MBN is responsive to both elastic and plastic stress, with pulse height distributions especially sensitive to dislocation formation and variations in dislocation configuration. Therefore, changes in pulse height distributions are used to analyze the early stages of plastic deformation (microyielding), while continuing to be used to monitor the stress state of the specimen after its macroscopic yielding point. The data in this figure was obtained on a steel specimen which had already entered the microyielding stage at 0.03% deformation

layers by monitoring variations in V_{rms}^2 for different magnetizing frequencies which penetrated several depths beneath the surface. The group was able to identify variations in relative permeability or specific electric conductivity which led to an evaluation of microhardness, microstructure, and residual stresses for different depths. MBN signals exhibited distinct signatures in these surface-hardened specimens. Studies by other groups also exist, related to MBN analysis based on the frequency dependence of the rms voltages under different dc bias fields. Hysteresis losses measured as area under the $B-H$ curve are known to increase with increasing dc field strength.

The popularity of using rms voltage pulses for MBN analysis notwithstanding, some researchers have focused instead on the magnetizing current amplitude [11]. The change in the latter is due to the fact that the MBN signal differs significantly for various magnetizing currents. But MBN emissions are suppressed at higher magnetizing current amplitudes; therefore, large amplitudes can decrease the level of Barkhausen activity. This is confirmed through studies of low-range current amplitudes that seem to give the most sensitive response to changes in stress, and are therefore commonly employed to detect any signs of stress in steel specimens. This is, however, also contingent on selecting the optimum magnetization level while minimizing the effects of other factors that affect Barkhausen measurements. In any event, regardless of parameters used, observations of MBN responses obtained under a variety of conditions display a common set of patterns. Thus, whether measurements are taken under stress, with varying microstructure or different crystallographic textures, the large variety of most advantageous input and output parameters is capable of revealing more about the underlying conditions related to MBN behavior. As different parts of the magnetization curve are swept by applying a dc bias field, different levels of micromagnetic activity will be detected. This is largely due to the change in the variety of mechanisms of these micromagnetic emissions at different locations on the magnetization curve while also choosing different dc bias fields. The principle contribution to MBN, the irreversible motion of 180° domain walls is no longer the dominant mechanism at higher strength fields. Non-180° domain walls, as well as creation and annihilation of domains near saturation become the dominant players.

3.1.3 MBN as a Reliable Stress Detector

We have seen in Chap. 2 that magnetic domains are influenced by stress and the resulting strain that builds up inside the material. When a ferromagnetic material is subjected to stress, changes occur in its magnetization, even in the absence of an applied magnetic field. An applied stress, just like an applied magnetic field, destroys the balance of the five energies which have to readjust to minimize their sum. As a consequence, domains experience stress-induced volume changes similar to those occurring under an external magnetic field, so that a strong magnetic easy axis becomes present in the stress direction. This is because domain magnetizations have a natural tendency to align parallel to [100] crystallographic directions which are closest to the direction of the principal stress in the material.

The magnetic anisotropy α in the specimen has been noticed to be strongly affected by elastic stress which has only a minor influence on the isotropic background signal β. As such, an applied elastic tensile stress increases the "MBN_{energy}." For this reason, a good indicator of the stress condition present in a ferromagnetic material are the above-mentioned polar plots of "MBN_{energy}" which change shape with stress-induced changes. Studies in the past have demonstrated a high magnetic response in steel to applied elastic tensile or compressive stress, with an easy axis development along the direction of the tensile, and away from the compressive stress direction [12–14, 16]. The magnetic easy axis develops further with increased elastic stress, sometimes changing orientation while following any directional variations in stress. On the other hand, an applied compressive elastic stress decreases the "MBN_{energy}" in the compression direction [17], and creates a magnetic easy axis perpendicular to applied stress. Some researchers have found that trapped tensile residual stresses increase the amplitude level of the voltage signal, whereas compressive residual stresses lower their level. This is in accordance with the trend observed in MBN_{energy} for externally applied stresses. Nevertheless, different stress conditions do not seem to affect the time delay of the measured voltage signal.

Example 3.1. Where do residual stresses originate?

Answer: Residual stresses are commonly introduced through cold deformation and forming of a material, machining and grinding, or welding. Heat treatment will often introduce residual stresses; however, heat treatment is also used to remove residual stresses when these are present due to a different factor. Nevertheless, the importance and consequences of residual stresses is often underestimated in practice until it is too late and the component fails. Numerous research groups have proven that residual stress profiles can be determined by Barkhausen noise investigations, while being subsequently verified by X-ray diffraction studies. It was found that MBN signals correlate particularly well with intergranular strain along magnetization directions lying closest to the stress axis. As MBN is affected by stress-induced reorientations of 180° magnetizations, increases in 180° wall populations, and pinning energy modifications due to stress, it can be used to unam-biguously determine the presence of residual stress. With adapted measurement and analysis parameters, as well as a careful selection of MBN parameters specifically sensitive to residual stress areas, it is possible to correlate Barkhausen noise signals to residual stress profiles and map them sufficiently accurate.

A magnetic easy axis becomes more pronounced with plastic deformation but only up to a certain point. It has been observed that "MBN_{energy}" values experience only a slight variation in the plastic regime, as opposed to the large increases observed during the elastic deformation of the material. In this case, an analysis of changes in pulse height distributions is more revealing of the undergoing stress deformations in the material when the first signs of plastic deformation show up. Note again the slight increase in the number of event pulses of intermediate amplitude in Fig. 3.4. After the specimen enters the microyielding regime, dislocations are generated in some grain boundary regions. Nevertheless, dislocations do not occur simultaneously in all grains, as some grains start to deform plastically at a later stage. As the specimen continues to deform plastically, more and more dislocations are formed extending to more and more grains, so the number of event pulses of intermediate amplitude rises [18]. When plastic deformation sets in, the crystalline lattice becomes permanently distorted. A phenomenon termed *slipping* takes place when the critical resolved shear stress is reached on a slip plane. When this happens, dislocations form that alter the spacing between slipped and unslipped planes. Dislocation-induced strain fields form as well, enhancing local variations in interplanar distance along a certain crystallographic direction within a grain.

As more strain fields are created, they affect pinning sites and induce volume changes in magnetic domains. With increased internal stresses, larger magnetic fields are required to move a domain wall across a pinning site which is also influenced by stress [19]. With dislocations developing their own strain fields, they redistribute the strain within the grain so that associated changes in domain configuration are observed in measured MBN data. As stated, these changes are best represented by changes in pulse height distributions which become especially noticeable along directions of maximum shearing stress. However, increased dislocation densities and dislocation entanglements that occur at higher levels of plastic deformation start to impede a simultaneous motion of domain walls [20]. As the critical value for slip increases, the material accommodates more and more elastic stress through increases in lattice spacing. Because the lattice can only accommodate so much elastic strain before slip systems are activated, plastic flow initiates, leading to work hardening. The number of event pulses will not increase further when the material becomes work hardened. It is believed that hard mechanical regions consisting of dislocation tangles are surrounded by a softer matrix [21]. The hard regions are built of small volumes assumed to be under tensile stress, while separated by larger volumes (soft regions) under compressive stress. This will increase slightly the threshold for plastic flow, allowing some elastic strain to continue to build up, but this build up will eventually reach its limits. At advanced levels of plastic deformation, competing effects between dislocation amalgamation and crystallographic texture formation leave their imprint on the material and its magnetic response to MBN. And as stated earlier, these changes in MBN depend strongly on the magnetizing field intensity [11]; therefore, several other MBN parameters are usually recommended for further analysis of plastic deformation.

When grains within a polycrystalline sample deform, they do so differently, depending on their crystallographic orientation with respect to the direction of applied stress. Nevertheless, every magnetic domain size modification results in wall movement, meaning that variations in MBN signal are expected. While deforming, grains can lock in elastic intergranular stresses [22], and their influence is visible in the MBN signal. As such, residual stresses are elastic in nature; however, they can be surrounded by nonuniform microstructural changes in the specimen. Studies have shown a correlation between residual stresses detected using X-ray diffraction and

MBN peak height in unloaded specimens. Furthermore, it was shown that residual stresses trapped in some parts of cold-rolled specimens can be separated from other plastic deformation effects. These residual stresses, whether axially compressive due to the rolling process, or built up at grain boundaries or between crystallographic planes, bring their own contribution to the magnetic anisotropy of the cold rolled specimens and as such can be detected through changes in "MBN_{energy}" or pulse height distributions [20]. It can therefore be seen that due to this high sensitivity to stress, magnetic Barkhausen noise can be used as an independent evaluation tool of elastic stresses and strains [23] present in the material. It can also reliably detect any signs of plastic deformation and its consequences [24]. Other factors can also contribute to altering a magnetic domain configuration and therefore are reflected in the MBN signal. These are, for instance, crystallographic texture as well as various microstructural inhomogeneities [25].

3.1.4 Effects of Compositional Microstructure on MBN

Materials at high temperatures can flow plastically and are in danger of fracturing when an external stress is applied. Micromagnetic Barkhausen emissions are very sensitive to variations in microstructure and stress states beneath the surface of magnetic materials. As such, the MBN technique lends itself to nondestructive investigations when it becomes necessary to detect early any kind of damage that can lead to costly failures. For instance, high strength materials used in landing gear components in the aerospace industry are in need of residual stress characterization underneath the surface. This is because mechanical and thermal treatments are often applied to materials and thereby residual compressive stresses are introduced to improve their fatigue strength. Also, engineering products such as steel pipelines underground or in power plants have creep damage due to microstructural changes that occur during their service life. Creep is a form of permanent deformation that occurs with time at a constant stress. Creep damage is associated with thermodynamic processes that result in strain, void nucleation and growth of voids, all leading to impending failure. Void nucleation occurs in regions of microscopic heterogeneous deformation such as grain boundaries and matrix interfaces. These microstructural changes inhibit movements of domain walls which contribute to Barkhausen transitions and therefore MBN measurements can offer some insight into these changes.

It has been suggested that the number of 180° domain walls increases under an applied tensile stress because of changes that occur in magnetoelastic energy. Nevertheless, these changes occur only above a threshold stress that would result in the addition of another domain wall to an existing domain configuration. Such a threshold stress depends on the grain size, and it increases with the number of existing domain walls. It is also believed that the interaction between domain walls and dislocation tangles results in different MBN profiles than the interaction between the walls and grain boundaries. This fact is attributed to the physical nature of the barrier which is assumed to dictate the restoring force acting on the wall. It is also known that grain size influences the number of defects in the specimen; therefore not surprisingly, its magnetic properties are affected as well [26]. Other researchers have independently confirmed [27, 28] that the boundaries of grains are sources for domain wall pinning, with a larger number of grain boundaries resulting in more intense signals. The larger number of boundaries is found in samples with smaller grains that have been annealed at lower temperatures. Small grains give rise to larger MBN signals because the boundaries act as pinning sites, and since their fractional volume is larger, more pinning sites need to be overcome when the walls move. Precipitates and segregation of phosphorus at grain boundaries can offer additional pinning sites for domain walls, increasing the number of MBN pulses even in large grained specimens. Nevertheless, it can be said that MBN activity is strongly linked to grain size and grain boundaries and many more studies support these findings. Grain size has therefore an effect on the magnetic Barkhausen emissions in a specimen.

A reorientation of grains can occur with plastic deformation which is likely to cause a redistribution of internal stresses. This potentially destroys the effects of other factors on the recorded MBN signal, such as for instance that of residual stresses. Fortunately, crystallographic reorientation only happens at high deformation levels, usually associated with mechanical processing. On the other hand, it is possible for magnetic domains of neighboring grains of similar crystallographic orientation to become simultaneously active under an applied field or stress. When this happens, the enhanced MBN signal may not be entirely due to crystallographic texture. Furthermore, an analysis of the MBN behavior observed in a material can be complicated by compositional variations which can bring their own contribution to signals, rendering the MBN analysis more problematical. Different signals have been obtained when phase changes such as carbide precipitation [29] were present in plain carbon steel. Sometimes, an intergranular impurity segregation is left behind after heat treatment, offering increased pinning of domain walls by carbide particles. Networks of lamella carbide have been observed to increase the signal by providing stronger pinning. Heat-treated carbon steel results in different phases, as pearlite, bainite, and martensite each have their own, distinct MBN signature. A comparison study in cementite, pearlite, and martensite specimens [30] showed that MBN signals are more pronounced in the magnetically soft spheroidized cementite, however, are reduced in pearlite or in the magnetically hard martensite specimens which were tempered at 400°C.

There is a need for calibration of the magnetic Barkhausen noise method as a nondestructive evaluation tool. At the moment, user knowledge and experience are the most important factors when using MBN data analysis. For this reason, and also because Barkhausen noise depends on so many influential variables, it is still necessary these days to resort to comparative studies. An undeformed specimen or a known microstructural state needs to act as a standard for a particular type of alloy. For instance, in spite of commonalities observed in different types of plain carbon steels, one should not directly compare MBN results. This is true for other materials, as well. The prior magnetic history of each sample, as well as processing treatment and ultimately the type of steel will determine the MBN response. The main advantage of magnetic Barkhausen noise as a nondestructive technique is that it reveals complex changes in materials structure that originate in microstructural properties, as well as stress configurations, together leaving their imprint on the material. But in the end, how the material responds to a magnetic field depends on multiple factors, each with a specific influence on the recorded signals. Nevertheless, these factors are difficult to dissociate and treat individually, requiring an experienced user to give the appropriate interpretation to the measured MBN data.

3.2   The Remote-Field-Eddy Current Nondestructive Testing Method

3.2.1   Purpose of Method

Environmental or age-induced stress corrosion cracking is a major concern in many technological pipeline applications because of material damage such as axial cracks that can be present both inside and outside of pipes or other tubular components. When these fail, it is often with catastrophic consequences; therefore, there is an urgency to avoid or at least predict such failure. Since there are different techniques for manufacturing tubular products or entire pipelines, as well as a variety of materials from which these components are made, testing is not always straightforward. Furthermore, during the operational life of a pipe or tubular product, the latter is subjected to a multitude of processes, each exerting its own influence. At some point during the use of the product, it is necessary to evaluate the state of the tubing both inside and outside, most of the time while it is still in service. Nevertheless, testing of tubing or pipelines while they are operational is difficult because one wants to minimize the impact of the testing itself, and allow the tube or pipe to continue to function. As such, a body of technical work, both theoretical and experimental, has developed over time in order to address the needs of the NDT industry which has grown around particular applications, just for the purpose of servicing their equipment infrastructure.

Most methods of testing require eddy currents, and therefore a careful calculation of test frequency is necessary so that material characteristics are taken into account in order to accurately predict the depth of a flaw. When evaluating ferromagnetic tubing, the testing is even more complicated, given that these respond to a magnetic field. In these cases, unpredictable magnetic permeability variations introduce noise in the measurement, making it less reliable. The Remote-Field-Eddy Current (RFEC) method is a popular and reliable way of testing conducting tubes, both ferrous and non-ferrous. In spite of the fact that both exciter and detector coils are located inside the tube, it allows a throughwall investigation with similar sensitivity to both internal and external defects. There are a variety of applications where only the inside of the tube is accessible, such as heat exchangers, gas mains, tubing in nuclear reactors, or oil well casings buried deep underground. In other situations, two or more metal tubes are arranged so that one is inside the other, making the accessibility of the inner tubes extremely difficult, if not impossible. However, RFEC is able to test these applications, and as such it has become a valuable investigation method when other inspection techniques have failed.

Example 3.2. What is stress corrosion cracking?

Answer: Stress corrosion cracking is a form of material failure due to a corrosive environment and a tensile stress state in the material. Specifically, if an engineering component such as a pipe is left in an environment where it corrodes over time, but is also subjected to a tensile stress due to how it is positioned or mounted, cracks will form in the material. Residual stresses may also contribute to cracking; therefore even if the pipe is not under an externally applied stress, its material will experience stress corrosion cracking in time. Stress corrosion cracking is a catastrophic type of corrosion. Cracks are initially very fine, penetrating into the material. They can form between grains, and are therefore *intergranular*, or they can be within grains in which case they are called *intragranular*. The detection of fine cracks is extremely difficult, while their propagation through the material cannot be easily predicted. Thereby, they can create disastrous damage in an unexpected way.

Stress corrosion cracking may be prevented or at least delayed by removing stress and environmental chemical species that lead to this failure. These include hydroxides, chlorides and oxygen. The nuclear industry is very concerned about chloride stress corrosion cracking and many inspection steps are taken to prevent failure. A low chloride and oxygen environment is ensured, as well as low carbon steels are employed as they are more resistant to this type of cracking. The three most important factors that need to be remembered that cause stress corrosion cracking are (a) a susceptible material, (b) an environment that will cause cracking of that material, and (c) the stress state of the material.

3.2.2 Basic Principles of Operation

The RFEC effect is observed when an exciter and a detector coil are introduced and moved inside a conducting tube while separated from each other by roughly two or even three tube diameters. As such, this sensor configuration created by these two coils is of crucial importance, and much effort has been invested over the years by many research groups to design probes with high accuracy, resolution, signal strength, and minimum distortion. A schematic of an RFEC probe is shown in Fig. 3.5. The exciter coil, placed coaxially inside the tube is excited by a low frequency ac current of typically 10–100 Hz for steel pipes. Two electromagnetic regions are created inside the tube, a so-called *direct-field zone*, and a *remote-field zone*. These regions are separated by a transition zone where the two electromagnetic fields interact. The indirect transmission path field traverses the wall,

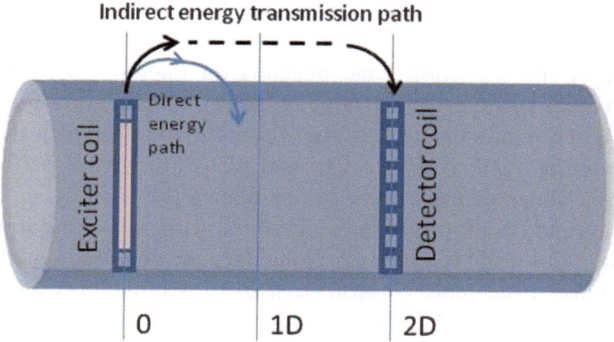

Fig. 3.5 Schematic illustration of a remote-field-eddy-current (RFEC) detection geometry. An internal solenoidal exciter coil is driven by a low frequency ac signal. The detector coil is placed at approximately two pipe diameters distance from the exciter coil and may contain an array of individual coils. Two distinct coupling paths exist between the exciter and the remotely spaced detector. The direct path inside the tube is attenuated rapidly by circumferential eddy currents induced in the tube's wall. Note the throughwall transmission of the indirect energy path. An electromagnetic signal traveling through the wall is affected by wall variations; therefore, any defects would alter the signal before it reaches the detector coil in the remote-field zone. There are two remote-field zones at approximately two pipe diameters, one on each side of the exciter coil

thereby signals traveling away from the exciter coil pick up any wall variations such as defects which are detected by the detector coil in the remote-field zone.

But where do these electromagnetic fields originate? The first electromagnetic field is created by the exciter coil, and is strong inside the wall in the vicinity of the coil. This is the *direct field*. Traditional eddy current testers operate in this zone and use an impedance plane analysis for the detected signals. The direct field sets up circumferential eddy currents inside the wall which migrate toward the outside of the wall where they spread in both directions along the exterior surface of the pipe. These eddy currents rapidly attenuate the direct field as it moves further away from the exciter. Nevertheless, these eddy currents form a circulating sheath of currents which acts as a second exciter that induces outside the pipe a second electromagnetic field, the *indirect field*. The indirect field travels along the pipe but also back through the wall, creating an electromagnetic field inside the pipe. In the meantime, the outside (indirect) field which is attenuated less rapidly than the inside (direct) field, is stronger than the other one in the transition zone. By the time the remote-field zone is reached, the inside (direct) field becomes negligible, so that the indirect field is the only significant transmitted energy which is then used for detection and analysis of wall defects [31]. Any detector devices that operate in this zone where the indirect field path through the wall dominates are remote-field instruments. Note that the signal received at the detector has traveled twice through the wall. This through wall double transit allows both external and internal defects to be detected with approximately equal sensitivity. Nevertheless, the nature of defect interactions with the electromagnetic field is not well understood and several theories, as well as experimental geometries, exist that try to account for the presence of these defects in the signal.

The width of the transition zone depends on operating frequency and the magnetic permeability of the pipe. A phase shift and change in the slope of the logarithmic signal amplitude indicate the transition zone between the direct-field zone and the remote-field zone. Background scans are always performed before a defect scan is undertaken, in order to distinguish the defect (i.e., abnormality) from the background. Unfortunately, background variations are not consistent, as bumps and ridges, as well the weld vary along the pipe are not symmetrical between different sections of the pipe. Nevertheless, these details depend on the type of pipe, its manufacturing and mode of employment, as heating tubing is different from oil pipes. In any case, the anomalous signal from the defect is calculated by vector subtracting the background signal from the recorded defect signal. The result indicates the change in the received signal caused by the defect. To be more specific, the defect signals are obtained by subtracting the phasor components of a defect-free background from those of a defect scan.

Both axial and circumferential sensor movements inside the pipe are employed in order to detect a defect with high resolution. Phase and amplitude response from a single, large defect is likely different from two closely spaced smaller defects. Furthermore, to improve the accuracy of the analysis, radial, axial, and circumferential field component comparisons are made. The combination of data from several field components, as well as a skilful analysis of the amplitude and phase may allow a separation of responses from different yet closely spaced defects. It seems that defects show more clearly in the radial field component than in the axial, nevertheless, this is dependent on sensor design [32]. Smaller defects seem to be indistinguishable from the background when circumferential field scanning data is used. A challenge is posed by axial slit-type defects which are parallel to each other, as quite often they cannot be easily differentiated from each other. Research indicates that the radial and circumferential field components may be the best solution for parallel slit comparison; however, more studies need to be performed. This information is useful when developing a variety of RFEC sensors [33], as it offers design possibilities for enhancing defect detection based on different field components.

3.2.3 Theoretical Considerations vs. Practical Experience

When the external indirect field diffuses back into the wall further along the pipe, it is attenuated and shifted in phase. The detector coil is placed at an axial position of approximately two or three inside diameters from the exciter coil (Fig. 3.5) where the remote-field zone begins. In that region, this indirect field is the dominant field inside the tube. The detector thus picks up magnetic fields that have traveled through the wall twice. There are software packages (e.g., TubeCalc) that have been developed to calculate the amplitude attenuation and phase shift in both direct and indirect fields in the different zones of the pipe [34]. An amplitude attenuation factor of ~150, and a phase shift of ~50° are common for an indirect field passing through the wall back into the pipe [34].

An initial theoretical description of the RFEC effect was based on the one-dimensional skin effect equation for a semi-infinite, plane-surfaced, homogeneous, and isotropic conductor. Although the skin-depth equation applies to a planar situation, while the electromagnetic field and the RFEC geometry are quite different from that case, the skin-depth equation was considered for a while the best approximation for thicker walls. A wall of this type is that for which its thickness is more than one skin depth. The following example discusses the basic ideas behind the relationship between skin depth and wall thickness.

Example 3.3.   Comment on the relationship between skin depth and wall thickness

Answer: In an RFEC experiment, the exciter and detector coil represent two induction coils which are placed inside a conductive pipe, centered within it. When a magnetic field is induced through the applied current, the magnetic moments that are created in the two coils become aligned with the axis of the pipe. Under the right circumstances, when the coils are spaced sufficiently apart and the frequency is low, the mutual impedance between the two coils is very sensitive to the thickness of the pipe wall. The phase of the mutual impedance will then be proportional to

$$\text{phase of mutual impedance} \sim 2\sqrt{\frac{\omega\mu\sigma}{2}}t, \qquad (E3.3.1)$$

while the magnitude of the mutual impedance will vary as

$$\text{magnitude of mutual impedance} \sim e^{-2\sqrt{\frac{\omega\mu\sigma}{2}}t}, \qquad (E3.3.2)$$

where t is the wall thickness and the significance of the other symbols is given at the end of the chapter. This is basically the skin-depth equation which in our case shows the desired relationship between skin depth and wall thickness. It is also the basis for using RFEC to detect wall thickness variations [35]. Note that the method can also be applied for two pipes, one inside the other. In this case, the mutual impedance has a magnitude that is proportional to the exponential in the combined wall thickness from both pipes, while the phase is linear in the combined wall thickness.

In time, more accurate theoretical investigations have been performed in the form of either finite element studies [36] or seeking analytical solutions [37, 38] which solve the Maxwell equations in the quasistatic case. In the latter case, solutions of the Bessel functions are sought, involving more than one "through wall equation," thus making the mathematics rather complex. TubeCalc, the software based on this theory allows calculation of the magnetic vector potential and the magnetic field at any point in or outside the tube, based on the input parameters such as excite coil dimensions, number of windings, current, as well as pipe dimensions and magnetic and electric properties (e.g., permeability and electrical conductivity). The detector coil characteristics are also taken into account. The assumptions in TubeCalc are those of an infinitely long piping tube without defects, and all input parameters allow a calculation of the detector signal. The two quantities of most interest that are obtained from this formulation are the amplitude attenuation factor a (3.2a), and the phase change Φ (3.2b) of the electromagnetic field, given by

$$a = \sqrt{\pi\mu_0}\frac{f\sigma\gamma_i\gamma_0}{\mu_\gamma\cosh(2\gamma) - \cos(2\gamma)}, \tag{3.2a}$$

$$\Phi = -\frac{\pi}{4} + \tan^{-1}\left(\frac{\tanh(\gamma)}{\tan(\gamma)}\right). \tag{3.2b}$$

The meaning of the other quantities is given at the end of the chapter.

When such an analysis is made, it is customary to graph both quantities a and Φ as a function of the wall thickness expressed in skin depths, while the amplitude attenuation factor is usually represented as a natural logarithmic quantity. If the outside radius of the tube approaches the dimensions of the inside radius, the amplitude attenuation factor a goes to infinity. Therefore, (3.2a) and (3.2b) are not valid for wall thicknesses below 0.15 mm [34]. Furthermore, defects are not included in a TubeCalc analysis. In general, theories that include defects in their analysis cannot take into account all possible defect geometries. This is due to the way the defect interacts with the electromagnetic field, an issue regarded in many theories as an electromagnetic scattering problem with a radiation field originating in the defect itself. However, in a scattering formulation not all possible defect shapes are accounted for. In practice, there are defects along the pipe that neither the probe nor the theories can record/analyze. Nevertheless, a qualitative understanding of the RFEC effect can be gained from these analyses.

3.2.4 RFEC and Defects

A variety of research laboratories in industry and academia have performed experiments with human-induced defects to learn more about RFEC signals and their correlation with real-life anomalies that occur in the pipe wall. This is because one cannot always rely only on theory to extract the necessary information about a physical defect. As such, evaluation RFEC scans are made in a laboratory environment by pulling a probe through the tube at a moderate speed which depends on the time constant of the lock-in amplifier. In industrial settings, the probe may be powered by the oil/gas it is immersed in, as the latter travels through the pipe. A scan resulting from a trip of the RFEC probe through the pipe contains information about normal regions in the pipe (considered background), as well as areas in which either the wall has thinned out or has cracks, pits, and other types of anomalies. In nonferrous pipes, defect signals arise from anomalous eddy currents which give rise to a distinct pattern in a defect area. A sketch of such an anomalous eddy current pattern in a slit formed along the axis of a nonferrous pipe is shown in Fig. 3.6. On the other hand, Fig. 3.7 shows an example of a pit defect in a ferrous pipe which creates a change in the magnetic field pattern in and around the pit. It should thus be noted that in contrast to nonferrous pipes, defects in ferrous pipes contribute to the signal through anomalous magnetic field patterns which form around defects, as well as in the defects themselves (Fig. 3.7). Hence, an analysis of RFEC scans differs for nonferrous and ferrous pipes; therefore, an experienced technician is required to perform it, in order to give the scan the correct interpretation.

Separating defect signals from the background is not an easy task given the tough reality of RFEC probing. Scans may not be able to detect all defects, or give enough information about those that get recorded. One reason for this shortcoming is the sensitive area of the detector coil where a field change above a certain threshold is normally detected. For instance, a defect can be long enough in the axial direction

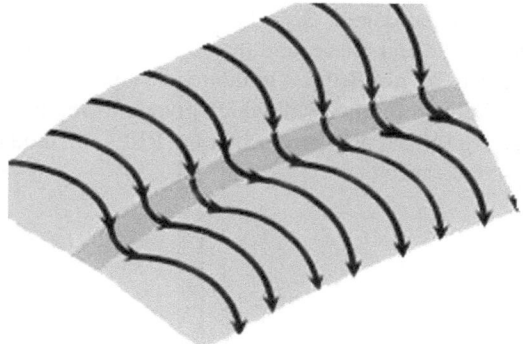

Fig. 3.6 Anomalous eddy current pattern at an axial slit defect present along a pipe. The axially diminishing eddy current pattern gives rise to a distinctive shape in the signal

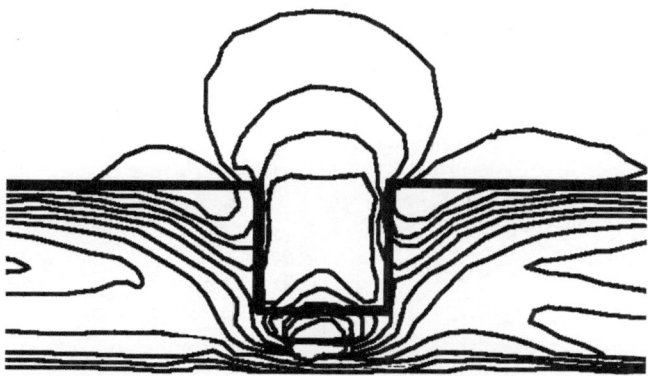

Fig. 3.7 Anomalous magnetic field pattern around and in a defect present in the pipeline wall. The magnetic flux is said to "leak out" of the defect. Hence, this effect is exploited in the magnetic flux leakage (MFL) nondestructive testing technique

to completely cover the sensitive area, a fact which complicates defect recording. Experience has shown that there is a significant amount of cross-talk between the exciter coil and detector coil circuits. This cross-talk needs to be measured and subtracted from the detector signal. The cross-talk is measured by inserting the probe in a thick steel block where the remote field signal is extremely small, and therefore undetectable. The measured signal is then due solely to the cross-talk and needs to be subtracted afterward from the defect data.

The detector coil usually has a number of windings at least an order of magnitude larger than an exciter coil to make the low field levels in the remote-field zone measurable. When a RFEC signal is obtained, the recorded data from a scanned defect contains two peaks measured as a function of distance along the pipe. Figure 3.8 shows examples of RFEC signals in the vicinity of a defect. The first peak is due to the exciter coil passing the defect. The second peak is the result of the wave traveling along the electromagnetic coupling path (Fig. 3.5) that is influenced by any anomalies in the wall [39]. Therefore, the second peak is recorded when the detector coil scans over the defect. Signals from these anomalies also contain an increase in both amplitude and phase, and this is usually an indication that a defect was successfully detected. This distinct signature in a signal occurs because a defect present in the wall implies a thinner wall which gives a lower amplitude attenuation and a smaller phase lag [40]. It also means that the RFEC method can thus measure wall thickness, as stated earlier.

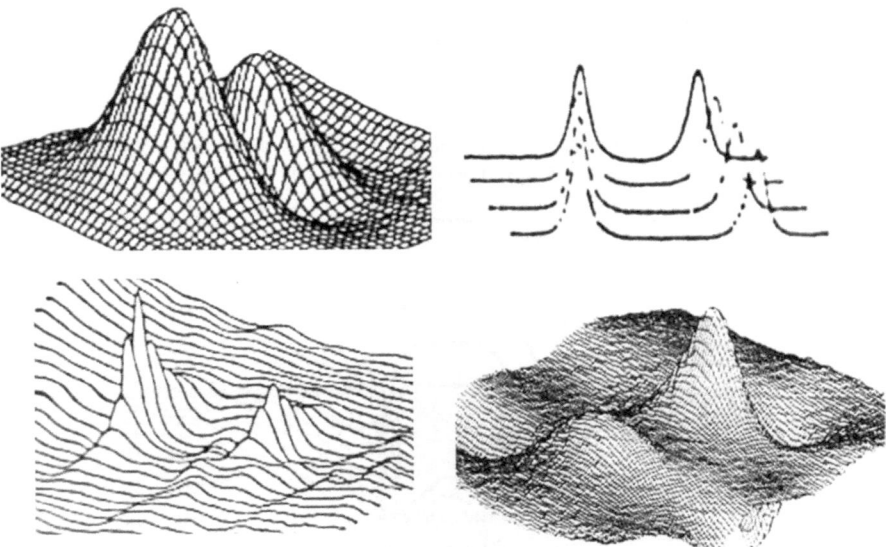

Fig. 3.8 Examples of RFEC signals in the vicinity of defects showing two peaks. The first peak is due to the exciter coil passing the defect, while the second peak is recorded when the detector coil scans over the defect. [Applied Magnetics Group Archive]

It was noticed that axial slit defects of different orientation do not interact the same way in ferrous and in nonferrous materials. Since stress corrosion cracks in ferrous pipelines are predominantly axial, the detection of slit defects in this direction is of great interest. The difference in signal strength is due to the interaction of the defect with the magnetic field component of the electromagnetic field, or the interaction with induced eddy currents. In nonferrous tubes/pipes such as aluminum, axial slits are perpendicular to the eddy currents which are induced circumferentially (Fig. 3.6). Hence, the signal is stronger for them than for circumferential slits. A full circumferential slit forming a groove around the pipe acts essentially as a small circumferential coil wound in opposition to the exciter coil. The anomalous radial field given by this type of defect is small, and has a shape similar to the background radial field from the exciter. On the other hand, axial slits in ferrous tubes/pipes do not produce such strong signals. Magnetic fields tend to align normal to the surface of a ferrous pipe (Fig. 3.7). For these pipes, circumferential slits which are perpendicular to magnetic fields give rise to the stronger signals [41]. As such, for nonferrous pipes it is the disruption of the eddy current paths by the defects that give rise to larger signals, whereas in ferrous pipes magnetic field interactions with the defects are mostly accountable for the measured signals [42].

3.3   Using Magnetic Flux Leakage for Defect Evaluation

3.3.1   Magnetic Flux Anomalies and the MFL Probe

Magnetic flux leakage (MFL) is a nondestructive inspection technique based on detecting anomalous magnetic flux patterns due to defects in pipelines. The magnetic flux "leaks" from pits and other surface indentations when a ferromagnetic material is magnetized preferably to saturation, thus creating anomalies in the magnetic flux configuration. An example of such an anomalous magnetic flux distribution due to a surface defect present in a ferromagnetic material was shown earlier in Fig. 3.7. That figure illustrates clearly how the flux "leaks" out of the pit, hence justifying the name of the technique specializing in detecting these flux leakages. A customized probe known as a "pig" is used for measuring such flux anomalies, while its construction and design may vary from one pipeline application to the next, depending on how the pipelines are used during regular service. An example of a typical MFL probe or tool is shown in Fig. 3.9. Testing by detecting leakages in the magnetic flux distribution is performed by pumping the specialized "pig" through the pipeline, usually from one compressor station to the next [44].

Generally, an MFL "pig" is fitted with a circumferential array of MFL detectors incorporating powerful permanent magnets (e.g., neodymium iron boron) that nearly saturate magnetically the ferromagnetic pipe wall. A typical MFL tool for a 36" diameter pipeline can have up to 16 detector tracks consisting of magnets and brushes circumferentially arranged around the wall to allow inspection of the entire

Fig. 3.9 *An example of a typical MFL probe.* Reprinted with permission from [43]

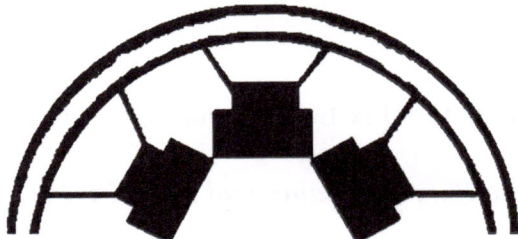

Fig. 3.10 Schematic of a circumferential array of magnetic flux leakage detectors incorporating powerful permanent magnets pushing against the pipe wall via brushes. The role of the permanent magnets is to nearly saturate magnetically the ferromagnetic pipe wall in regions between the brushes

pipe. Figure 3.10 shows a rough schematic of circumferentially distributed magnets pushing against the pipe wall through strong brushes.

The distribution of magnetic flux will show anomalies in its pattern in the vicinity of defects such as corrosion pits present in the pipe wall. These fringing magnetic fields are then recorded with Hall probes or sensor coils that move with the detector "pig." Axial or radial components of the magnetic field density B are detected by the Hall sensors. A sketch showing the main components of an MFL setup is shown in Fig. 3.11.

The MFL inspection tool also contains an odometer that measures the distance traveled within a pipeline. These position measurements are periodically checked against signals from beacons which are placed above ground next to buried pipelines. The beacons consist of magnetic coils functioning at 60 Hz.

3.3.2 When Does MFL Work?

The MFL technique works best for in-service pipelines that need to be inspected for interior and exterior corrosion pits, hard spots and many similar types of defects. However, flux leakage tools do not respond well to fine cracks that appear due to stress corrosion for which RFEC inspection is much better suited. Figure 3.12 serves as a practical example of how magnetic patterns recorded using MFL can distinguish between different sorts of surface imperfections. It illustrates a comparison between two controlled defect types obtained in a laboratory MFL study where researchers [45] recorded the signal due to a grooved (Fig. 3.11a,b), and a hollowed rhombic defect (Fig. 3.11c,d). An analysis of the magnetic flux density distribution around the defects was also performed for completeness. The defects were intentionally produced in the surface of an 8-in. diameter test pipe. The experiment indicates that the correct type of defects was successfully identified from the recorded signals based on the MFL patterns they produce. A "back yoke" type of sensing system had to be developed and employed behind the Hall sensors in order to increase the signals more effectively. The latter were increased up to 200% as a result of concentrating the leakage fields on the high permeability "back yoke." Additionally, a nonlinear finite element analysis was undertaken in order to verify the MFL results.

Example 3.4. Give an example of a theoretical method of estimating defects in a pipeline that is based upon principles on which MFL inspection functions

Answer: Modeling a moving inspection vehicle in the vicinity of defects is a transient problem incorporating nonlinearity and velocity effects. An example of how this could be implemented theoretically is given by the Leismann–Frind method [46] which offers an improved accuracy, and is therefore considered more reliable [47]. A motion related term $(v \times B)$ is included in Maxwell's equation

$$\nabla \times \frac{1}{\mu} \nabla \times \vec{A} = \vec{Js} - \sigma \frac{\partial \vec{A}}{\partial t} + \sigma \vec{v} \times \nabla \times \vec{A}, \qquad \text{(E3.4.1)}$$

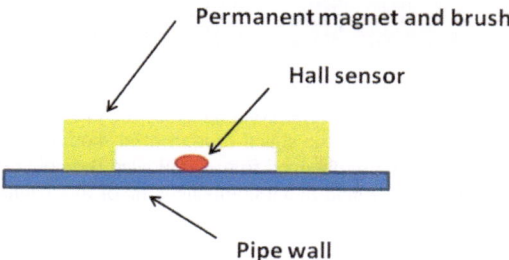

Fig. 3.11 Sketch of an MFL setup showing the Hall probe sensor against the pipe wall surrounded by the powerful magnet. The Hall sensor detects axial or radial components of the magnetic field

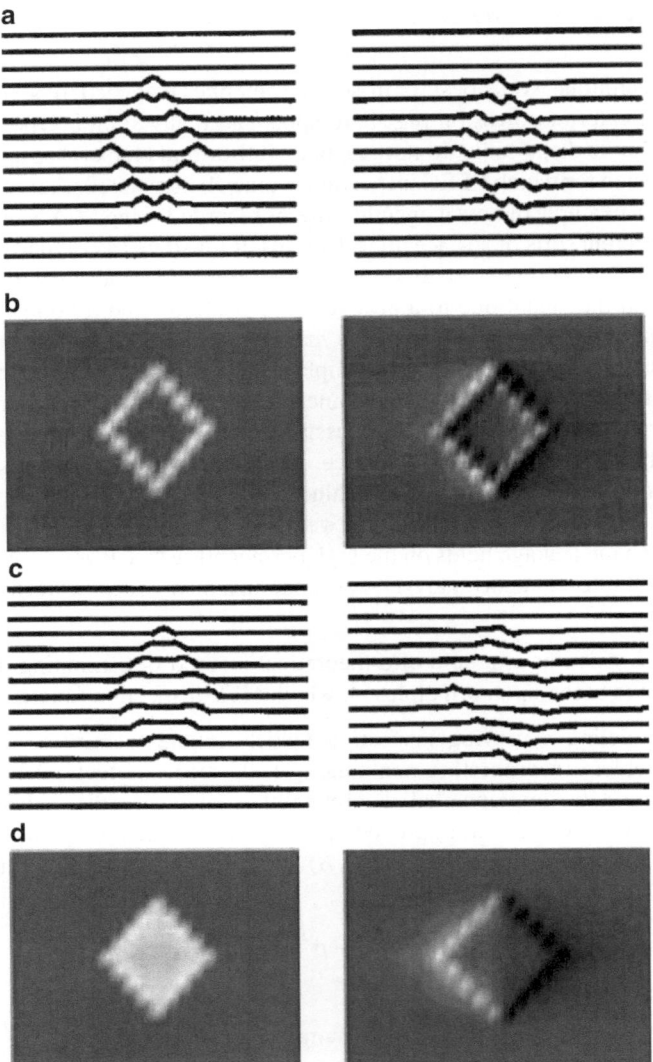

Fig. 3.12 An example of how different types of defects affect the magnetic flux pattern: (**a**) signals from the z-component B_z and radial component B_r of the magnetic field density in a grooved rhombic defect; (**b**) images of the signal in a grooved rhombic defect. These magnetic field patterns are compared with those obtained from a hollowed rhombic defect as illustrated in: (**c**) signals from the z-component B_z and radial component B_r of the magnetic field density in a hollowed rhombic defect, and (**d**) images of the signal in a hollowed rhombic defect. Reprinted with permission from [45]

where v is the velocity of the MFL "pig," μ magnetic permeability, A magnetic vector potential, σ electrical conductivity, J_s surface current density and t is time. The surface current represents the permanent magnet as an equivalent electromagnet. An axisymmetric geometry is considered, thereby simplifying the equation. Also, the probe is seen as stationary, while the pipe wall is moving from left to right such that the overall magnetization and sensor motion takes places from right to left. Leismann and Frind introduce: (a) an artificial diffusion term, and (b) time-weighting factors for each individual term in the governing equation. These are then computed during the process of error minimization. The resulting equation is expanded in a Taylor series for which the one-dimensional case is shown below [47]

$$\left[\frac{\sigma}{\Delta t} - \frac{1}{\mu} \frac{\partial^2}{\partial z^2} - \theta_a \frac{\sigma v^2 \Delta t}{2} \frac{\partial^2}{\partial z^2} \right] A^{n+1}$$

$$= Js + \left[\frac{\sigma}{\Delta t} + (1 - \theta_a) \frac{\sigma v^2 \Delta t}{2} \frac{\partial^2}{\partial z^2} - \sigma v \frac{\partial}{\partial z} \right] A^n. \qquad (E3.4.2)$$

The solution at iteration $n + 1 (A^{n+1})$ is calculated based on the solution obtained previously at step A^n. A symmetric matrix is obtained because of the individual weighting factors. A finite element code is used by researchers in [47] to update the magnetic vector potential at every step. If a defect is present in the wall, the defect edge would force a magnetic field component perpendicular to the direction of motion. Furthermore, due to this radial field component, currents are induced at the defect edges, altering the usual MFL signal.

3.3.3 Sensitivity of the MFL Technique

Classification methods exist for the many types of defects based on fracture mechanics considerations. In general, defects are seen as either non-injurious or potentially hazardous. MFL can only detect some of these; nevertheless recorded defects are assessed for both depth and width in order to determine their threat to the integrity of the pipeline. A defect width can be estimated with MFL by analyzing the signal duration, whereas information about a defect depth is contained in the signal amplitude which is compared to calibration runs in pipelines with known defects. Near and far-side corrosion pits give different signals, so it is important to distinguish between these two types of defects before a test run is performed. The distinction can be made by comparing responses of slightly different types of detectors that react in a distinct manner to various defects.

It has been observed, for instance, that a special Hall probe configuration is needed for near-side corrosion pits. Also, the sensitivity of the sensor system depends on the operating point in the magnetic saturation curves of the investigated material, as well as the position of the Hall sensor. If the operating point on the magnetic saturation curve is too low, the material is not magnetically saturated in the defect region, and the recorded signals are too weak. A high operating point on the saturation curve is not desired either, as the change in magnetic flux (given by the slope) around the defect area is in this case too small, and therefore the sensor signal too weak. As such, the operating point of the magnetic system needs to be optimized in order to maximize the change in the magnetic flux in the region of a defect. MFL sensitivity depends on the change in field, not just the magnitude of the field.

Signal sensitivity in MFL testing is dependent on a number of other factors, as well, which include the velocity of the sensor, and the magnetic permeability variations in the pipeline wall. Raw MFL signals should not be used for a direct evaluation of defects, as magnetization level, scan velocity, remanent magnetization, and material stress level will all contribute to the shape and magnitude of the recorded signals. Most experimental parameters will vary from one scan to another, altering MFL signals, thereby resulting in misleading information if used inappropriately. As such, various processing techniques have been developed through which MFL signals are made invariant to these factors in order to obtain a more reliable defect detection.

To better visualize these experimental effects and their influence on the MFL signal, consider Example 3.4. Notice that the last two terms of relationship (E3.4.1) become significant at high sensor velocities. In particular, the last one determines the shape of the MFL signal because of the leakage field which depends not only on material characteristics, but also on scan velocity and overall geometry of the experiment. When the probe moves, currents are induced around the defect that distort the shape of the defect signal. Also, an induction due to motion is affecting the signal dc level. Furthermore, operational variables are not constant and vary along the magnetization curve which in turn is determined by the type of ferromagnetic material, as well as any residual stresses that have been trapped.

All things considered, various techniques had to be developed to compensate for these disparities to render the signal more or less invariant to test parameters, prior to characterization. Some methods employ a feature extraction system, while others rely on a signal restoration scheme. Which method is used depends on the details of the experiment, on what parameters are known. If the magnetization curve is not known, it is necessary to extract information that is permeability invariant due to the lack of details about the latter. However, this can only happen if the sensors move with known velocity so that the invariance algorithms compensate only for the those variables that are not desired, while staying sensitive to defect details [48]. Such invariance algorithms rely on shape or statistical descriptions of the signal, nevertheless they work for MFL signals if the shape is compensated for, and the statistical behavior is preserved. If a signal restoration scheme is employed, a restoration filter is constructed by computing a filter kernel or filter coefficients dependent on the value of the distortion parameter. For instance, experimentally obtained low velocity (basically, static), and velocity-distorted signals are compared so that the restoration filter will deconvolve the effects of tool velocity. Such details go beyond the scope of this book; therefore, interested readers should consult [48] and references therein.

Example 3.5. Comment on another method of modeling velocity effects on MFL signals, and compare to that described in Example 3.4

Answer: As another example of a theoretical prediction of the influence of tool velocity on the recorded MFL signals from defects, the method employed by D. Zhiye et al. [43] is described briefly. The theory starts from the acknowledged fact that sources, fields, velocity, and defects in a ferromagnetic material to be inspected interact together in an intricate way and therefore give rise to complicated signals. As MFL probes move at rather high velocities, the eddy currents that are induced in the steel pipe wall because of the relative movement between the exciter and the wall play a significant role. These eddy currents change the profile of the electromagnetic field, increasing the difficulty of MFL signal interpretation, and correct identification of defects. Conventional magnetostatic models are therefore unsuitable for these dynamic conditions, because the distribution of the two field components used for defect analysis (radial or axial *B*) are distorted by the eddy currents.

The MFL method is based on detecting anomalies in the pipe wall. Thereby, the cross section of the pipe changes in time during the journey of the MFL probe. This calls for a time-transient analysis. A reference frame needs to be placed on the moving tool by employing a moving boundary. In this reference frame, the relative velocity is set to zero. A magnetic scalar potential φ is used for the whole domain, while a magnetic vector potential *A* defines the iron region. This splits the magnetic field *H* into four parts

$$\boldsymbol{H} = \boldsymbol{H}_{\mathrm{s}} + \frac{1}{\mu}\nabla \times \overset{\mathrm{r}}{\vec{A}} + \nabla\,\varphi + \sum \left(\frac{1}{\mu}\nabla \times \overset{\mathrm{r}}{\vec{A}} \right), \qquad (E3.5.1)$$

where $\boldsymbol{H}_{\mathrm{s}}$ is the source field due to known total currents, and $\sum (\frac{1}{\mu}\nabla \times \vec{A})$ is the total source field due to unknown currents in the voltage-driven solid conductor loops of the coil winding. To this relationship (E3.5.1), Ampere's law, Faraday's law, and the condition of the solenoidality of the flux density are applied, yielding a differential equation in the conducting region. For the non-conducting region, a much simpler expression is obtained. The problem region is then meshed and the magnetic scalar field is represented by node-shape functions, while the magnetic vector potential field is represented by edge-shape functions.

By applying Galerkin's scheme, the field equations are expressed in matrix form and solved. Results indicate distributions of eddy currents produced at the inside wall under probe moving conditions (e.g., 10 m/s), and in the presence of a defect (L 20 mm × W 10 mm × D 4 mm). They also show that the maximum of eddy currents occur in the pipe wall in the vicinity of the defect, a fact which is not surprising. Furthermore, velocity-induced eddy currents are also significant at the location of the coils terminal. Eddy currents exist due to the moving probe even when the excitation current is dc, and are dependent on the velocity itself. With increased velocity, the eddy currents tend to concentrate more on the outside of the wall due to the skin-depth effect. The profile of the magnetic field is noticed to be distorted because of the induced eddy currents. MFL signals are affected in both shape and amplitude by the eddy currents. This simulation compares to Example question 3.4 in as much as it accounts more for the eddy currents set up in the steel wall. Nevertheless, both examples acknowledge the influence of velocity effects on the MFL signal, and stand in contrast to other calculations which deal only with the magnetostatic case.

3.3.4 Effects of Stress on MFL Signals

It seems that quite a few people ignore the important consequences of stress when it comes to giving the correct interpretation of an MFL signal. This is particularly noticeable in theoretical assessments and analysis that involve computer simulations without an experimental comparison. Stress is rarely taken into account when estimating defect sizes from MFL measurements. Most corrosion pits occur in pipelines while these are operational, and therefore under stress. There are numerous studies by the Applied Magnetics Group at the request of the Gas Research Institute (Columbus, Ohio) that emphasize detectable differences in MFL signal under conditions of stress. Studies show that stress can change the MFL signal by more than 50%, depending on the magnetization of the pipe wall [49]. For instance, a circumferential hoop stress changes the direction of the easy axis toward the hoop direction. Therefore, the permeability increases in the circumferential direction and diminishes toward the pipe axis. In this case, more magnetic flux will flow through the steel around circular defects in axially magnetized pipes, than will leak out of the pipe wall. As such, an MFL signal decreases with increased hoop stress. Conversely, and axial tensile stress increases the permeability along the axis of the beam while reducing it in a direction perpendicular to the beam axis. An axial stress thus enhances MFL signals.

It should also be noted that the effect of stress on the MFL signal is at a maximum for a particular value of the magnetic flux [49]. A steep increase in MFL peak-to-peak value is seen above a critical flux density, fact associated with a decreased strength of the detour flux dipole that develops at the bottom of the defect at low flux densities. The dipole is believed to generate an opposing magnetic field in the defect region. As such, at low magnetic fields, this dipole reduces the flux diverted into the leakage field. On the other hand, at high fields, the permeability of the region at the bottom of the pit decreases due to the high flux concentration there. The strength of the dipole is decreased and more flux is diverted into the air causing a steep increase in MFL signals.

Changes in MFL signals have been particularly observed to occur when the applied stress is high enough to create plastic deformation in the regions of stress concentration near corrosion pit type of defects. A defect that penetrates more than half into the thickness of the pipe wall develops almost twice the magnitude of applied stress at the edge perpendicular to the direction of applied stress. These edges around the stress concentration areas become plastically deformed and the number of dislocations increases in those regions. When tensile plastic deformation is reached, residual compressive stresses develop parallel to the deformation axis. In this case, microstresses are created that impede domain wall motion and domain rotation, increasing magnetic hardness. Magnetic hardness at defect edges causes more flux to leak out, and less flux to flow through the material around the defect. As such, magnetic Barkhausen noise evaluations can offer complementary information on the stress state near a defect in a steel pipe. The technique is therefore used in some laboratory studies to assist interpretation of stress-dependent MFL results.

Another argument in favor of why another magnetic nondestructive testing technique such as MBN can aid in MFL research is given by the change in magnetic properties with stress, a change that can be detected indirectly. It is a well known fact that magnetic coercivity, permeability and other magnetic properties of steel are significantly affected by stress. This is because the latter increases the magnetoelastic energy, which in turn changes the 180° domain volume and domain wall population, modifying the pinning energies of wall movement. A rapid and quite large increase in MBN_{energy} is noticed near the defect edge when an external hoop stress is applied. However, if plastic deformation has occured in the defect area, MBN_{energy} results are smaller than before plastic deformation. Therefore, MBN_{energy} variations with stress can even be used to construct calibration curves for estimations of stress concentrations near defects.

The climatic conditions to which transmission pipelines are usually exposed range from sub-zero temperatures to hot desert environments. Unstable ground, excavation, and construction equipment can also leave an imprint on the walls of pipelines. Corrosion pits, manufacturing, operating, and environmental stresses as well as human-induced mechanical damage all add up to the number of factors affecting these pipeline structures while in use over many decades. Quite often these factors accumulate in densely populated areas, leading to catastrophic failure and the loss of human lives. Over the years, a number of nondestructive inspection techniques have been developed for detection of damage in pipelines. MFL has established itself as the most widely-used in-line inspection technique for the evaluation of wall defects in pipes, measuring a leakage field that contains significant information about the condition of the wall. Therefore, it is likely that the interest in this technique will continue to exist over the coming years, even if newer inspection techniques are bound to be developed and put into practice.

3.3.5   The Future of Magnetic NDT

The aging infrastructure and raised public consciousness regarding the quality of pipelines in service have placed higher demands on the accuracy and reliability of magnetic inspection techniques. But pipelines are not the only steel products that necessitate defect inspection. Cans made of alloys that contain iron and are magnetic are often inspected by MFL, in order to detect changes in permeability brought about by nonmagnetic defects. These can be oxide inclusions such as Al_2O_3, MgO, $CaO–Al_2O_3$ which cause cracks in the can due to the drawing process of the sheet. Rates of thickness reduction of 70% can be obtained by drawing; hence, the stress factor on these sheets is very high. Quite often, flux-sensitive sensors such as Hall probes or magnetodiodes are used for leakage flux inspection before the sheets are drawn. Nevertheless, no magic single NDT method can satisfy these days the ever-increasing need for defect detection and characterization, thereby requiring multiple evaluation approaches in order to obtain more information. Any industrial NDT method must be at least capable of detecting signals produced due to inter-actions with the defect, while being sensitive enough to small variations in defect dimensions. An increased defect sensitivity, faster testing speeds, longer scanning distances, and results trustworthiness are among the minimum expectations that the magnetic NDT industry is facing. Thus, researchers are taking advantage of every possibility of investigation that alternative modalities associated with magnetic phenomena are offering in order to obtain further defect data.

Standard magnetic evaluation techniques are frequently supplemented with other forms of information gathering related to the condition of the pipe. More knowledge is gained by exploiting additional physical processes initiated by a particular inspec-tion technique. New transducer designs, or stronger excitation sources are only a couple of examples of high priority development areas in research laboratories. Usually, a wealth of information is obtained after a scan; however, not all of it is processed optimally. If more details about possible defects could be extracted from the usual recordings, an existing inspection method could be tapped extra and taken further without major hardware modifications. But no matter what modality of obtaining a more comprehensive picture of the state of the inspected pipeline is chosen, the fact remains that all magnetic NDT techniques are experiencing an increased pressure of producing more with less and in a shorter amount of time. The future of magnetic NDT looks bright, if not by choice, at least out of necessity.

Tests, Exercises, and Further Study

P3.1 Compare MBN signal profiles observed in spherodized cementite, hard martensite, and pearlite specimens. *See for instance* [30, 50].

P3.2 Comment on observed MBN activity in spike or residual domains. *See for instance* [51].

P3.3 What are the effects of fatigue and overstressing on the MBN response of case-carburized En36 steel? *See for instance* [52].

P3.4 Name at least three sources of noise that can interfere with defect detection in RFEC testing.

P3.5 Design guidelines for RFEC sensors were developed from observations of various sensor resolution, recorded signal strength, as well as distortion of signal. Name at least four (4) guidelines for sensor design acquired through experience. Note that sometimes a compromise is required, as improving one aspect of sensor design may worsen another one. *See for instance* [33].

P3.6 Name three types of RFEC sensors that would be optimum for measuring: (a) the radial, (b) axial, and (c) circumferential field component. *See for instance* [33].

P3.7 Give a brief description of the RFEC method for defect inspection.

P3.8 What are some advantages and disadvantages of the MFL method for defect inspection as compared to RFEC?

P3.9 RFEC responses from axial defects in aluminum pipes are easily observable, whereas circumferential slits go mostly undetected. On the other hand, in steel pipes, the circumferential slits give significant responses, while axial slits result in much smaller signals. What are these differences attributed to?

P3.10 Describe in summary how the "prime frequency formula" can be used to determine an eddy current test frequency for a homogeneous, non-ferromagnetic tube of relative magnetic permeability ~ 1.

P3.11 Why is eddy current (not RFEC) testing of ferromagnetic materials not straightforward?

P3.12 Comment on so-called magnetically saturated windows in RFEC testing, their primary advantage, and the effect these have on overall tool length. What is the effect of using only one magnetic saturation window?

P3.13 Summarize the principles behind MFL NDT.

P3.14 Circumferential pipe wall stresses due to line pressure of the oil or gas can be as high as 70% of the yield strength. There can also be axial stresses in the pipes due to localized bends or natural, environmental conditions. How does stress, in particular hoop and axial tensile stress present in the ferromagnetic pipe wall change the MFL signal? Are corrections necessary?

P3.15 Comment on modeling velocity effects on MFL signals in [53], and compare to those described in Examples 3.4 and 3.5. Are corrections necessary?

P3.16 Comment on the effects of magnetic anisotropy on MFL detector signals. *See for instance* [54]. Are corrections necessary?

P3.17 How are MFL "pigs" propelled through the pipeline, and what equipment is necessary to ensure they function appropriately?

P3.18 Give an example of a finite element model that predicts the size, shape, and location of a defect. *See for instance* [55].

P3.19 Unlike corrosion pits which are usually associated with climatic and environmental hostile conditions, mechanical damage to the steel pipe wall may occur due to human manipulation. This can actually happen more often, especially in densely populated areas where excavation is more common. Metal loss, gouging, and denting are among the usual forms of mechanical damage experienced by pipelines. Describe a theoretical model that looks specifically at mechanical damage and its signature in the MFL signal. How can this damage be distinguished from corrosion pits influences on the MFL signal? *See for instance* [56].

P3.20 How can finite element analysis predict the effect of residual stresses on the MFL signal? Discuss such an analysis. *See for instance* [57].

Hints and Partial Answers to Problems in Chap. 3

The answers to these problems are in many instances just a hint as to where further information is to be found. They contain the main idea, but the reader needs to find the references that are sometimes recommended for each exercise and complete the answer through further bibliographic reading and study. Reference numbering follows the one used in that particular chapter.

P3.1 For spherodized cementite particles, a single peak in the undeformed state was followed by peak broadening with three overlapping peaks. It is assumed that cementite lamellae provide strong directional pinning to domain wall motion, as previous reports indicate. Hard martensite also displayed multiple MBN peaks with strain, whereas pearlite and a martensite specimen tempered at 180°C showed only a single peak. However, that peak increased systematically with strain. It should be noted that the MBN parameters employed in these studies are quite different from the ones described above in earlier sections. Instead, magnetization curves were recorded and it is on these curves that one or more peaks were noticed. All MBN signal changes were reversible when loads were removed while still in the elastic stage of deformation, which is expected MBN behavior.

P3.2 Three peak MBN profiles were observed in the magnetization curves of unstrained mild steel and were attributed to spike (or residual) domains. These are usually nucleated and annihilated in the knee region of the hysteresis loop, but some are retained in the specimen upon saturation. Because some spike domains are retained, reversal of the field causes abrupt nucleation of new domains when the knee region is reached again. Around saturation, the incomplete annihilation of these spike domains causes some MBN activity, as other researchers pointed out when they observed a second peak at higher fields in the magnetization curve of low carbon steel.

P3.3 Given the type of steel and its mechanical processing, these specimens had a sharp change in microhardness with depth level, so that a crack initiated at the surface propagated very fast into the material. Furthermore, dislocations had a small chance to form prior to crack formation, as smaller stress levels did not allow dislocation initiation, while crack formation preceded larger stresses necessary for developing dislocations. Magnetic behavior was therefore observed to vary under these circumstances. MBN peak height increased in the vicinity of the crack; hence this technique was able to detect the crack location. Nevertheless, crack growth represented only a small fraction in the fatigue life of the case-carburised steel, with specimens failing soon after crack initiation. On the other hand, detection of residual stresses proved to be a more useful indicator of impending failure, since the MBN technique was able to better assess the maximum level of bending stress prior to crack initiation. Alerting users before cracks have a chance to form is definitely a more favorable alternative to finding out that the component is about to fail as soon as a crack is detected.

P3.4 Sources of noise in RFEC can be from vibration, elevators, electronic equipment, background variation, and other interfering electromagnetic fields.

P3.5 A few guidelines are: (a) it is recommended to use a shield around a sensor which measures the axial field component in order to improve the resolution and decrease the signal strength; (b) the smaller the sensor is by design, the better the resolution and the smaller the signal strength; (c) the coil and core length of a sensor that measures the axial field component has a direct influence on the resolution of the sensor; (d) using a core in the sensor may have an adverse effect on the resolution; however, a soft iron core will boost signal strength causing little or no distortion in the detected field.

P3.6 (a) radial field: coil with soft iron core, made as small as possible; (b) axial field: very short shielded "U" core sensor; (c) circumferential field: a good axial sensor, rotated 90° about the radial axis.

P3.7 The RFEC NDT is based on an electromagnetic interaction with surface or through wall defects in conductive pipes. It takes place in the vicinity of a solenoidal excitation coil, although different zones in the pipe can be distinguished. The most useful zone in the pipe is that in which there is no direct coupling from the exciter, but there is coupling from the flux that has penetrated the pipe walls. Detector signals are attenuated and shifted in phase and can have a distinct signature in the vicinity of a defect. RFEC is a reasonably reliable technique for inspecting ferrous and nonferrous metal tubes when the tube can be accessed from the inside only.

P3.8 Because MFL is primarily a magnetic anomaly detector, it is effective for detecting pits and certain cracks in steel pipes. However, it is relatively insensitive to gradual wall thinning, as well as to cracks aligned with the magnetic field. For these purposes, RFEC is a better option.

P3.9 The differences in RFEC responses are due to the fact that steel is a ferromagnetic material with a high relative permeability, whereas aluminum is not ferromagnetic with a relative permeability approximately equal to that of air.

P3.10 An eddy current test frequency can be established for a non-ferromagnetic, homogeneous tube with relative magnetic permeability ~ 1 by using the "prime frequency formula"

$$f_{\text{test}} = \frac{10\rho}{t^2}, \tag{P3.10}$$

where f_{test} is the test frequency, ρ is the electrical resistivity, and t is the material thickness. This formula allows the application of a calibration curve that can accurately predict the depth of a flaw based on its measured phase angle.

P3.11 When using eddy currents (not RFEC), testing of ferromagnetic materials is not straightforward due to the variation in magnetic permeability noise and its lack of predictability. These are important factors that influence the accuracy of the electromagnetic test. Due to permeability variations, a formula for calculating a test frequency by using solely eddy currents cannot be reliably used.

P3.12 During RFEC inspection, the regions where the two coils are located (exciter and detector) could be magnetically saturated. Magnetic saturation would allow for higher operating frequencies and scanning speeds during testing. The higher frequencies would suppress any unwanted signals caused by local variations in the permeability of the pipe wall. However, for best results, it would be better to saturate just the regions near the two coils while leaving the pipe wall in between them unmagnetized. The saturated windows at the exciter and detector coils are in the energy transmission path. A high magnetic permeability guides the flow of energy on the outside of the tube closer to the wall, and thereby decreases the attenuation in the axial direction. The direct field coupling is also attenuated, for the same reason. This allows the two coils to be brought closer together so that the overall tool length could be correspondingly reduced. Nevertheless, the primary advantage of using magnetically saturated windows is that it allows the use of higher frequencies which permit higher scanning speeds. If only one saturation window is used such as near the exciter, the dominant interaction with the pipe wall is near the detector. Therefore, it is more beneficial to have the window near the exciter, to allow the detector to record defects as it passes them.

P3.13 An MFL tool called a "pig" is powered through a pipeline. The "pig" contains powerful permanent magnets held against the pipe wall with the help of brushes. This inspection "pig" magnetizes regions of the wall between the brushes. If a defect is present in the pipe wall, it will distort the magnetic flux distribution around and within the defect. Some magnetic flux will therefore "leak out" of the defect. The leakage field, in particular the axial or radial component, is detected using Hall probes. Recorded signals can distinguish between defects of different width and depth.

P3.14 Many magnetic properties of steel are affected by stress. For instance, a 330 MPa hoop or axial tensile stress is common in many underground pipelines. Significant variations in the MFL signal are observable when the stress in the pipe is high enough to create plastic deformation near pit defects. This is because a pit changes the stress distribution in the surrounding area of the pit, and can result in high local stress concentrations. Therefore, the magnetic permeability of the wall in the pit area, as well as other magnetic properties, changes. For this reason, stress can significantly change the MFL signal. Yet, stress is not often taken into account when estimating defect sizes from these MFL measurements. The signal will also change depending on the magnetization state of the pipe wall, such as while magnetically saturating the wall during inspection. Remember, the MFL probe magnetizes the pipe wall to near magnetic saturation in order to detect the flux leakage in and around a defect. Reference [49] reports that a tensile stress in a composite steel beam increases the peal to peak MFL signal, while hoop stress on a hydraulic pressure vessel decreases the same type of signal.

Consequently, corrections are necessary to account for such factors *See* [49] *for further details.*

P3.15 In the mentioned reference, a 2D finite element model and a time step algorithm are used to investigate MFL signal variations due to different specimen velocities with respect to the sensor. This particular model uses the Leismann–Frind time-step algorithm similar to the one used by the researchers in Example 3.4; however, a 2D finite element model is preferred over the simplified 1D. It was found that changes in velocity affect base signal levels and peak-to-peak values, but these variations are not proportional to the velocity itself. Motional induction currents are produced inside the conducting material which set up their own magnetic fields. As such, the sensor detects not only leakage fields caused by defects, but also velocity induced fields. In contrast to this approach, Example 3.5 emphasized the effect on the MFL signal of induced eddy currents. From the analysis in [53], it is concluded that optimal operating conditions that yield the maximum defect signal can be predicted, including the most appropriate velocity and sensor location. All three examples/models acknowledge velocity influences on the MFL signals. Thus, corrections must be made for velocity.

P3.16 Changes in the *amplitude* of the experimentally measured MFL signals are believed to be due to stress-induced changes in the bulk magnetic anisotropy, also known as the magnetic easy axis of the steel pipe wall. On the other hand, changes in MFL signal *patterns* are attributed to defects acting as local stress raisers, in particular surface stress concentrations. 3D finite element calculations undertaken in [54] confirm that this is the case. The calculations were performed using both isotropic and anisotropic pipe wall permeabilities. These model the effects of global or bulk changes in magnetic anisotropy without variations due to any local stress concentrations around defects. The results are in good agreement with experimental measurements and indicate that it is important to correct for these effects in order to interpret MFL signals accurately. Additionally, signals depend not only on the easy axis and the defect, but also on velocity, line pressure, and other running conditions.

P3.17 The MFL "pig" is propelled through the pipe by oil (or some other pipe product) acting on elastomeric cups which ensure that there is a small differential pressure across the "pig." The "pig" is free swimming and for this reason carries battery power supplies, electronics for signal processing, as well as high density magnetic tape data storage systems. Modern "pigs" are fitted with hundreds of sensors in order to acquire detailed MFL signals.

P3.18 From [55]: A large number of uncertainties are expressed as "noise" in the MFL signal. The finite element method may help distinguish this noise from the actual defect signal. One approach is to reconstruct the crack shape by finding correct values of specific parameters such as length, depth, and size given certain performance criteria. This is known as an inverse problem where information is extracted from the field distribution measured by the sensor. The problem is solved iteratively until the system output matches the input MFL signal. The authors in [55] use topology design and shape optimization for defect identification. Specifically, a topological sensitivity analysis explores the parametric search space, predicting an appropriate initial structure as an initial starting point in the optimization. Shape optimization is used based on a "continuum design sensitivity analysis" (CDSA) in order to obtain smooth boundaries of the resulting structure. The solution is then compared to the noise signal to provide useful information related to improved sensor design. See [55] for further details.

P3.19   During MFL inspection, wall irregularities are detected and stored within the "pig." Signals are dependent on the magnetic permeability of the material, which in turn is dependent on material chemistry, the level of magnetization, and stress. Intelligent algorithms for defect characterization require large amounts of acquired MFL data; therefore, it is necessary that a reliable numerical model exists to simulate the effects of all forms of defects on the recorded signal, not only those due to corrosion pits. As such, mechanical damage leaves its own signature on an MFL signal. The model presented by the authors in [56] contains two stages: (1) a structural analysis is carried out and the stress distribution associated with mechanical damage is found, and (2) a stress distribution model is incorporated in a magnetic finite element representation used to predict MFL signals caused by mechanical damage. For instance, stress distributions resulting from a gouge are obtained from a static elastic model. On the other hand, a finite element model is built for the magnetic properties of pipe wall by taking into account stress, and separately for comparison, the lack of stress. The stress distribution from step (1) is mapped onto the elements of the pipe section geometry. See [56] for further details.

P3.20   From [57]: A 3D magnetic finite element analysis was used to simulate the effect of localized residual stresses on MFL signals from a steel plate that contained no geometrical defects. Simulating a residual stress pattern involved a structural modeling of dents. Nevertheless, the effect of residual stresses on MFL signals is not limited to dents. Residual stresses can be created by a variety of factors. In their work, the authors needed a residual stress pattern to incorporate into a magnetic finite element model. Residual stresses change the local magnetic anisotropy, perturbing the magnetic flux path. The magnetic model took into account appropriate directional permeability values in magnetically anisotropic materials. Locally assigned, anisotropic permeability functions were incorporated into the magnetic finite element model. A close collaboration with Infolytica software manufacturers was necessary to accommodate these complex requirements. The model has produced a successful application of finite element magnetic simulations to the examination of stress effects on MFL signals. See [57] for further details.

References

1. K. Tiito, *Solving Residual Stress Measurement Problems by a New Nondestructive Magnetic Method*, Nondestructive Evaluation: Application to Materials Processing, ed. by O. Buck, S.M. Wolf (ASM, Materials Park, OH, 1984), p. 161
2. X. Kleber, A. Vincent, NDT&E Int. **37**, 439 (2004)
3. B.D. Cullity, *Introduction to Magnetic Materials*, 2nd edn. (Addison-Wesley, New York, 1972)
4. M. Blaow, J.T. Evans, B. Shaw, Mater. Sci. Eng. **A386**, 74 (2004)
5. D. Utrata, M. Namkung, Rev. Progr. Quant. Nondestr. Eval. **2**, 1585 (1987)
6. R.S. Tebble, Proc. Phys. Soc. Lond. B**68**, 1017 (1955)
7. T.W. Krause, L. Clapham, D.L. Atherton, J. Appl. Phys. **75**(12), 7983 (1994)
8. V. Babbar, B. Shiari, L. Clapham, IEEE Trans. Magn. **40**(1), 43 (2004)
9. T.W. Krause, L. Clapham, A. Pattantyus, D.L. Atherton, J. Appl. Phys. **79**(8), 4242 (1996)
10. C. Jagadish, L. Clapham, D.L. Atherton, IEEE Trans. Magn. **25**(5), 3452 (1989)
11. L. Piotrowski, B. Augustyniak, M. Chmielewski, Z. Kowalewski, Meas. Sci. Technol. **21**, 115702 (2010)
12. C. Jagadish, L. Clapham, D.L. Atherton, NDT Int. **22**(5), 297 (1989)
13. C. Jagadish, L. Clapham, D.L. Atherton, J. Phys. D Appl. Phys. **23**, 443 (1990)
14. C. Jagadish, L. Clapham, D.L. Atherton, IEEE Trans. Magn. **26**(1), 262 (1990)
15. P. Zerovnik, J. Grum, G. Zerovnik, IEEE Trans. Magn. **46**(3), 899 (2010)
16. T.W. Krause, K. Mandal, C. Hauge, P. Weyman, B. Sijgers, D.L. Atherton, J. Magn. Magn. Mater. **169**, 207 (1997)
17. H. Kwun, G.L. Burkhardt, NDT Int. **20**, 167 (1987)
18. C.-G. Stefanita, L. Clapham, D.L. Atherton, J. Mater. Sci. **35**, 2675 (2000)
19. C.-G. Stefanita, D.L. Atherton, L. Clapham, Acta Mater. **48**, 3545 (2000)
20. C.-G. Stefanita, L. Clapham, J.-K. Yi, D.L. Atherton, J. Mater. Sci. **36**, 2795 (2001)
21. H. Mughrabi, Acta Metall. Mater. **31**, 1367 (1983)
22. L.E. Murr, Met. Trans. **6A**, 427 (1975)
23. J.-K. Yi, *Nondestructive Evaluation of Degraded Structural Materials by Micromagnetic Technique*, Ph.D. Thesis, Department of Nuclear Engineering, Korea Advanced Institute of Science and Technology, Taejon, Korea (1993)
24. H. Kwun, G.I. Burkhardt, *Electromagnetic Techniques for Residual Stress Measurement*, in Metals Handbook **17**(9), 159 (1989)
25. W. Six, J.I. Snoek, W.G. Burgers, De Ingenier **49E**, 195 (1934)
26. H. Sakamoto, M. Okada, M. Homma, IEEE Trans. Magn. **23**, 2236 (1987)
27. G. Bertotti, F. Fiorillo, A. Montorsi, J. Appl. Phys. **67**(9), 5574 (1990)
28. D.H.L. Ng, K.S. Cho, M.L. Wong, S.L.I. Chan, X.-Y. Ma, C.C.H. Lo, Mater. Sci. Eng. **A358**, 186 (2003)
29. J. Kameda, R. Ranjan, Acta Metall. **35**, 1515 (1987)
30. M. Blaow, J.T. Evans, B.A. Shaw, Acta Mater. **53**, 279 (2005)
31. T.R. Schmidt, Mater. Eval. **42**(2), 228 (1984)
32. B.J. Mergelas, D.L. Atherton, Br. J. NDT **35**(10), 568 (1993)
33. J.C. Winslow, *High Resolution Detectors for Remote Field eddy Current Probes*, M.Sc. Thesis, Dept. of Physics, Queen's University, Kingston, ON (1995)
34. C.S. Stellema, *Remote Field Eddy Current Inspection of Small Bore Ferromagnetic Tubes*, M.Sc. Thesis, Dept. of Physics, Queen's University, Kingston, ON (1993)
35. S.M. Haugland, IEEE Trans. Magn. **32**(4), 3195 (1996)
36. D.L. Atherton, W. Czura, D.A. Lowther, IEEE Trans. Magn. **26**(5), 2863 (1990)
37. C.V. Dodd, W.E. Deeds, J. Appl. Phys. **39**(6), 2829 (1968)
38. C.V. Dodd, C.C. Cheng, W.E. Deeds, J. Appl. Phys. **45** (2), 638 (1974)
39. T.R. Schmidt, Mater. Eval. **42**(2), 225 (1984)
40. D.L. Atherton, W. Czura, Electrosoft **2**(2/3), 157 (1991)
41. D. L. Atherton, W. Czura, Res. NDE **5**(2) (1993)

42. D.L. Atherton, O. Klink, T.R. Schmidt, Mater. Eval. **49**(3), 356 (1991)
43. D. Zhiye, R. Jiangjun, P. Ying, Y. Shifeng, Z. Yu, G. Yan, L. Tianwei, IEEE Trans. Magn. **44**(6), 1642 (2006)
44. P.C. Porter, H.A. French, D.L. Atherton, Proc. Int. Conf. Pipeline Inspection, Edmonton, June 1983, p455
45. G.S. Park, E.S. Park, IEEE Trans. Magn. **38** (2), 1277 (2002)
46. H. Leismann, E. Frind, Water Resour. Res. **25**, 1133 (1989)
47. G. Katragadda, W. Lord, Y.S. Sun, S. Udpa, L. Udpa, IEEE Trans. Magn. **32**(3), 1581 (1996)
48. S. Mandayam, L. Udpa, S.S. Udpa, W. Lord, IEEE Trans. Magn. **32**(3), 1577 (1996)
49. K. Mandal, D. Dufour, D.L. Atherton, IEEE Trans. Magn. **35**(3), 2007 (1999)
50. M.G. Hetherington, J.P. Jakubovics, J. Szpunar, B.K. Tanner, Philos. Mag. B **56**, 561 (1987)
51. D.G. Hwang, H.C. Kim, J. Phys. D Appl. Phys. **21**, 1807 (1988)
52. V. Moorthy, B.A. Shaw, P. Hopkins, NDT&E Int. **38**, 159 (2005)
53. Y-K Shin, IEEE Trans. Magn. **33**(2), 2127 (1997)
54. S. Leonard, D.L. Atherton, IEEE Trans. Magn. **32**(3), 1905 (1996)
55. M. Li, D.A. Lowther, IEEE Trans. Magn. **46**(8), 3221 (2010)
56. P.A. Ivanov, Z. Zhang, C.H. Yeoh, L. Udpa, Y. Sun, S.S. Udpa, W. Lord, IEEE Trans. Magn. **34**(5), 3020 (1998)
57. V. Babbar, B. Shiari, L. Clapham, IEEE Trans. Magn. **40**(1), 43 (2004)

Chapter 4
Basis of Magneto-Optical Applications and Materials

Abstract The Faraday effect, as well as the magneto-optical Kerr effect (MOKE), dates back to the 1800s when it was noticed that optical phenomena can be influenced by a magnetic field. Although distinct, the two effects are commonly treated together. Furthermore, given their versatility they have resulted in applications that span over a century while involving a variety of magnetic materials. Magneto-optical recording and MOKE systems are just two examples of such applications, enjoying widespread use at different points in time, and continuing to be currently available in some technological form. As such, MOKE systems are nowadays sophisticated instruments that are prevalently employed in the characterization of thin magnetic films of a certain type. These MOKE systems are based on polarization changes experienced by light beams traveling under a magnetic field. Thus, magnetic–optic interactions allow us to gain some valuable information about, for instance, magnetization orientation or magnetization reversal. For this reason, we need an understanding of the basis of these effects where magnetic and optical phenomena coexist, as well as the type of materials to which such a magnetization direction analysis can be applied. This insight enables us to uncover important contemporary issues in magnetism, such as spin direction in different domains, spin reversal, or the highly desired spin reorientation transition. The latter is a significant effect that occurs in magnetic thin films that have a strong perpendicular magnetic anisotropy, belonging to a class of materials that are today's promising candidates for ultrahigh-density recording media. Nevertheless, before expanding on the subject of today's uses of magneto-optical phenomena, it is equally noteworthy to elaborate on time-honored subjects such as magnetic recording and optical readout by the Faraday–Kerr effect.

It is a known fact that magnetic materials with intermediate values for magnetic anisotropies have historically been employed as magnetic recording media. As such, traditional magnetic tapes are made of ferromagnetic solid solutions of $\gamma - Fe_2O_3$–Fe_3O_4, a substance that usually presents itself in the form of acicular particles with certain magnetic characteristics. Aside from these characteristics being measurable with MOKE systems, this unique acicular shape gives them a uniaxial magnetic anisotropy and consequently a good coercivity which is strongly

C.-G. Stefanita, *Magnetism*, DOI 10.1007/978-3-642-22977-0_4,
© Springer-Verlag Berlin Heidelberg 2012

dependent on the ratio of ferrous to ferric ions. It was noticed that higher coercivity values can be obtained when the ratio Fe^{2+}/Fe^{3+} is greater than 0.04; however these better values are also due to the type of binder in which the particles were dispersed, such as paraffin or epoxy resin. Nevertheless, some of the first magnetic recording materials to which it was possible to apply optical readout by the Faraday–Kerr effect were intermetallic compounds such as MnBi. As such, the Faraday–Kerr effect was initially used to study magnetic domains in MnBi films [1]. MnBi has a hexagonal nickel arsenide crystal structure with a very high magnetic anisotropy constant where the direction of easy magnetization is along the c-axis [2]. Due to the method of preparation, this c-axis was oriented normal to the surface with a spread of only 15°. This detail was particularly important in early experiments of optical measurement, as the domain magnetization vectors needed to have components along the direction of propagation of light. Combined with the fact that the optical setup was made such that domains of either polarity appeared significantly dark, while those of opposite polarity remained noticeably light, the idea of using this high contrast for magnetic recording and optical readout emerged. In fact, the first reported MO recording was in 1957 [3].

Since those early days of magneto-optical recording, the field of magneto-optical applications has expanded, and at the same time shifted toward uses that include in situ vectorial analyses of magnetization measurements in various geometries at different light scattering planes. As such, we are encountering these days a highly developed area of magneto-optical instruments allowing not only precise investigations at different radiation wavelengths, but also high-throughput screening and optimization of magnetic materials with applications beyond magnetic recording.

4.1 Magneto-Optical Measurement Methods

4.1.1 Magnetic Characterization through Faraday Rotation and Kerr Effect

The *Faraday rotation* or *Faraday effect* is a magnetic field influenced optical occurrence, first reported by Michael Faraday in 1845. The effect is observed mostly in optically transparent dielectric materials, in *transmission* of polarized light through a medium, and under an applied magnetic field. As such, a *Faraday rotation* takes place when the plane of polarization of the incident light is rotated proportionally to the component of the magnetic field intensity in the direction of light propagation, as the light passes through the material (Fig. 4.1). The Faraday rotation is an experimental evidence of the interaction between light and a magnetic field. Similarly, the *MOKE* describes changes in polarization of light that is *reflected* from a magnetized material, and takes its name after John Kerr who reported it in 1875 [4]. Quite often, people refer to the two effects, Faraday and Kerr as one, the *Faraday–Kerr* effect.

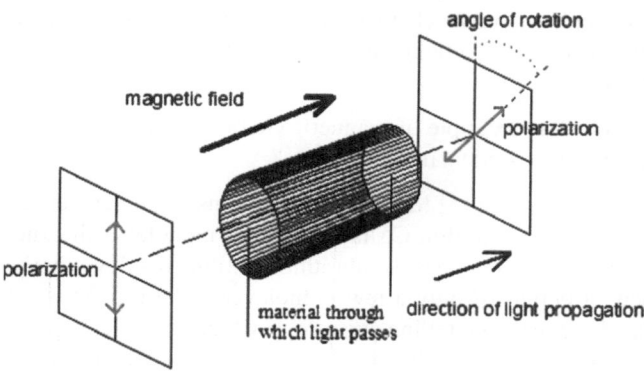

Fig. 4.1 Schematic representation of the Faraday rotation which occurs when polarized light is transmitted through an optically transparent medium while a magnetic field is acting. In this case, the plane of polarization of the incident light rotates proportionally to the intensity of the magnetic field component aligned with the direction of light propagation

Some readers may already be familiar with the more popular applications of the Faraday–Kerr effect, which is particularly expected since the effect enjoys widespread use in magnetic recording and optical readout. For this reason, parts of this chapter are dedicated only to this topic. Nevertheless, we also review other aspects of the Faraday–Kerr effect, such as its ability to be applied to highly accurate magneto-optical characterization techniques. The latter are extensively employed nowadays, and are known as MOKE methods, usually consisting of vectorial analyses of magnetization. In this section, we examine specifically the magneto-optical characterization abilities of the Faraday–Kerr effect, in view of the many developments in more advanced fields of magnetics where MOKE are successfully applied. For instance, many contemporary magnetic systems such as those used in spintronic devices consist of hybrid structures, generally ferromagnetic-semiconductor. Among the most commonly known are epitaxial ferromagnetic thin films grown on semiconductors such as GaAs and InGaAs. However, half metallic magnetic oxides which are not ferromagnetic have gradually gained usage in spintronic devices as well, given the exceptional spin polarization at the Fermi level that is found in these materials. Although we do not concern ourselves in this chapter with the specifics of different types of magnetic materials, we nevertheless reveal the capabilities of magneto-optical techniques to characterize their magnetic properties.

Consider the fact that many spintronic and magnetic recording applications rely nowadays on different spin directions in magnetic domains, or in layers of magnetic ultrathin films. As such, the highly desired property of spin reorientation can be found in devices based on giant magnetoresistance (GMR), tunneling magnetoresistance (TMR), or ultrahigh-density recording media. For this reason, the example below illustrates how magneto-optical measurements can be helpful in observing spin reorientation away from a given direction in ultrathin magnetic films, with an emphasis on the characterization aspect of MOKE methods, and less on the specific type of magnetic material, or the actual experimental setup which is reviewed in another section.

Example 4.1. Give an example of magnetic characterization of ultrathin magnetic films through magneto-optical measurements?

Answer: Recently reported [5] MOKE investigations have uncovered details about magnetic properties of ultrathin Co layers of 5–6 monolayer thickness. This particular material was chosen because ultrathin Co films display a broken symmetry at the surface, at depths of only a few monolayers. The broken symmetry results in an enhanced magnetocrystalline anisotropy which is sufficient to overcome the demagnetizing energy of the Co films. As such, the spontaneous magnetization reorients out of plane as the film thickness decreases. In this study, the ultrathin Co films were grown by electron beam evaporation on a Pd (111) single crystal. Growth of the films was performed in an ultrahigh vacuum chamber that was specially designed to allow in situ magneto-optical measurements, without the necessity of sample transportation. Kerr rotation and ellipticity measurements were carried out with a sensitivity of 0.001° through phase-sensitive detection, by using a photoelastic modulator and crystal polarizers [5]. A rotation of the sample stage by ±60° in the film plane was facilitated by a precision manipulator which allowed various measurement geometries at different scattering planes.

By measuring the rotation of the three-dimensional magnetization vector under an applied magnetic field, it was possible to perform a full vectorial magnetization analysis, as follows. Magneto-optical Kerr rotation signals were measured at three configurations shown schematically in Fig. E4.1: Polar measurements in the yz and xz scattering planes with a magnetic field applied normal to the film, and a longitudinal measurement in the yz scattering plane for a magnetic field much larger than the coercivity, applied parallel to the y-axis. The incident angle of the light beam was 45°. Measured complex Kerr angles relate to the magnetization components along the three axes, so that all magnetization components can be determined. By employing MOKE measurements, the study proves that these are capable of delivering detailed observations of spin reorientation. The spin reorientation was thereby determined to be away from the normal of the film, a behavior that was dependent on the Co thickness. Interpretation of MOKE data also revealed that multidomains of perpendicular magnetization are formed, and switch via collective spin rotation when the magnetic field reaches values near the coercive field.

Most often, researchers are interested in the magnetization direction in a material in order to determine magnetization reversals, or even spin reorientation. However, what is perhaps equally intriguing is that magneto-optical measurements can detect superparamagnetic behavior. A recent study [6] reports on a superparamagnetic phase that was observed to form in a narrow thickness range of 3.5–4.8 monolayers of Fe grown by molecular beam epitaxy (MBE) on GaAs. Similarly, the superparamagnetic phase was also noticed to form in the narrow thickness range of 2.5–3.8 monolayers of Fe on InAs substrates. Through magneto-optical hysteresis loops,

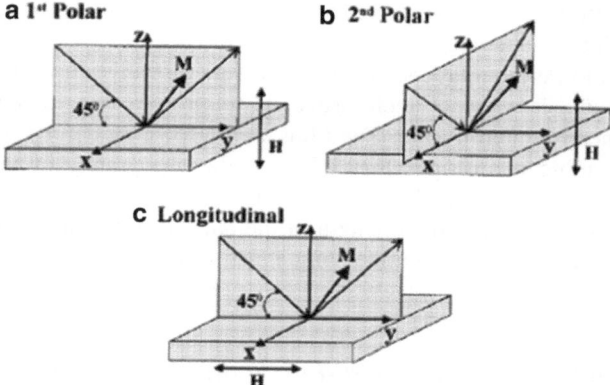

Fig. E4.1 Magneto-optical Kerr effect measurements of the three-dimensional magnetization vector under an applied magnetic field in ultrathin films of Co. The three measurement configurations are as follows: (**a**) polar measurement in the yz plane, (**b**) polar measurement in the xz scattering plane, and (**c**) longitudinal measurement in the yz scattering plane for a magnetic field much larger than the coercivity, applied parallel to the y-axis. Reprinted with permission from [5]

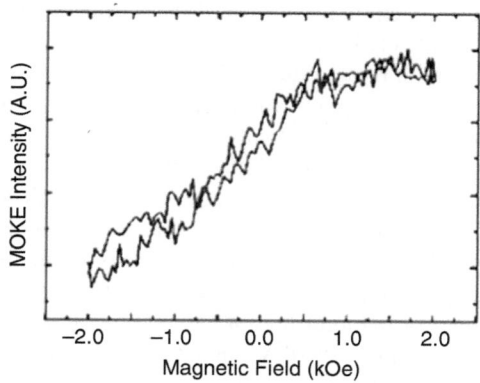

Fig. 4.2 Magneto-optical Kerr effect (MOKE) hysteresis loop obtained in four monolayers of Fe grown by molecular beam epitaxy (MBE) on GaAs. Reprinted with permission from [6]

researchers were able to obtain more information on the exchange interaction in the grown Fe. Specifically, the exchange interaction led to internal ferromagnetic ordering within the three-dimensional Fe clusters formed on the semiconductor substrates, thereby resulting in the observed superparamagnetic behavior. The superparamagnetic response was determined in the measured magneto-optical hysteresis curves obtained for different monolayer thicknesses of Fe, and is exemplified in Fig. 4.2 for the sample with a 4 monolayer thickness. Given that interface properties are an important issue for a successful spin injection between different materials, these results are significant in view of the capability of MOKE techniques to characterize interface effects expected to occur between ferromagnetic monolayers and semiconductor substrates.

In addition to Fe monolayers, Fe_3O_4 films were also investigated in the above-mentioned study, in view of their usage in spintronics due to their high spin polarization near the Fermi level. The films were grown on deformed GaAs (100) substrates, forming nanostripes that followed the topography of the substrates. MOKE measurements revealed that a uniaxial magnetic anisotropy developed with an easy axis perpendicular to the length of the nanostripes, confirming once more the magnetic characterization abilities of magneto-optical techniques.

4.1.2 Optical Contrast Readout of Magnetization Orientation

Consider magnetic thin films that have a certain domain structure. The domain configuration of these thin films can then be determined by optical contrast readout of their magnetization orientation. This type of characterization is possible because an optical contrast is obtained through MOKE in magnetic domains of opposite magnetization orientation. The experiment is set up so that a plane-polarized beam of light is sent perpendicularly toward the surface of the specimen. When the polarized light is reflected back from the magnetized material, the plane of polarization of the beam is rotated due to MOKE. The measurement is especially successful when the domain magnetization is perpendicular to the thin film surface, and magnetic domains appear as optically distinct. This is actually a necessary requirement given that the rotation of the polarization plane occurs when the magnetization vector of each domain has a component along the direction of propagation of the light. In this manner, optical readout can occur by detecting the plane of polarization which has rotated due to reflecting domain regions. With the help of an optical analyzer or a rotation compensator, domains will be seen as either reflecting back the light or extinguishing it, depending on the magnetic polarity of the domain. The *magnetic polarity* of a magnetic domain is given by the *direction of total domain magnetization* in that particular domain.

Specifically, when the light is reflected back by the magnetic domains at the surface of the film, the plane of polarization is rotated either clockwise or counterclockwise, as determined by the reflecting domains. Consequently, when the optical analyzer is rotated, the light from domains of one polarity is extinguished (background of Fig. 4.3a), while the light from domains of opposite polarity is not (background of Fig. 4.3b). This makes the domains where light is extinguished appear as dark, while those domains that reflect light appear as bright. A Kerr–Faraday rotation of only a few degrees already gives a good visual contrast. If the optical analyzer is rotated to the other position of maximum light contrast, the dark and bright regions are reversed. Also, if the domains have an in-plane magnetization instead of the required magnetization perpendicular to the surface, oblique illumination must be used. In this case, the magnetization vector will have a component along the direction of light propagation and the polarization rotation

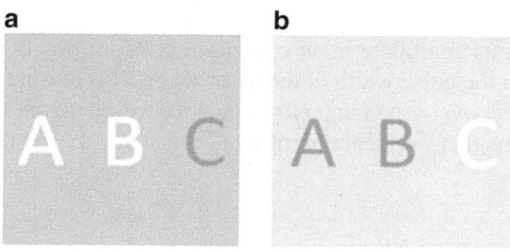

Fig. 4.3 Optical contrast readout of magnetic domains where the magnetic domains are represented as rectangular backgrounds (**a**) and (**b**). Upon rotation of the optical analyzer, (**a**) light from domains of one polarity is extinguished while the background (domain) on the left appears as dark, and (**b**) the background (domain) on the right of opposite magnetic polarity reflects light, appearing as bright. This allows writing on magnetic domains by using the two opposite magnetic polarities so that letters "A" and "B" are written with the tip of a magnetic pen of one magnetic polarity, while letter "C" is written with the other end of the tip of the pen, of opposite magnetic polarity

can occur. Oblique illumination has been applied for the investigation of magnetic domain structure in single crystals of SiFe, and evaporated films of NiFe. On the other hand, direct perpendicular illumination has been historically used for magnetic domain observation in single crystals of cobalt, or intermetallic compounds such as MnBi [3]. Furthermore, with some changes in illumination and overall experimental setup, the technique can be extended to other types of magneto-optical materials. For instance, thinner and more transparent films can be analyzed in transmission utilizing the Faraday effect, while the light beam is inclined at 45° (see [3] and references therein).

The MOKE can also be used for magnetic writing on the film, as exemplified in Fig. 4.3a, b. As such, the film can be written with a magnetic probe tip where each end of the magnetic probe has a different magnetic polarity, so that writing with one end of the probe appears as dark, while writing with the other end appears as bright, determined by the direction of magnetization in the respective region. Of course, the whole procedure occurs under the same experimental conditions as magneto-optical observation of magnetic domains, meaning that a polarized light is sent perpendicular to the surface of the film while an optical analyzer is used for light detection. It should also be noted that in this example, the entire film was magnetized perpendicular to the surface, and that it is necessary for the films to have a high coercive force, in order for the writing not to be easily disturbed by stray fields. The writing on the film can be erased by either saturating or demagnetizing the film perpendicular to the surface.

An interesting proposal for erasure of magnetic information from a thin film involves the absence of a magnetic field [7]. In this case, the laser-induced erasure process is possible due to angular momentum compensation which has an effect on domain wall motion. Angular momentum compensation occurs at a specific temperature when the angular momentum becomes zero and the mobility of the wall increases significantly. It is believed that under these circumstances, the wall of one domain moves inward, while the wall of the adjacent domain moves outward, so that the two domains annihilate in an outside-in collapse process. Domain erasure depends critical on the pulse width of the laser beam. This erasure effect is observed in single layer direct overwrite magneto-optical recording media which differs from the conventional magneto-optical thin films.

4.1.3 Magneto-Optical Kerr Effect Systems

MOKE systems are a class of magneto-optical characterization equipment that are nowadays increasingly used for the investigation of the vectorial direction of magnetization at the surface of thin magneto-optical films, or of magnetic nanostructures in different growth configurations. Depending on the details of the system design, these instruments are at least capable of in situ qualitative evaluations of magnetization direction, or even quantitative measurements during a magnetization reversal process. MOKE systems have evolved in time allowing to be used simultaneously in ultrahigh vacuum chambers, displaying an increased sensitivity to monolayer magnetization, magnetic phase transitions, and magnetization switching in the thickness range of the spin reorientation transition. The basic principles of operation of a MOKE system consist of shining polarized light onto a magnetic sample while under an applied magnetic field, and then passing the reflected and rotated polarized light through an analyzer, before collecting it by a photodetector. A simplified schematic of such an analyzing setup showing only the basic experimental components is displayed in Fig. 4.4. The Kerr rotation is measured by the relative change in intensity, and is proportional in the linear approximation to

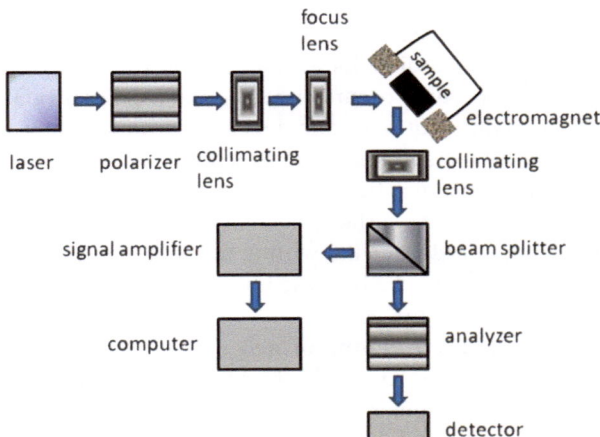

Fig. 4.4 Simplified schematic of a magneto-optical Kerr effect measurement system in longitudinal geometry

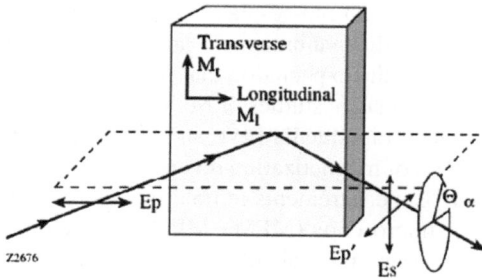

Fig. 4.5 Schematic representation of MOKE measurement geometries. Polar MOKE geometries reveal information about the magnetization component perpendicular to the sample surface, while longitudinal or transverse measurements detect in-plane magnetizations. Reprinted with permission from [10]

the magnetization in the sample which in this example is longitudinal. A second photodetector may be used to subtract ambient light fluctuations and common laser drift.

The output of the analyzing setup consisting of the Kerr rotation measurements in either longitudinal, transverse, or polar geometry vs. applied magnetic field can be plotted on a graph in order to obtain a MOKE hysteresis loop. It should be noted that hysteresis loops in longitudinal or transverse geometry involve an in-plane magnetizing field, perpendicular to the light plane of incidence, and are therefore sensitive to an in-plane magnetization. On the other hand, polar Kerr rotation hysteresis loops are acquired when the magnetic field is perpendicular to the sample surface and incident light is linearly polarized, responding to an out-of-plane sample magnetization. Figure 4.5 shows these measurement geometries. For each measurement configuration, the complex Kerr rotation angles Θ are recorded, their magnitude given in terms of Kerr rotation θ and ellipticity ε

$$|\Theta| = (\theta_k^2 + \varepsilon_k^2)^{\frac{1}{2}}. \tag{4.1}$$

The complex Kerr rotation angles Θ are determined for increasing and decreasing applied magnetic field values, while a *MOKE hysteresis loop* is drawn. This hysteresis loop consisting of a Θ vs. applied magnetic field H plot is very similar to the M vs. H or B vs. H hysteresis loops discussed in Chaps. 1 and 2. Furthermore, MOKE hysteresis loops of Θ vs. H are obtained for each measurement configuration. For situations where the magnetic film has an arbitrary magnetization orientation and the light beam is obliquely incident to the film, there are published, simplified analytical expressions for magneto-optical characterization.

MOKE systems have the ability to perform a 3D vector magnetometry and monitor magnetization reversal in conventional magneto-optical recording films, ultrathin films for high-density recording media, or specialized materials for spintronic applications. Through magneto-optical measurements researchers are also able to quantitatively characterize magnetization reversals in single nanodots, especially when these structures exhibit an out-of-plane magnetization. It is well known that nowadays large-scale periodic nanomagnets are of interest in ultrahigh-density magnetic data storage, with the expectation that one magnetic nanodot will retain a single bit of information. As such, studies of MOKE techniques applied to the characterization of potential storage media have already been published. Consider for instance, the investigation of magnetization reversal in multilayered Co/Pt nanodots [8] that compares MOKE measurements in polar geometry with another technique such as magnetic force microscopy (MFM). MFM has been previously applied for the characterization of films and magnetic structures of sub-100 nm size, and as such was successful in visualizing the stray field distribution above the magnetic surface of the sample, as well as the domain structure beneath the film surface, or even domain wall substructure. For this particular study of the multilayered Co/Pt nanodots [8], MFM data analysis allowed the acquisition of conventional magnetic hysteresis loops from MFM images for various external magnetic fields, from which coercivity was then determined. Corresponding MOKE hysteresis loops were also measured, and not surprisingly, MFM and MOKE results were in good agreement. Thus, MOKE measurements were validated in spite of the MFM image analysis being only semiquantitative due to tip effects. Specifically, the samples in question were large-scale periodic arrays ($\sim$20 cm^2) of magnetic nanostructures of Co/Pt nanodots that were obtained by optical interference lithography with Ar + ion lasers operating at wavelengths of 457 nm and 244 nm, respectively. Array periodicities ranged from 125 nm to 1,100 nm with dot diameters between 70 nm and 740 nm [8]. To further reveal the capabilities of magneto-optical investigation techniques, we introduce the example below that illustrates how MOKE systems can help characterize the exchange coupling between magnetically hard and soft layers. The emphasis in this example is on the usefulness of MOKE hysteresis loops acquired with these systems.

Example 4.2. Magnetic recording applications involving spin-based devices use magnetic structures composed of exchange-coupled soft and hard magnetic layers. Describe how MOKE systems can characterize the exchange coupling between hard and soft magnetic layers in spin-based structures?

Answer: The functionality of many spin-based magnetic devices relies on the exchange interaction that takes place at the interface between alternating soft and hard layers. This interaction plays an important role in the magnetization process due to resulting pinning effects explained by some theoretical formulations as follows: Bloch walls form in thick, but soft magnetic layers so that the magnetization of the soft layer pinned to the hard layer rotates reversibly during the magnetization process. As this happens, the magnetization in the remaining part of the material does not rotate until the coercivity of the hard layer is reached, initiating the reversal. These assumptions were put to the test in an experimental study [9] of FeTaN–FeSm–FeTaN multilayers investigated through MOKE hysteresis loops measured at room temperature for both longitudinal and transverse MOKE geometries, while the external field was applied in plane along the easy axis direction. By acquiring minor longitudinal hysteresis loops it was possible to determine the exchange field at the

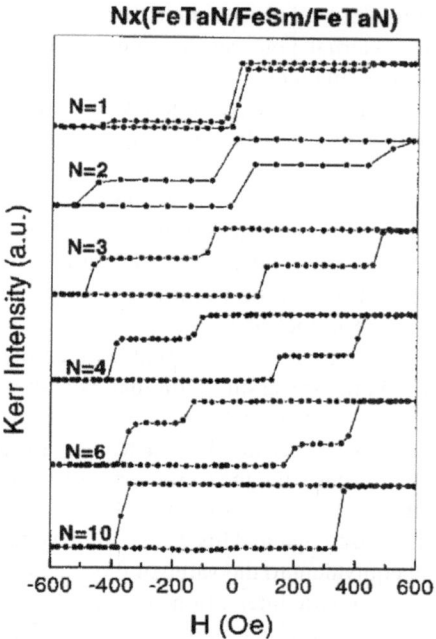

Fig. E4.2 Longitudinal MOKE hysteresis loops for FeTaN–FeSm–FeTaN multilayers with different N number of basic units. Reprinted with permission from [9]

interface between soft and hard magnetic layers. Furthermore, the dependence of the interlayer exchange coupling on soft (FeTaN) and hard (FeSm) layer thickness was also desired. In order to obtain these results, FeTaN and FeSm multilayers were prepared at room temperature by dc magnetron sputtering from alloy targets, so that films were deposited onto oxidized (001) Si wafers. An AlNiCo permanent magnet supplied a magnetic field sufficiently high to induce a uniaxial in-plane magnetic anisotropy. Hard and soft multilayers of 150 nm and 38 nm respective thickness were obtained, with a basic unit consisting of a soft–hard–soft material sequence. X-ray diffraction characterization revealed an amorphous structure for the FeSm layers, while the FeTaN were nanocrystalline.

To facilitate the study of the thickness dependence of the interlayer exchange coupling, several samples needed to be prepared, each sample containing a stack of predefined N number of basic units with varying hard and soft layer thickness. The magnetization reversal process was then examined in all samples using a MOKE system so that in-plane hysteresis loops either parallel or perpendicular to the applied field were acquired and plotted. Light penetrated a limited depth of only a few tenths of nanometers in these samples. For instance, Fig. E4.2 shows longitudinal MOKE hysteresis loops for FeTaN–FeSm–FeTaN multilayers with different N number of basic units. It is observed from this figure that with the exception of $N = 10$, the FeTaN and FeSm layers undergo magnetization reversals separately, at two different field values. As such, the magnetization in the soft layers reverses as soon as the external field reaches the soft layer coercivity, an event marked by a sudden drop in magnetization. We also notice that between the field value where the soft layer and hard layer coercivities are reached respectively, a plateau in the field region is observed that increases with increasing N. Beyond this plateau, saturation is rapidly reached for all multilayers.

The MOKE hysteresis loops and the results that can be inferred from them also show an easy axis direction along the measurement direction, in view of the highest remanence values that were revealed in Fig. E4.2 graphs, as compared to those acquired in other magnetization directions. Furthermore, we can see that for $N > 6$, the soft layer magnetization is locked to the hard layer, corresponding to FeTaN layer thicknesses of 1.9–3.2 nm. This is remarkable given that the exchange length of Fe is 4 nm, thus invalidating the assumption that exchange coupling is observed only when the magnetically soft material can form a Bloch wall.

The exchange coupling was measured from MOKE hysteresis loops acquired only for the soft layer. In this case, the measurements on the loop start at the positive saturation point while field values are lowered just beyond the coercivity of the soft layer, before they are increased again. This minor Kerr hysteresis loop appeared as displaced by an amount equivalent to the value of the exchange field, corresponding to the spatially averaged magnetic interaction between the two layers [9]. For large N, both hard and soft layers nucleate and reverse simultaneously. The study of the dependence of the exchange field on soft layer thickness reveals that a smaller fraction of the material volume is free to rotate in thinner layers due to enhanced interface pinning effects. As such, the interlayer coupling increases with decreasing soft layer thickness, indicating a fully coupled system for a soft layer thickness below the exchange length in the soft layer. Also, it was noticed that for thinner soft layers, coercivity and exchange field values are very similar. The power dependence of the nucleation field on thickness does not follow the trend predicted by theoretical formulations, invalidating some of the assumptions made, as stated at the beginning of this example.

Thus, MOKE systems have established themselves as high-resolution spatial tools for measuring the three magnetization components in many magnetic systems. As seen, polar MOKE geometries reveal information about the magnetization component perpendicular to the sample surface, while longitudinal or transverse measurements detect in-plane magnetizations. This characterization ability of magneto-optical measurements led to an increased application of the technique to contemporary spin-based devices. For instance, in some advanced technologies such as magnetic random access memories (MRAM), the binary information is stored in the form of a parallel or antiparallel orientation of magnetic moments of two adjacent ferromagnetic layers. Hence, in order to read or write the information in a timely manner it is necessary to achieve a fast magnetization reversal also known as *magnetization switching*. The act of fast magnetization switching in these technologies, as well as in spin valves or other spintronic devices is accomplished through short duration high current density electrical transients in the form of magnetic field pulses applied to the switching element. Two aspects are particularly

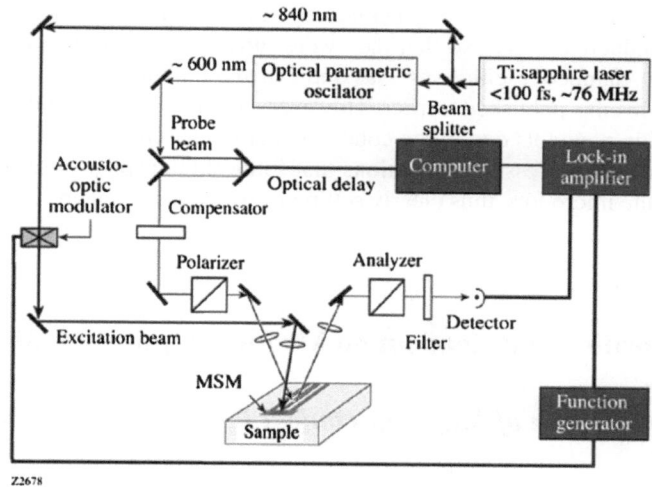

Fig. 4.6 Sketch of the experimental setup used for the MOKE characterization of Co dots of nano- and microsize. Reprinted with permission from [10]

considered, namely how fast can a spin-based device switch between two stable magnetization states, and how reduced in size does the device need to be in order to achieve the picosecond switching dynamics.

Spin dynamics which occurs on a picosecond timescale and is triggered by in-plane magnetic pulses is of great interest for these new spintronic technologies, and has therefore been recently investigated [10] using a MOKE technique. A sketch of the experimental setup used in this study is contained in Fig. 4.6. The investigated magnetic system consisted of closely spaced rectangular Co dots ranging in size from $2 \times 6\,\mu m^2$ down to $100 \times 300\,nm^2$. The dots were fabricated using MBE, and were studied for their switching capabilities. Oblique incidence was used, with a light beam inclined at 30° to which both transverse and longitudinal magnetization components responded. Nanosized magnetic structures contain only a single domain, as opposed to microsized, multidomain configurations. For this reason, the former are expected to change their magnetic state much faster. The variation in size for the studied magnetic structures was deliberately meant to confirm this assumption. As such, room temperature time-resolved spin dynamics Co dots of various sizes were exposed to in-plane picosecond magnetic pulses. This occurred in the absence of an external static magnetic field, given that the goal was to observe the time evolution of the magnetization rotation and relate its dynamics to the physical size of the tested dots. Thus, magnetic transients of ~25 ps were generated using a photoconductive switch fabricated on a low-temperature grown GaAs substrate in the vicinity of the tested structures, at the end of a coplanar waveguide transmission line. The time-resolved spin dynamics was obtained by measuring the MOKE signal with the help of femtosecond optical probe pulses. The pulses were generated by the same laser as the one used for the photoconductive switch so that they were fully synchronized. An initial spike followed by damped oscillations was indicative of a magnetization evolution, confirming results published earlier. However in this case, the spike was more pronounced in nanodots due to the coherent small-angle magnetization excitation. Nevertheless, the precession oscillations that followed the spike were significantly stronger in the microdots, thus clearly setting apart results for nano- vs. microsized structures.

4.2   Recording and Readout on Magneto-Optical Media

4.2.1   Emergence of Magneto-Optical Recording and Readout

It did not take long for engineers to apply the magnetic writing explained above to magnetic recording. As such, they used finely grained magnetic films for writing, where the dark/bright changes in background were due to the many fine magnetic domains within the film that had an opposite domain magnetization orientation (Fig. 4.7). These grains ranged in size from 3,000 to 6,000 Å, and were a high technological achievement for the magnetic recording industry. The films were engineered such that the width of the magnetic domain boundaries of a certain magnetization orientation, for instance, up or down coincided with the grain boundaries, allowing a very fine magnetic domain structure to be obtained among grains with adjacent boundaries. An estimated 10^6 bits were recorded on one square centimeter of magnetic film, with the width of the finest line assessed at 0.001 cm. However,

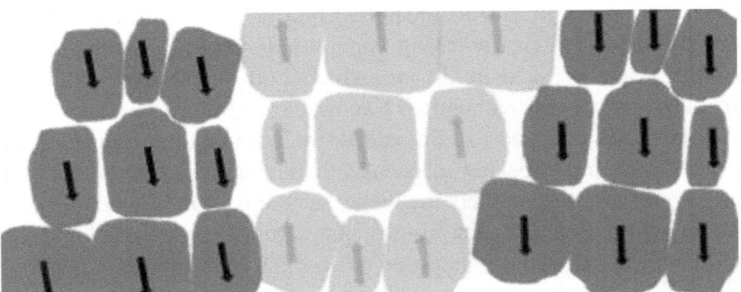

Fig. 4.7 Sketch of finely grained magnetic films for magneto-optical writing and recording. A domain appearing as bright has an opposite domain magnetization orientation from a domain with a dark appearance. The films are engineered such that the width of the magnetic domain boundaries of a certain magnetization orientation (up or down) coincides with the grain boundaries, allowing a very fine magnetic domain structure to be obtained. For clarity, the size and spacing of grains are exaggerated, as the actual grains are closely packed without empty spaces. Magneto-optical thin films are continuous, and magnetic grains are not separated

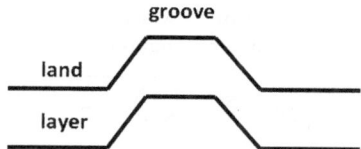

Fig. 4.8 Sketch of land and grooves on magneto-optical disks. Marks were then recorded on these land areas that were tracked using grooves obtained on a polycarbonate substrate using a stamper. Several layers of different materials cover the substrate

the read/write equipment had limited capabilities, providing some impediments in magnetic recording. For instance, the writing was sometimes erased by either saturating or demagnetizing the film perpendicular to the surface. Therefore, in spite of the Faraday–Kerr angle of rotation being fairly large, the magneto-optical recording method had the disadvantage of large media noise mostly due to the grain boundaries themselves. These shortcomings initiated decades of experimentation combined with theoretical studies of spin–orbit interaction and exchange coupling in search for better magnetic films as well as recording techniques [11].

With the advent of the 1970s, a diversity of magneto-optical media for magnetic recording emerged. For instance, amorphous GdCo films were used [12] because they had no grain boundaries. Other materials that were considered were the amorphous rare-earth transition metals [13]. Nevertheless, the newly developed semiconductor lasers offered immediately improved tracking and focusing servo-techniques, so that both materials and read/write equipment experienced significant progress. Magneto-optical disk drives were developed in the 1980s based on amorphous TbFe [14] and GdFeCo as the recording medium [15], and by the end of the decade 130 mm MO disks with a two-side capacity of 650 MB were already commercially available. At the beginning of the 1990s, single-sided 90 mm MO disks of 128 MB for data recording and 64 mm rewritable Mini Disks for digital audio recording were put on the market [16]. For instance, Sony produced in the 1990s a magneto-optical disk with marks recorded on land areas that were tracked using grooves (Fig. 4.8) obtained on a polycarbonate substrate using a stamper, usually made of Ni [17]. The layers were deposited onto the substrate by sputtering such that for example, a 20 nm TbFeCo acted as the magnetic layer while two dielectric layers of Si_3N_4 were deposited on either side to protect against corrosion [18]. TbFeCo is a magnetic material with a high perpendicular magnetic anisotropy, a large coercivity (800 kA/m), low noise, and a reasonable Kerr rotation angle.

Example 4.3. What are magnetic bubble domains?

Answer: Magnetic bubble domains were used for a while for writing of bits, and were part of an alternative magneto-optical recording and readout technique. A magnetic bubble domain is an entity that is generated in certain materials when a laser beam is acting, heating the material so that the coercivity is lowered [19]. The coercivity is brought down to the value of the bias field which is necessary for writing, and for the maintenance of the magnetic bubbles as written bits [20].

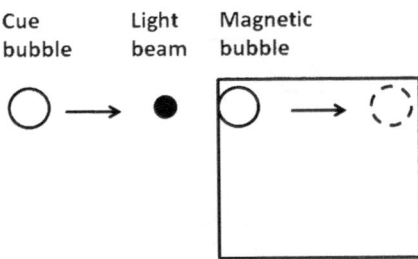

Fig. E4.3 A magnetic bubble is generated at the upper left corner of the etched square. This bubble is a bit of memory. Another bubble known as a "cue bubble" is generated in order to move the bit bubble to the opposite corner. The cue bubble follows the light beam, as cue bubbles prefer heat. The two bubbles repel each other by the interbubble repulsive force, so that the repelled bubble is shifted to the upper right corner

In the absence of the laser beam, the coercivity of the magnetic medium is kept higher than the bias field so that the material retains the magnetic bubbles, becoming nonvolatile [21]. Only a few mW of power are necessary for writing on a film with low coercivity [22]. However, in order for the writing to be stable, a requirement is that these bubbles nucleate only at the center of the irradiating beam, in a region with higher temperature. For instance, a bubble constituting a bit of memory is generated at a corner of a square formed on the magnetic film by etching or ion implantation. Another bubble, called a "cue bubble," is also generated in order to move the bit bubble to the opposite corner. This occurs by having the cue bubble follow the movement of the light beam, as cue bubbles prefer heat. Due to their low coercivity typically less than 1 Oe, magnetic bubbles move along the temperature gradient to the region heated by the light beam. As the light beam approaches the bit bubble so does the cue bubble. The bit bubble is repelled by the interbubble repulsive force and is subsequently stabilized at the opposite corner of the square (Fig. E4.3). The size of the written bit is the size of the magnetic bubble domain, and it depends only on the bias field, regardless of the light beam intensity, pulse width, or beam diameter. The bit is circular in shape, irrespective of the shape of the light beam [23].

Bubble formation is expected to occur easier at higher temperatures, because if bubbles were to form in the cooler area of the beam, they would escape the center and writing would become unstable. Materials that satisfy these requirements are Bi containing liquid phase epitaxial garnet films that have a high Faraday rotation angle so that a written bit or magnetic bubble can be read out with a high contrast ratio [24]. These days, Bi substituted garnets are still being studied in detail for technological applications of the MO effect, in particular in optical storage devices [25]. Enhanced Faraday rotation angles obtained in these materials are attributed to different effects, such as mixing of the 3d Fe^{3+} orbitals with the 6p Bi^{3+} orbitals [26], or the large splitting of the excited state induced by the large spin–orbit coupling of the Bi^{3+} ions [27].

4.2.2 Challenges Encountered with Magneto-Optical Recording and Readout

Magneto-optical writing was done on magnetic film disks through light pulses of $\sim$10 mW reaching approximately 1 μm diameter area on the magnetic layer. The temperature within the light spot rose by heat absorption so that when it reached a critical point of for example 170°C for TbFeCo, the coercivity of the material became equal to the applied magnetic field. The field was applied in order to write a mark. Since the magnetization within the light spot resulted in redirecting the magnetization upward [28], readout was possibly through the magneto-optical effect. As such, a second light of approximately 1.5 mW was bounced off the disk, just weak enough to raise the temperature only slightly [29]. Because the light was linearly polarized, the polarization of the reflected light rotated in opposite directions for oppositely directed spins so that Kerr rotations occurred. The rotation angles were then picked up by a beam splitter, and divided into s-polarized and p-polarized light before entering two detectors. The disk was rotating in the meantime, allowing recorded marks that were advancing to be read out by the light beam focused on the media [30].

Unfortunately, there were resolution limitations in the magneto-optical detection method, given by the recording wavelength p of the marks (Fig. 4.9a). This wavelength needed to be longer than the optical limit given by $\lambda/(2NA)$, where λ is the readout light wavelength and "NA" is the numerical aperture of the objective lens. Therefore, some solutions were adopted in order to increase the resolution of the optical readout process. For instance, the microscope was equipped with an optical system containing a pinhole smaller than the size of the spot shining on the object. This allowed detection of areas smaller than the limit imposed by conventional detection [31]. However, this method also had its limitations, and in order to overcome them, the pinhole was produced right in the media, at the intersection of the readout light spot with a heated area produced by the laser. In this manner, a portion of the focused spot was optically masked [32], leaving

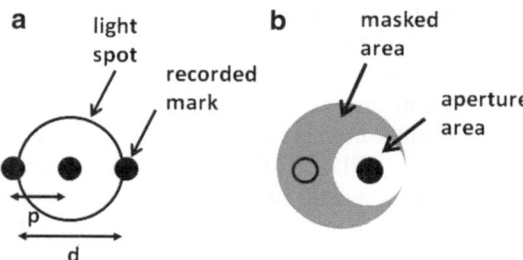

Fig. 4.9 (a) Limit of conventional detection reached when the recording wavelength p is greater than the optical limit $\lambda/(2NA)$, where d is the diameter of the first Airy ring, and (b) an example of rear aperture detection in magnetically induced super resolution

an effective aperture much smaller than the focused spot. Thus, the signal was optically detected within the effective aperture when a magnetic field higher than the coercivity of the heated region was applied during readout [33]. The magnetization of the heated region aligned with the field, corresponding to an erasure process [34]. This method expanded the readout resolution limit of the magneto-optic disk, and was known as the *magnetically induced super resolution* (MSR) method [35]. Two types of detection were obtained with MSR, a front aperture detection [36], and a rear aperture detection [37] (Fig. 4.9b). This readout resolution obtained by the MSR method exceeded [38] the conventional optical limit without the need for a shorter wavelength or a larger numerical aperture of the objective lens [39].

4.2.3 Improvements in Magneto-Optical Recording and Readout Techniques

The MSR method is, unfortunately, a destructive optical readout technique because it erases the recorded mark after readout [40]. For this reason, Sony developed a nondestructive alternative that used an exchange-coupled magnetic triple layer film. The layers had three functions, namely *readout*, *switching*, and *recording* [41] as shown schematically in Fig. 4.10. The readout layer had a low coercivity [42], while the coercivity of the recording layer was high [43]. The magnetically stored information was thus still retained in the recording layer, even after the magnetic layers cooled down, and the recorded mark in the readout layer was erased by the applied field in the heated area [44]. The retained information was then copied onto the readout layer thanks to the exchange coupling between the readout and recording layers [45], allowing the recorded marks to be read out several times [46]. Through this readout method, it was possible to resolve marks with a recording wavelength of 0.76 mm, for a heating power of 2.5 mW known as the *resolving power*. Sony also developed in the mid-1990s Pt/Co multilayer [47] and GdFeCo/TbFeCo double layer [48] magneto-optical disks that are used for readout of a high power green laser that gave an improved signal-to-noise ratios, as well as areal densities three times higher than other MO disks available at the time [49]. The shorter wavelength of the green laser of only 532 nm compared to the previously used 780 nm laser [50] allowed increased recording densities in magneto-optical disks due to higher coercivities and Kerr rotation angles.

An optical readout technique used in the mid and late 1990s was based on *domain wall displacement detection* (DWDD) [51], which also used a magnetic triple layer film embedded in a magneto-optical disk [52]. The advantage of this technique was that its resolving power was not determined by an optical limit, but rather by the driving force of a wall displacement which was temperature dependent [53]. The magnetic triple layer film was exchange coupled so that information recorded on the memory layer was copied onto the displacement layer for readout [54]. However, the exchange coupling was cut off when the switching layer reached the

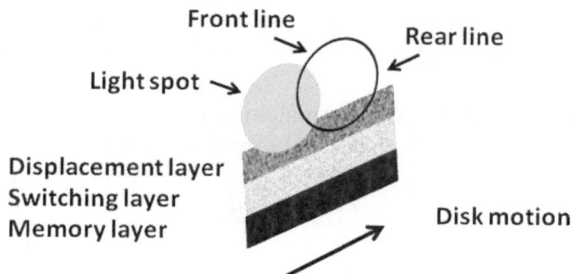

Fig. 4.10 Domain wall displacement detection (DWDD) using a magnetic triple layer film such as for instance, TbFeCo (memory layer), TbFe (switching layer), and GdFe (displacement layer). The magnetic triple layer film was exchange coupled so that information recorded on the memory layer was copied onto the displacement layer for readout. The mark copied onto the displacement layer passed through a critical temperature contour. When the leading edge of the copied mark passed the front line of the critical temperature contour, a waveform of the signal rose steeply, and then fell down rapidly as the trailing edge of the mark passed the rear line

critical temperature, almost as high as its Curie temperature. The copied mark on the displacement layer was then read. A magnetic domain wall within the mark was displaced to the maximum temperature point when the wall passed through a critical temperature contour (Fig. 4.10). As such, when the leading edge of the copied mark passed the front line of the temperature contour, a waveform of the signal rose steeply, and then fell down rapidly as the trailing edge of the mark passed the rear line. The annealed areas on both sides of the track ensured that no side magnetic walls were present [55], a characteristic that was in sharp contrast to the size dependence of magnetic bubble domains used in those times, domains that had circumferential walls [56]. Thus in DWDD, the gradient of the Bloch wall energy, basically independent of domain size, acted as a driving force and was responsible for wall displacement.

Example 4.4.   Give an estimate for the driving force of domain wall displacement?

Answer: In order to estimate the driving force for wall displacement in DWDD, consider the velocity of the wall. The linear velocity v_w of the domain wall can be calculated according to the formula [52]

$$v_w = \mu_w(H_\sigma + H_M + H_{ext} - H_{cw}), \tag{E4.4.1}$$

where μ_w is the wall mobility, H_σ is the magnetic field caused by the gradient of Bloch wall energy, H_M is the demagnetizing field from the magnetic triple layers, H_{ext} is the external magnetic field, and H_{cw} is the wall coercivity. The latter has a small value for GdFe in the displacement layer when there is no exchange force. Also, the external magnetic field H_{ext} of about 300 Oe is only applied to cancel a ghost signal observed when a longer mark (>0.25 μm) passes the rear line. Based on the above equation, and using some additional theoretical considerations

[57, 58] as well as experimental data [59], the velocity of the wall displacement was estimated to be 17 m/s, which compares reasonably well to the experimental value of 28 m/s [52].

The magnetic wall velocity v_w determines the available data transfer rate because the minimum mark length L that can be read out by DWDD is given by the relationship

$$\frac{L}{v} > \tau = \frac{d}{v_w}, \tag{E4.4.2}$$

where v is the linear velocity of the rotating disk ($\sim$1.5 m/s), and d is the displacement of the wall. Equation (E4.4.2) means that the wall displacement time τ has to be shorter than the time L/v for the domain wall to reach the terminal, before the next wall reaches the front line. Since v/L gives the data transfer rate, it indicates that the latter is limited by the linear velocity v_w of the domain wall. In the referenced DWDD experiments [52], the maximum data transfer rate was calculated to be 100 Mbits/s. At the same time, an estimate of the minimum mark length that can be read out by DWDD was $v\tau = 50$ nm [60]. It should also be noted that a large wall displacement time τ results in a large jitter, where the jitter in nm is given by the jitter in ns times the disk velocity. The jitter in the time domain is inversely proportional to the maximum slope of a signal wave. But the slope is determined by the wall velocity which becomes a quality factor for the jitter. However, the main reason for the jitter in DWDD remains the fluctuation of the front line.

4.3 Magneto-Optical Thin Films

4.3.1 Characteristics of Conventional Magneto-Optical Thin Films

Conventional magneto-optical thin films were usually fabricated by X-ray lithography and ion beam etching [61], so that regular micrometer-sized regions were achieved on the film as stated earlier. A perpendicular easy magnetization axis was frequently obtained through this process; however, this particular magnetization orientation was also dependent on the material composition as shown in the next section. In order to attain well-separated magnetic regions on the film, a mask was used during X-ray lithography patterning illustrated schematically in Fig. 4.11, followed by removal of the unprotected parts of the film by ion beam etching [62]. MOKE measurements were made to inspect the quality of the film, due to the necessity of ensuring a perpendicular magnetization direction. This magnetic characterization was possible because the direction of magnetization in these films showed a strong dependence on the direction of light polarization, as well as a high ratio of film to background reflectance. Other complementary characterization techniques such as *reflection high-energy electron diffraction* (RHEED) offered further insight into the optical qualities of the film, by indicating epitaxial growth

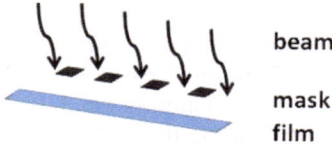

beam

mask

film

Fig. 4.11 Sketch of fabrication of magneto-optical thin films by a lithography method where a mask is used during irradiation. The unprotected parts of the film are subsequently removed by ion beam etching. The goal is to achieve regular micrometer-sized regions patterned on the film. The spacing on the mask is exaggerated for clarity, while the masked regions are actually closely packed

Fig. 4.12 Schematic of a magnetic thin film displaying perpendicular magnetizations for storing "1" and "0" bits that are read out magneto-optically

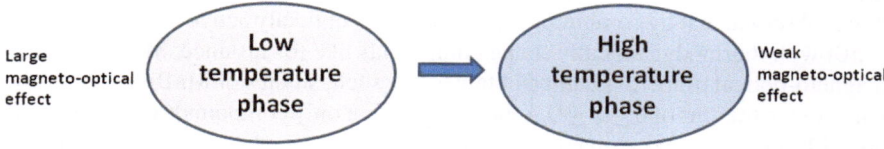

Structural phase transition

Fig. 4.13 A disadvantage of some magneto-optical materials is the structural phase transition that occurs during thermomagnetic writing, causing a significant decrease in the magneto-optical signal. As such, the locally written region of the film transforms from a low-temperature phase with a large magneto-optical effect to a high-temperature phase with a weak magneto-optical effect

obtained in these films. X-ray diffraction studies showed whether the material was crystallographically textured, as a highly textured material with an easy magnetization axis perpendicular to the film was usually desired (Fig. 4.12).

It should be emphasized at this point that magnetic properties of conventional magneto-optical thin films are continuous, in contrast to other magnetic films such as those used in traditional magnetic recording tapes where the magnetic grains that form a bit are separated from each other due to exchange coupling. Conventional magneto-optical thin films allowed thus higher storage densities, packing more bits on the film. For instance, during the prime years of magneto-optical recording, the data storage marks on the thin film of magneto-optical disks consisted of magnetic domains separated by a magnetic wall of about 8 nm, where the mark width spanned about 170 nm in typical areal densities of 100 Gbits/in.[2]. Another advantage of magneto-optical films was that they had an improved thermal stability, in contrast to traditional magnetic recording tapes that suffered from thermal stability issues due to magnetization reversals of single-domain particles that occurred spontaneously under thermal activation [63]. Therefore, the benefit of using magneto-optical media was that the energy barrier for the wall motion was 320 times larger than the thermal activation energy kT, meaning that the thermal stability of the magnetic walls in magneto-optical media was higher than that of conventional recording tapes with perpendicular magnetization films [64].

In spite of their advantages, it was noticed that there were several inherent problems with conventional magneto-optical materials, preventing them from continued use over the last decades and into the present day. For instance, it was noticed that during thermomagnetic writing, a structural phase transition seemed to occur when a domain was written. When this happened, the locally written region of the film transformed from a low-temperature phase with a large magneto-optical effect, to a quenched high-temperature phase with a weak magneto-optical effect [65] (Fig. 4.13). Unfortunately, this metastable quenching process caused a significant decrease in the magneto-optical signal after just a few read/write cycles. Therefore, in order to avoid a structural transition to a high-temperature phase, more thermodynamically stable magneto-optical materials were necessary. For instance, it was observed that by combining highly magneto-optically active materials such as MnBi with thermodynamically stable compounds like for instance, MnSb, superior magneto-optical properties can be obtained. As such, studies of $MnBi_{1-x}Sb_x$ alloys with concentrations of $0 \leq x \leq 0.4$ showed [65] that only compounds with more than 50% Mn displayed good magneto-optical properties, emphasizing the importance of having this particular element as part of the mixture. On the other hand, it was necessary to maintain the Sb concentration rather low, 4–8% substrate dependent, in order to obtain a perpendicular anisotropy and Kerr rotation angles larger than 0.5°.

The MnBi–MnSb system merged the perpendicular anisotropy, and therefore the considerably high Kerr effect of MnBi, with the thermodynamic stability of MnSb which fortunately did not undergo a structural transition to a high-temperature phase. The structural similarity of the two ferromagnetic compounds allowed this opportune combination in either polycrystalline bulk form or thin film samples. Whether the obtained samples were polycrystalline or epitaxially grown depended to a large extent on the choice of substrate; nevertheless their magneto-optical properties were determined mostly by the capability of having a perpendicularly oriented magnetization. On the other hand, the coercivity of the magnetic material had a strong dependence on substrate choice, as well as microstructural details of the film. Many laboratories performed over the years systematic studies of structural, magnetic, and magneto-optical properties of alloys, in order to obtain the optimum compound. Also, with improved fabrication techniques such as for instance, sputter deposition, special properties particulates dispersed in polymer matrices were obtained, paving the way for new types of magneto-optical materials that distanced themselves from more conventional thin film approaches. Some research results of improved conventional magneto-optical alloys, as well as new and enhanced present day magneto-optical materials, which often occur in the form of nanosized particles are discussed in the sections that follow.

4.3.2 Conventional Magneto-Optical Materials

It was noted that MnBi is an intermetallic compound that enjoyed widespread use in magnetic recording and optical readout by the Faraday–Kerr effect due to its high magnetic anisotropy constant. A common method of obtaining MnBi magnetic films in the early days was by vapor depositing on a glass substrate a layer of Mn, followed by a layer of Bi. The obtained film was a composite that was baked in successive steps at temperatures above 200°C, while sealed off at pressures of 1.5×10^{-7} mm Hg. The film was then further baked at temperatures ranging from 225 to 350°C [3]. The films were usually not perfectly uniform nor exactly continuous. Nevertheless, it was possible to obtain areas of approximately a square inch with crystallographically arranged c axes normal to the surface, which was also the bulk uniaxial direction of easy magnetization. If magnetization vectors were in plane, they required oblique illumination, and reports exist about magnetic domain observation experiments carried out using oblique illumination on single crystals of silicon iron and evaporated films of NiFe [66, 67]. For films that were thinner and more transparent, magnetic domain observation was made with the transmitted light inclined at an angle of 45° to the surface of the film [68].

Sony used for a while TbFe magneto-optical films because their large coercivity had proven to be a good choice for the writing layer [69]. On the other hand, GdFe displayed a high Kerr rotation angle and was therefore very suitable for a readout layer in Sony magneto-optical disks [70]. These exchange-coupled magnetic double layers showed improved carrier levels [71], encouraging further exploration and subsequent development of magneto-optical disks with slightly different element combinations: GdFeCo and TbFeCo. As such, the magneto-optical films used by Sony in the mid-1990s had a combined coercivity higher than 10 kOe. The layers were successively deposited by sputtering on photo-polymerized glass substrates [72], while a specially prepared stamper left sharp grooves with a track pitch of 0.9 μm on the substrate. Scanning tunneling microscopy investigations revealed improved surface smoothness and sharper boundaries between lands and grooves leading to reduced disk noise. The thickness of the layers was optimized for the green laser wavelength of 532 nm. Simultaneous investigations were also performed on Pt/Co multilayers successively deposited by sputtering on the same kind of substrate. The Pt/Co films displayed a relatively high Curie temperature of 400°C and a lower coercivity of approximately 1 kOe. Through layer thickness optimization it was possible to obtain at least as good magneto-optical properties in these Pt/Co layers as in the GdFeCo/TbFeCo films.

The enduring legacy of rare-earth and transition metal amorphous alloy films made them the materials of choice throughout the 1990s. Among accomplishments achieved with these films were the 0.3 μm diameter bits that were recorded using a magnetic field modulation method. Nevertheless, their small size made them hard to read considering that their diameter was less than that of the readout beam. Unfortunately, the minimum bit pitch of an optical recording system was still limited in many cases by the "optical transfer function" (OTF) given by $\lambda/(2NA)$, which required precise arrangement of optical parts and shorter wavelength lasers. For this reason, and as stated above, other approaches for high-density recording were adopted such as the MSR which relied on the temperature distribution in the medium irradiated by the readout beam spot, allowing readout of those smaller bits beyond the OTF limit. However, the MSR method required at least three magnetic layers in the recording medium, and magnetic fields for readout. Therefore, an alternative to using magnetic fields for readout was to use an in-plane magnetization film that exhibited the optical Kerr effect only at high temperatures. The temperature of the medium at the center of the readout beam is usually higher than in neighboring regions. If the readout film displays an in-plane magnetization at lower temperatures and a perpendicular magnetization direction at higher temperatures, it would not require an additional magnetic field for readout, thus simplifying the readout process.

It seems that the unique properties of rare-earth and transition metal amorphous alloy films allowed significant accomplishments. These alloys are ferrimagnetic materials, and as such have a *magnetization compensation point* at a certain temperature, and within a certain composition range where the magnetic moments of the rare-earth metal and those of the transition metal cancel each other. This is called the *magnetization compensation point* where the overall effect of the magnetic moments is annulled, and the material displays a net magnetic moment of zero. This point is achieved for a particular compound composition range that varies with temperature and material type, and is due to rare-earth magnetic moments decreasing more rapidly with temperature than those of transition metals. Furthermore, there are certain rare-earth and transition metal amorphous alloys that at that compensation temperature, and for a particular composition percentage, show a perpendicular magnetization, which comes in handy in a magneto-optical readout process [73]. Remember that the incident laser beam needs to be normal to the surface of the film for the optical Kerr effect to occur and be detected. Therefore, a rare-earth and transition metal amorphous alloy film with this property of exhibiting an in-plane magnetization at room temperature and a perpendicular magnetization at higher temperatures can be chosen as a readout layer. On the other hand, there are other rare-earth and transition metal amorphous alloy films that exhibit a perpendicular magnetization from room temperature to their Curie point, allowing these materials to be used as recording layers.

Research showed that two rare-earth and transition metal amorphous alloys were suitable for this particular magneto-optical recording and readout method: GdFeCo and DyFeCo (Fig. 4.14). As such, the films displayed the following magnetic properties. The magnetization of the readout layer was in plane at room temperature while coupled to the recording layer with perpendicular magnetization. Upon reaching higher temperatures after spot laser heating, the magnetization of the readout layer became perpendicular and followed the direction of the recording

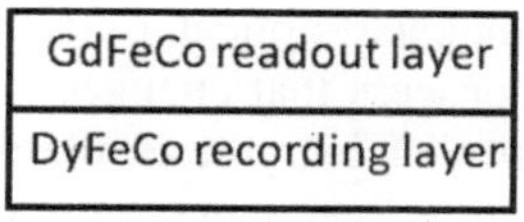

| GdFeCo readout layer |
| DyFeCo recording layer |

Fig. 4.14 Sketch of the concept of a bilayer arrangement of two rare-earth and transition metal amorphous alloy films with particular magnetic properties suitable for a magneto-optical recording and readout method that does not necessitate a readout magnetic field. GdFeCo is chosen as the readout layer because it exhibits an in-plane magnetization at room temperature, and a perpendicular magnetization at higher temperatures. On the other hand, DyFeCo shows a perpendicular magnetization from room temperature to its Curie point, allowing it to be used as a recording layer. Remember that the incident laser beam needs to be normal to the surface of the film for the optical Kerr effect to occur and be detected

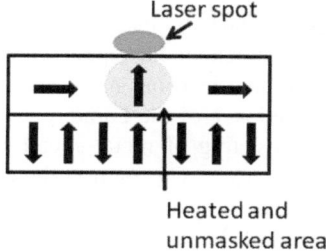

Laser spot

Heated and
unmasked area

Fig. 4.15 Sketch of optical readout using a magnetic bilayer without a readout magnetic field. The magnetization of the top readout layer is in plane at room temperature while it is coupled to the bottom recording layer with perpendicular magnetization. Upon reaching higher temperatures after spot laser heating, the magnetization of the top readout layer becomes perpendicular, and follows the direction of the bottom recording layer. All areas except around the center of the laser spot are masked by the in-plane magnetization of the top readout layer

layer (Fig. 4.15). The thickness of both layers was roughly 40 nm. It should be noted that all areas except around the center of the laser spot were masked by the in-plane magnetization of the readout layer, in conformity with the requirements of the MSR technique.

4.3.3   New and Enhanced Magneto-Optical Materials

Magneto-optical compounds with MnSb have been increasingly fabricated, given their thermodynamic stability due to the lack of a structural transition to a high-temperature phase. As such, the Heusler alloy PtMnSb stands out as having enhanced magneto-optical properties, especially if annealed at temperatures above 400°C. The magneto-optical Kerr rotation angle was measured [74] for wavelengths between 400 and 1,000 nm under an applied magnetic field of 10 kOe. It was noticed in bulk specimens that the magneto-optical Kerr rotation angle is larger if the PtMnSb alloy is annealed, a process that changes the magnetic anisotropy at the surface of the specimen. A similar increase in the magneto-optical Kerr rotation angle was observed in thin film specimens, confirming that this enhancement is not an isolated phenomenon, but rather related to the PtMnSb itself. Surface analyses by X-ray photoelectron spectroscopy (XPS) and glancing angle X-ray diffraction revealed details about the changes in the surface state of the material after it undergoes annealing. Overlayers were noticed to form at the surface due to oxidation, and they displayed various refractive indices. However, they were not considered responsible for the enhancement of the magneto-optical Kerr rotation angle.

The field of magneto-optical materials has seen significant advances over the years due to improvements in fabrication methods for multilayer films, especially those for magnetic superlattices. The latter has shown to be particularly timely, since the ability of layering different elements has opened up the possibility of not only developing new materials, but also learning more about magnetism at interfaces. As such, the wavelength dependence of the Faraday–Kerr rotation angle has been studied in multilayer films consisting of metal–metal combinations, for instance ferromagnetic Fe and Co with Au or Cu, or multilayer films of semiconductors paired with metals. The most employed semiconductor for multilayer films has historically been Si, due to its well-known electrical and optical properties. Silicon was combined with a ferromagnetic metal such as Fe, and its magneto-optical properties were thoroughly investigated particularly around the bandgap of 1.1 eV for Si, which corresponds to a wavelength of 1,100 nm.

In one study, multilayered films of Fe/Si were prepared by electron beam evaporation, while X-ray diffraction confirmed a superlattice structure in films with large lattice constants, or an amorphous structure in both Si and Fe monolayers [75]. It was noticed that polar Kerr rotation angles were small for wavelengths shorter than the corresponding wavelength of the Si bandgap. Also, the wavelength for the maximum Kerr rotation angle shifted toward higher values with increasing Si monolayer thickness. The multiple reflections due to the multilayers, as well as interface phenomena, have made the interpretation of results more difficult; however with proper experimental design, persistent measurements, and the aid of computer calculations, it was possible to reach some interesting conclusions regarding their magneto-optical properties. As such, the wavelength dependence of the Kerr rotation angle was attributed to multiple optical reflections that were not observed in bulk Fe or FeSi alloys. Furthermore, calculations showed large fluctuations in reflectivity as a consequence of multiple reflections, with a minimum in reflectivity where the Kerr rotation angle is near zero. However, experimental results for the Fe/Si multilayer films were not able to confirm these theoretical predictions, in spite of the fact that they had agreed for previous Fe/Cu multilayer experiments. It was therefore assumed that the discrepancy between theory and experiment in the Fe/Si results is due to the formation of an FeSi alloy at the interfaces that does not happen for Fe and Cu which form nonsoluble solid phases. With this assumption, experimental results agreed with their theoretical counterpart.

Similar to many other technological developments, devices with magneto-optical thin films have inevitably turned nowadays to nanosized materials. As such, bismuth-substituted yttrium iron garnet particles dispersed in a polymer matrix have emerged as potential candidates for magneto-optical thin films, becoming likely competitors for sputter deposited thin films. The most promising particles were prepared by a coprecipitation reaction in an aqueous solution with a high pH value due to added cations, followed by heat treatment [76]. It had been observed that the pH value of the reaction solution strongly affects particle size, with higher pH values leading to smaller particle sizes. Precise control of pH value and coprecipitation conditions during the reaction process was therefore important and necessary, in order to ensure optimum magneto-optical characteristics. The particles were mixed with an epoxy binder and milled using a planetary milling machine, while the resulting compound was placed on glass substrates and dried for several hours in a warm environment. After fabrication, magnetic properties of the particles were measured with a superconducting quantum interference device (SQUID) magnetometer, while the crystallographic structure was examined by X-ray diffraction. A room temperature evaluation of the absorption coefficient was made with an ultraviolet-visible spectrophotometer at 520 nm, while the Faraday–Kerr rotation was measured also at room temperature, by using a cross polarizer technique, a 3 kOe magnetic field, and a 520 nm wavelength. This specific fabrication method [76] resulted in the formation of crystallites of only ∼20 nm, the reduced size allowing suppression of light scattering, and an increased Faraday–Kerr rotation. Investigations of the relationship between particle size and magnetic properties of the particles, as well as magneto-optical properties of the thin films made with these particles, revealed that lower pH values of the reaction solution, as well as reduced heat treatment temperatures, were indeed responsible for the enhanced characteristics, yet no explanation was given further as to why these particular conditions were the only optimum ones.

Investigations are nowadays undertaken on what are other materials of interest, such as spinel ferrites. A recent study [77] examined the magneto-optical properties of epitaxial films of spinel ferrite solid solutions by using a composition spread approach to examine magneto-optical performance as a function of concentration. A variety of $(Zn,Co)Fe_2O_4$ solid solutions were grown on a $MgAl_2O_4$ substrate by combinatorial pulsed laser deposition for various concentrations of Zn and Co. The method of obtaining composition spread thin films involves growing wedge-shaped layers using moving masks and alternatively ablated targets during deposition, as shown schematically in Fig. 4.16. Film thicknesses of only ∼70 nm were obtained, while atomic concentrations of Fe, Co, and Zn were measured by *energy-dispersive X-ray spectrometry*. Magnetic circular dichroism and Faraday–Kerr rotation angle measurements revealed a significant change in magnetic properties with compositional variations, thereby allowing an evaluation of magneto-optical performance as a function of elemental concentrations. It was observed that magneto-optical performance is enhanced especially in the ultraviolet region by substituting Zn for Co in $CoFe_2O_4$ leading to possible short-wavelength applications.

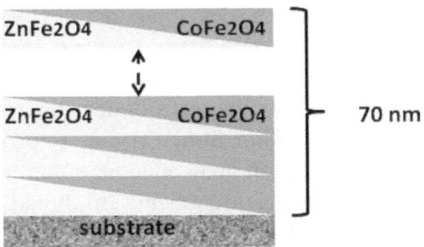

Fig. 4.16 Schematic illustration of a growth sequence in a composition spread thin film. Wedge-shaped $ZnFe_2O_4$ and $CoFe_2O_4$ layers are alternatively deposited until a film thickness of $\sim 70\,nm$ is reached [77]

It has been long assumed that spinel ferrite solid solutions have special magneto-optical properties. This is explained as follows [77]: $CoFe_2O_4$ is a ferrimagnet known for its magneto-optical properties. It has an inverse spinel structure with Co^{2+} ions located at octahedral sites and Fe^{3+} at both octahedral and tetrahedral sites. Conversely, the antiferromagnet $ZnFe_2O_4$ has a normal spinel structure with Zn^{2+} ions located at tetrahedral sites and Fe^{3+} ions at octahedral sites. It was predicted that magneto-optical properties of solid solutions of these two compounds occur because the tetrahedral Fe ions substitute for Zn ions, while octahedral Co ions substitute for Fe ions. Also, because of the presence of Zn in the solid solution, the optical absorption is reduced, in view of the reduction of charge transfer from Co^{2+} to Fe^{3+} ions, and crystal field transitions of Co^{2+} ions. Furthermore, these solid solutions were thought to display an increased magnetization and a reduced Curie temperature attributed to the reduced portion of antiparallel Fe^{3+} ions between tetrahedral and octahedral sites [77].

Magneto-optical phenomena are still important nowadays in magneto-optical information technology where a large magneto-optical performance is required with minimum optical absorption, as for instance in storage media. In order to obtain a faster magnetic recording speed, it is desired to reduce the Curie temperature, and therefore thermal demagnetization. It is hoped that by introducing nonmagnetic elements into magneto-optical compounds, higher magneto-optical performances are obtained. Nevertheless, it is important to know the optimum amount of nonmagnetic material that can be introduced before room temperature magnetic ordering is destroyed and magneto-optical effects disappear. Fortunately, new fabrication and characterization techniques have become available, as shown above. As such, through sequential deposition and a composition spread approach, it is possible to rapidly draw a phase diagram of various properties while the composition is varied continuously. This technique allows both a high-throughput screening and an optimization of materials, making it a powerful tool for materials exploration.

Tests, Exercises, and Further Study

P4.1 MSR is an improved magneto-optical readout technique that expands the readout resolution limit of an optical disk system. Two types of detection methods of MSR have been developed, front aperture detection and rear aperture detection. Briefly describe the principles of magneto-optical readout using MSR and one of the aperture detection methods. See also Problem P4.7 below. *Note: See for instance, Fig. P4.1, Sect. 4.2.2, and references therein for further details.*

P4.2 Name a few techniques for characterizing magneto-optical thin films.

P4.3 Describe briefly how magnetic bubbles are generated. What kind of material is necessary for these bubbles to form? See Fig. P4.3 and referenced publications in main text.

P4.4 Describe briefly what the purpose is of using magnetic bilayers or even multilayers (Fig. P4.4) in magneto-optical disks. Why is exchange coupling between layers important?

P4.5 Magnetoplasmonic materials find uses in optical switching, light guiding at the subwavelength scale, far-field optical microscopy, or as biological sensors. These materials are actually systems that combine the properties

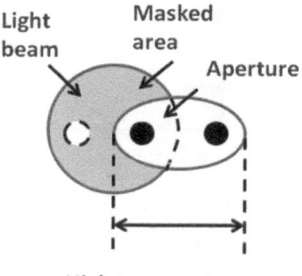

Fig. P4.1 Schematic of a magneto-optical readout using magnetically induced super resolution and rear aperture detection

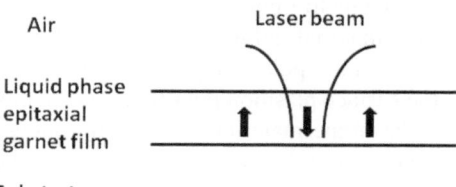

Fig. P4.3 Irradiation of a liquid phase epitaxial garnet film in order to generate a magnetic bubble. The laser is focused in this figure on the film itself; however, focusing the beam on the substrate is also possible

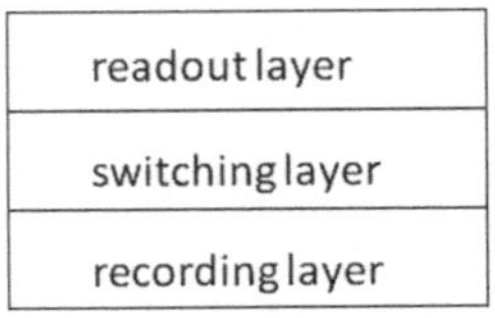

| readout layer |
| switching layer |
| recording layer |

Fig. P4.4 An example of a trilayer system of magnetic film for magneto-optical disks

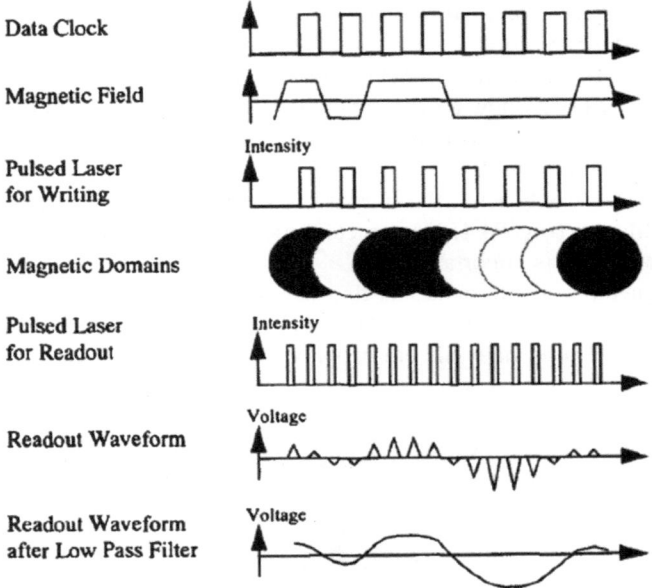

Fig. P4.6 Principles of laser-pumped magnetic field modulation recording (LP-MFM) and laser-pumped readout (LP-R). Reprinted with permission from [79]

of magneto-optical materials with the capability of plasmonic materials of setting up surface plasmon polaritons. Surface plasmon polaritons are electromagnetic waves set up at the interface between two media with dielectric constants of opposite sign. In a magnetoplasmonic system, both the plasmonic and the magneto-optical properties become interrelated so that the magnetoplasmonic material exhibits an enhanced magneto-optical activity when the surface plasmon polariton resonance is excited, or conversely, the wavevector of the surface plasmon polaritons can be modified by an external magnetic field. Plasmonic materials are usually noble metals such as gold and silver characterized by low absorption losses in the optical range, with silver exhibiting narrower and more intense plasmon resonances. The chemical instability of silver is overcome by capping it with platinum, making it a better candidate than gold for sensing and telecom applications. An example of a magnetoplasmonic material system is the Pt capped Ag–Co–Ag trilayers

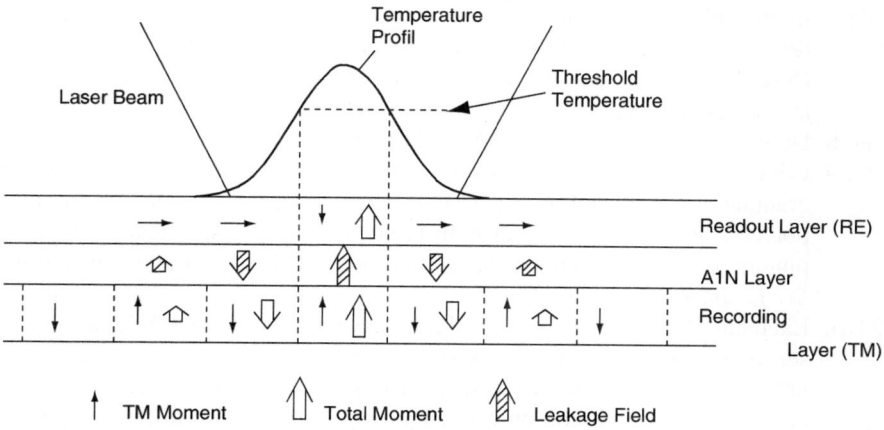

Fig. P4.7 Magneto-optical readout using magnetically induced super resolution and center aperture detection. Reprinted with permission from [79]

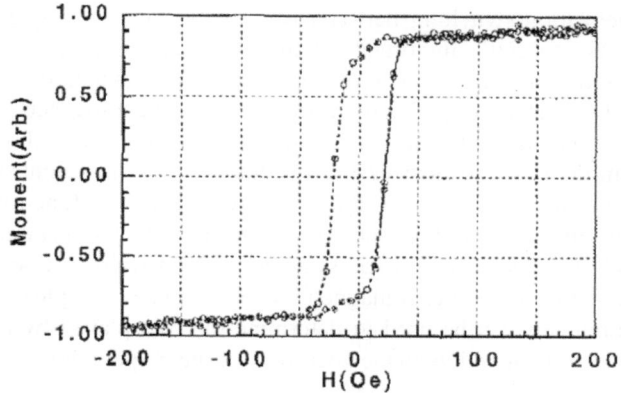

Fig. P4.9 Typical hysteresis loop of as deposited granular Co nanoparticles in a polymer thin film matrix. Reprinted with permission from [80]

grown by magnetron sputtering. Describe how magneto-optical properties of such a magnetoplasmonic material system can be determined by MOKE measurements. *See for instance* [78].

P4.6 Several types of high-density recording technologies based on magneto-optical effects have been historically tried out, in order to satisfy the demands of storing an increased amount of digital information. At the end of the twentieth century, a variety of such magneto-optical techniques were put on trial. Describe the principles of *laser-pumped magnetic field modulation recording (LP-MFM). See also Fig. P4.6 for guidance and* [79] *or a similar reference.*

P4.7 In this chapter, we have talked about a *front aperture detection*, or a *rear aperture detection* as magneto-optical readout techniques using MSR. Describe a *center aperture detection* method using MSR. *See also Fig. P4.7 for guidance and* [79] *or a similar reference.*

P4.8 Describe how you can control MOKE in an MSR disk.

P4.9 Comment on the differences observed with MOKE techniques, between granular cobalt crystallites dispersed in a polymer matrix, and a continuous polycrystalline film of cobalt with granular crystallites. Domain boundary pinning is expected to play a role in the characteristics of the two materials. *See for instance* [80]*and Fig. P4.9.*

P4.10 Exchange bias is an effect of significant importance in GMR and TMR sensor systems, as well as MRAM. Exchange bias helps define the magnetic reference direction. As seen in this chapter, MOKE techniques are able to give information on exchange bias field strength in ferromagnetic/ antiferromagnetic systems. Give another example of the magnetic characterization abilities of MOKE techniques applied to exchange bias in ferromagnetic/antiferromagnetic systems. *See for instance the permalloy/FeMn system described in* [81].

P4.11 Magnetization switching between two stable states in single-domain elements is especially important for the latest data storage applications. As such, ferromagnetic wires have served as part of switching devices due to their shape anisotropy which allows them to have a magnetization parallel to the wire. Magnetization switching in the wire is done through domain wall motion. These devices move domain walls in a controlled manner at propagation fields below the nucleation field. The functionality of the device relies on a well-defined direction of domain wall motion. In order to study these spintronic logic systems, domain walls can be injected into a junction formed by ferromagnetic wires. Give an example of how MOKE measurements can be used to determine the role played by domain walls injected at the junction on the coercivity of the output wire. *See for instance the experiment described in* [82].

P4.12 Double-layered perpendicular magnetic recording media consists of a recording layer and a soft underlayer. The magnetic properties of these layers can be evaluated by various techniques. Given that this chapter discusses magneto-optical methods of investigation, comment on how MOKE can be applied to the study of double-layered perpendicular magnetic recording media. *See for instance the measurements described in* [83].

Hints and Partial Answers to Problems in Chap. 4

The answers to these problems are in many instances just a hint as to where further information is to be found. They contain the main idea, but the reader needs to find the references that are sometimes recommended for each exercise and complete the answer through further bibliographic reading and study. Reference numbering follows the one used in that particular chapter.

P4.1 Consider for instance, *rear aperture detection* where two films play a key role. The recording layer has a high coercivity in order to allow magnetic storage of information, whereas the readout layer has a low coercivity so that it can be read. A magnetic field acts on the readout layer before readout, so that the stored information on this layer is selectively erased. A laser beam then shines on the readout layer creating a high-temperature area located within a fixed distance from the center of the laser spot due to a thermal delay induced by the rotating disk (Fig. P4.1). Because there is an exchange coupling between these two layers, the information from the recording layer is copied onto the readout layer in the high-temperature area where the coercivity of the layer has been lowered due to heat. Thus, only the high-temperature area inside the focused area (light beam) forms an effective aperture where the information is read out. With MSR, a portion of the focused area is optically masked, so that the effective aperture after masking is considerably reduced as compared to the focused area. The advantage of using MSR is that there is no need to use a laser with shorter wavelength, or increase the numerical aperture of the objective lens. MSR has been found to double the readout resolution limit along the track of the conventional optical disk system. *See also Sect. 4.2.2 and references therein for further details.*

P4.2 For instance, Rutherford backscattering was used in early magneto-optical thin film developments for determining chemical composition, while X-ray diffraction revealed the crystallographic structure of the deposited film material. For measuring the hysteresis curve of the magnetic material, vibrating sample magnetometry was commonly employed, while polar Kerr rotation, normal reflectance, and ellipticity hysteresis curves were used for magneto-optical characterization.

P4.3 Magnetic bubbles are generated on liquid phase epitaxial garnet films by focusing a laser light on the film (Fig. P4.3 film incidence), or on the substrate. In the latter case, the beam reaches the film through the substrate, and the method is called substrate incidence. For instance, an Ar laser beam is used for irradiation of the film, such that heat is absorbed in the film at the point of incidence over an area $<2\,\mu$m. Under these heating conditions, the coercivity of the film is lowered to the value of the magnetic field which is necessary for nucleating a bubble at the center of the beam. Bubble nucleation occurs through this method at higher temperatures, avoiding the cooler area of the beam where they would escape the center, destabilizing magnetic writing on them. The magnetic material of the film must satisfy the requirement of having a high coercive force at room temperature, higher than the bias field, in order to ensure it is nonvolatile. The room temperature bias field is necessary to maintain the magnetic bubble domains which constitute the written bits. See relevant references for further details.

P4.4 Magnetic bilayer films form a system of a readout layer and a writing layer. For instance, GdFe has a high Kerr rotation angle and as such can be used as an optical readout layer. On the other hand, TbFe has a high coercivity and is therefore suitable as a writing layer. Nevertheless, MSR requires magnetic triple and even quadrilayers to ensure transfer of magnetic information from the writing to the readout layer (see for example Fig. P4.4). This transfer of magnetic information occurs through the spin interface between the magnetic multilayers which allows interface control, and hence the existence of an exchange coupling between these layers. See relevant references for further details.

P4.5 See indicated reference for details. The magnetoplasmonic material system consisting of Pt capped Ag–Co–Ag trilayers is taken as the example in this problem. The system is grown by magnetron sputtering, while the morphology is studied by AFM and SEM. The magnetization reversal as well as magnetic anisotropy is investigated by acquiring MOKE hysteresis loops. Loops in transverse geometry (i.e., in-plane field, perpendicular to light plane of incidence) are sensitive to in-plane magnetization, while polar Kerr rotation loops (i.e., field perpendicular to sample, linearly polarized light) respond to an out-of-plane magnetization. By comparing the two types of hysteresis loops, it was concluded that the samples exhibited an in-plane magnetization, as revealed by the low saturation field and the high remanent magnetization. Also, the magnetic data were indicative of shape anisotropy dominance. Magneto-optical activity was determined by measuring polar Kerr rotation and through ellipticity spectra, showing a dependence on Co layer thickness. The transverse Kerr effect signal with and without plasmon excitation was measured in order to study the influence of surface plasmon polariton excitation. It was observed that an optimum excitation of the surface plasmon polaritons is obtained for a specific Co layer thickness, which is then diminished when the layer thickness is increased. Overall, an enhanced transverse Kerr effect signal is obtained with surface plasmon polariton excitation, as compared to the case without excitation.

P4.6 In LP-MFM, a pulsed laser synchronized with the data clock is used to irradiate the magneto-optical disk in order to cause it to heat up. A magnetic field modulated with the data is applied at the same time with the laser. Once the temperature of the recording layer has risen above the Curie temperature, a direct overwrite can be achieved and a mark recorded. Clear edge marks can be formed by synchronizing the time of cooling down of the magnetic layer with the transition time of the magnetic field. The heat distribution on the magneto-optical disk is confined to a small area due to the pulsed laser irradiation, as compared to those cases of continuous laser irradiation. Hence, the track pitch can be reduced, and a stable high-density recording can take place.

P4.7 As we have seen in this chapter and Problem P4.1, in order to use magnetically induced super resolution also known as MSR, a special MSR magneto-optical disk is used. This disk has at least two magnetic layers, a recording and a readout layer. During the readout process, a laser beam irradiating the readout layer heats up an area so that the temperature distribution becomes like a Gaussian distribution. When this happens, recorded magnetic domains on the recording layer are copied to the readout layer within a limited range of temperature. The area on the layer over which the domains are copied (or rather the information stored in the domains is copied) is called an "aperture," while the area with the hidden domains is called a "mask." The aperture is smaller than the spot of the laser beam used for readout, in order to ensure a high readout resolution. In center aperture detection, the disk has a nonmagnetic layer between the readout layer and the recording layer. At room temperature, the magnetic domains in the recording layer are hidden by the readout layer which has an in-plane magnetization. The magnetization of the readout layer changes from in plane to perpendicular when higher temperatures are reached. When the temperature rises above a certain threshold, the perpendicular magnetization "follows" the leakage field of the recording layer. In the area where this happens, also known as the aperture area, the recording and readout layers are magnetostatically coupled. These are the principles behind magneto-optical readout using MSR and center aperture detection.

P4.8 The recording layer of an MSR disk is the reflective layer, while the non-magnetic layer between the readout and recording layer is the interferential layer. As such, by controlling the thickness of the nonmagnetic layer, Kerr rotation angles can be enhanced while Kerr ellipticity can be reduced.

P4.9 Nanosized magnetic materials are of great interest nowadays for a variety of applications. However, quite often, the medium in which these materials are contained plays a significant role in the properties that are observable for these materials. As such, whether the matrix is electrically insulating or magnetically active is expected to make a difference in the characteristics of the nanostructures embedded within the matrix. Highly crystalline Co nanoparticles randomly dispersed in an insulating polymer matrix displayed ferromagnetism due to percolation effects, evidenced by electrical resistivity measurements and observations of magnetic domains. By using MOKE magnetometry, in-plane magnetic hysteresis loops of the type depicted in Fig. P4.9 were obtained. The coercivity of Co nanoparticles dispersed in the polymer matrix was lower than that of polycrystalline granular Co films deposited under similar conditions. Results confirmed earlier reports of lower coercivities obtained in magnetic nanostructures of different types dispersed in other matrices. Subsequent observations of domain structure for the Co nanoparticles in the polymer matrix and, respectively, the granular Co film helped understand the magnetic behavior of these two materials. The higher coercivity of the granular Co film is due to stronger pinning of the magnetic domain walls. The strong pinning of the domain walls was not present in the Co/polymer system and hence the lower coercivity. It was also noticed that the direction of magnetization of the film is in plane. Furthermore, the

Co/polymer system displayed changes in its domain structure, not otherwise present in the granular Co film.

P4.10 The study in the indicated reference describes the characterization procedure of a 20 nm $Ni_{81}Fe_{19}$ (permalloy)/8 nm FeMn system deposited on a Si wafer, on top of a 5 nm Ta buffer. An externally applied magnetic field of 50 Oe ensured an easy axis in permalloy and a pinning direction. MOKE hysteresis loops were obtained for longitudinal and transverse field directions, in order to determine the exchange bias field, known to be in the longitudinal direction of the specimen. Additional anisotropic magnetoresistance measurements were made to pinpoint specific exchange bias behavior. It was observed that the onset of switching from spring to full rotation of magnetization depends on the local value of the exchange bias field. This transition occurs within a small field range of about 1 Oe, such as between 76.5 and 77.5 Oe for a clockwise rotation, or 78.5 and 79.5 Oe for counterclockwise rotation. This dependence of the behavior on the direction of field rotation is attributed to the coexistence of uniaxial and unidirectional anisotropies with small misalignment. See [81] for further details.

P4.11 A spintronic system consisting of a three-terminal wire junction (two input and one output wire) was fabricated from $Ni_{80}Fe_{20}$. The switching of this submicrometer size ferromagnetic device was investigated by injecting domain walls into the junction, in a controlled manner. The domain walls were introduced using one or two domain wall injection pads with low switching fields. MOKE magnetometry was employed for the acquisition of hysteresis loops on the device output wires. These measurements revealed that the coercivity of the output is strongly dependent on the number of domain walls that are being injected at the junction. As such, it was determined that by injecting either one or two domain walls, the output wire can be switched at two distinct field values significantly lower than the nucleation field of the junction. See [82] for further details.

P4.12 In this investigation, it was of interest to measure the magnetic properties of the recording layer in the presence of the magnetically soft underlayer. Most magnetometry studies characterize the recording layer on its own due to the difficulties associated with separating the magnetic influence of the soft underlayer which dominates the signal. Hall effect techniques have proven to be successful in separating the magnetic characteristics of the recording and soft layers, because the Hall voltage comes from three distinguishable sources. As such, the first source is the anomalous Hall effect which depends on the perpendicular component of magnetization, present only in the recording layer. On the other hand, the second source in the form of the planar Hall effect depends on the in-plane component of magnetization. Thirdly, the contribution from the normal Hall effect is often negligible in magnetic films due to the high carrier density in metallic materials at room temperature. Based on these facts, Hall measurements were undertaken where the first two voltage contributions were separated on account of their symmetries with respect to the applied magnetic field. In order to test the validity of their results, the researchers used MOKE techniques for comparison with the anomalous Hall effect measurements, and a good correlation was found between the respective hysteresis loops. Nevertheless, MOKE measurements had their own limitations when used for the ultrathin films, as the incident laser penetrated through the recording layer into the soft magnetic underlayer. Furthermore, overestimates of coercivity were obtained through Kerr hysteresis loops, as higher magnetic field sweep rates were used. The hysteresis loops showed good agreement at low and intermediate fields. See [83] for further details of the magnetic characterization of double-layered perpendicular magnetic recording media.

References

1. B.W. Roberts, C.P. Bean, Phys. Rev. **96**, 1494 (1954)
2. C. Guillaud, thesis, University of Strasbourg, Strasbourg, France (1943)
3. H.J. Williams, R.C. Sherwood, F.G. Foster, E.M. Kelley, J. Appl. Phys. **28**(10), 1181 (1957)
4. A. Yariv, *Optical Electronics*, 3rd edn. (Holt, Rinehart and Winston, New York, 1985)
5. J-W Lee, S-K Kim, J-R Jeong, J. Kim, S-C Shin, IEEE Trans. Magn. **37**(4), 2773 (2001)
6. Y.B. Xu, E. Ahmad, Y.X. Lu, J.S. Claydon, Y. Zhai, G. van der Laan, IEEE Trans. Nanotech. **5**(5), 455 (2006)
7. R.S. Weng, M.H. Kryder, IEEE Trans. Magn. **29**(3), 2177 (1993)
8. A. Carl, S. Kirsch, J. Lohau, H. Weinforth, E.F. Wassermann, IEEE Trans. Magn. **35**(5), 3106 (1999)
9. G. Gubbiotti, G. Carlotti, M. Madami, J. Weston, P. Vavassori, G. Zangari, IEEE Trans. Magn. **38**(5), (2002)
10. D. Wang, A. Verevkin, R. Sobolewski, R. Adam, A. van der Hart, R. Franchy, IEEE Trans. Nanotech. **4**(4), 460 (2005)
11. D.K. Misemer, J. Magn. Mater. **72**, 267 (1988)
12. Y. Togami, IEEE Trans. Magn. **18**(6), 1233 (1982)
13. P. Chaudhari, J.J. Cuomo, R.J. Gambino, Appl. Phys. Lett. **22**(7), 337 (1973)
14. Y. Mimura, N. Imamura, Appl. Phys. Lett. **28**(12), 746 (1976)
15. N. Imamura, C. Ota, Jpn. J. Appl. Phys. **19**, L731 (1980)
16. E. Ikeda, T. Tanaka, T. Chiba, H. Yoshimura, J. Magn. Soc. Jpn. **17**(S1), 335 (1993)
17. M. Kaneko, K. Aratani, Y. Mutoh, A. Nakaoki, K. Watanabe, H. Makino, Jpn. J. Appl. Phys. **28**(28), 27 (1989)
18. M. Kaneko, IEEE Trans. Magn. **28**(5), 2494 (1992)
19. J.P. Krumme, H.J. Schmitt, IEEE Trans. Magn. **11**(5), 1097 (1972)
20. Y.S. Liu, G.S. Almasi, G.E. Keefe, IEEE Trans. Magn. **13**(6), 1744 (1977)
21. H. Callen, R.M. Josephs, J. Appl. Phys. **42**(5), 1977 (1971)
22. A.H. Bobeck, R.F. Fischer, A.J. Perneski, J.P. Remeika, L.G. van Uitert, IEEE Trans. Magn. **5**(3), 544 (1969)
23. M. Kaneko, T. Okamoto, H. Tamada, T. Yamada, IEEE Trans. Magn. **19**(5), 1763 (1983)
24. W.F. Druyvesteyn, J.Appl. Phys. **46**(3), 1342 (1975)
25. J. Yang, Y. Xu, F. Zhang, M. Guillot, J. Phys. Condens. Matter **18**, 9287 (2006)
26. K. Matsumoto, S. Sasaki, K. Haraga, K. Yamaguchi, T. Fujii, IEEE Trans. Magn. **28**, 2895 (1992)
27. G.F. Dionne, G.A. Allen, J. Appl. Phys. **75**, 6372 (1994)
28. A. Okamuro, J. Saito, T. Hosokawa, H. Matsumoto, H. Akasaka, J. Magn. Soc. Jpn. **15**(S1), 447 (1991)

29. Y. Mutoh, T. Shimouma, A. Nakaoki, K. Suzuki, M. Kaneko, J. Magn. Soc. Jpn. **15**(S1), 311 (1991)
30. J. Saito, M. Sato, H. Matsumoto, H. Akasaka, Jpn. J. Appl. Phys. **26**(4), 155 (1987)
31. M. Kaneko, K. Aratani, M. Ohta, Jpn. J. Appl. Phys. **31**(1), 568 (1992)
32. M. Ohta, A. Fukumoto, K. Aratani, M. Kaneko, K. Watanabe, J. Magn. Soc. Jpn. **15**(S1), 319 (1991)
33. A. Fukumoto, S. Kubota, Jpn. J. Appl. Phys. **31**(1), 529 (1992)
34. G. Bouwhuis, J.H.M. Spruit, Appl. Opt. **29**, 3766 (1990)
35. S. Yoshimura, A. Fukumoto, M. Kaneko, H. Owa, Jpn. J. Appl. Phys. **31**(1), 576 (1992)
36. F. Maeda, M. Arai, H. Owa, H. Takahashi, M. Kaneko, Proc. SPIE **1499**, 62 (1991)
37. A. Fukumoto, K. Aratani, S. Yoshimura, T. Udagawa, M. Ohta, M. Kaneko, Proc. SPIE **1499**, 216 (1991)
38. S. Yoshimura, I. Nakao, A. Fukumoto, M. Kaneko, H. Owa, IEEE Trans. Consum. Electron. **38**(3), 660 (1992)
39. M. Kaneko, K. Aratani, A. Fukumoto, S. Miyaoka, Proc. IEEE **82**(4), 543 (1994)
40. A. Fukumoto, S. Yoshimura, T. Udagawa, K. Aratani, M. Ohta, M. Kaneko, Proc. SPIE **1499**, 216 (1991)
41. K. Aratani, M. Kaneko, Y. Muto, K. Watanabe, H. Makino, Proc. SPIE **1078**, 258 (1989)
42. A. Nakaoki, S. Tanaka, T. Shimouma, M. Kaneko, Jpn. J. Appl. Phys. **31**(1), 596 (1992)
43. K. Aratani, A. Fukumoto, M. Ohta, M. Kaneko, K. Watanabe, Proc. SPIE **1499**, 209 (1991)
44. A. Nakaoki, Y. Mutoh, S. Tanaka, T. Shimouma, K. Asano, K. Aratani, M. Kaneko, K. Watanabe, Proc. SPIE **1316**, 292 (1990)
45. T. Fukami, Y. Nakaki, T. Tokunaga, M. Taguchi, K. Tsutsumi, J. Appl. Phys. **67**, 4415 (1990)
46. S. Tanaka, T. Shimouma, A. Nakaoki, K. Aratani, M. Kaneko, J. Magn. Soc. Jpn. **15**(S1), 331 (1991)
47. S. Hashimoto, A. Maesaka, Y. Ochiai, J. Appl. Phys. **70**, 5133 (1991)
48. W.B. Zeper, A.P.J. Jongenelis, A.J. Jacobs, H.W. van Kesteren, P.F. Garcia, IEEE Trans. Magn. **28**, 2503 (1912)
49. M. Kaneko, Y. Sabi, I. Ichimura, S. Hashimoto, IEEE Trans. Magn. **29**(6), 3766 (1993)
50. M. Takahashi, J. Nakamura, M. Ojima, K. Tatsuno, Proc. SPIE **1663**, 250 (1992)
51. T. Shiratori, E. Fujii, Y. Miyaoka, Y. Hozumi, J. Magn. Soc. Jpn. **22**(S2), 47 (1998)
52. M. Kaneko, T. Sakamoto, A. Nakaoki, IEEE Trans. Magn. **35**(5), 3112 (1999)
53. T. Shiratori, E. Fujii, Y. Miyaoka, Y. Hozumi, Digest Int. Symp. Optical Memory, Tsukuba, Japan, p 46 (1998)
54. T. Shiratori, E. Fujii, Y. Miyaoka, Y. Hozumi, J. Magn. Soc. Jpn. **23**, 145 (1999)
55. S. Kobayashi, S. Senshu, S. Iwaasa, S. Igarashi, S. Wachi, T. Funahashi, M. Ogawa, Proc. SPIE **1663**, 15 (1992)
56. M. Kaneko, T. Okamoto, H. Tamada, Y. Tamada, IEEE Trans. Magn. **22**(1), 2 (1986)
57. O.W. Shih, J. Appl. Phys. **75**, 4382 (1994)
58. E. Schloemann, J. Appl. Phys. **44**, 1837 (1973)
59. M. Tabata, T. Kobayashi, S. Shiomi, M. Masuda, Jpn. J. Appl. Phys. **36**, 7177 (1997)
60. J.C. Slonczewski, J. Appl. Phys. **44**, 1759 (1973)
61. C. Cesari, J.P. Faure, J. Nihoul, K. Ledang, P. Veillet, D. Renard, J. Magn. Magn. Mater. **78**(2), 296 (1989)
62. F. Rousseaux, A.M. Haghirigosnet, B. Kebabi, Y. Chen, H. Launois, Microelectron. Eng. **17**(1–4), 157 (1992)
63. H.J. Richter, IEEE Trans. Magn. **35**, 2790 (1999)
64. M. Kaneko, S. Yoshimura, G. Fujita, S. Kobayashi, IEEE Trans. Magn. **36**, 2266 (2000)
65. Z. Celinski, D. Pardo, B.N. Engel, C.M. Falco, IEEE Trans. Magn. **31**(6), 3233 (1995)
66. C.A. Fowler Jr., E.M. Fryer, Phys. Rev. **86**, 426 (1952)
67. C.A. Fowler Jr., E.M. Fryer, Phys. Rev. **100**, 746 (1955)
68. C.A. Fowler Jr., E.M. Fryer, Phys. Rev. **104**, 552 (1956)
69. Y. Mimura, N. Imamura, Appl. Phys. Lett. **28**(12), 746 (1976)
70. S. Tsunashima, H. Tsuji, T. Kobayashi, S. Uchiyama, IEEE Trans. Magn. **17**, 2840 (1981)

71. T. Kobayashi, H. Tsuji, S. Tsunashima, S. Uchiyama, Jpn. J. Appl. Phys. **20**, 2089 (1981)
72. S. Hashimoto, Y. Ochiai, J. Magn. Magn. Mater. **88**, 211 (1990)
73. A. Takahashi, J. Nakajima, Y. Murakami, K. Ohta, T. Ishikawa, IEEE Trans. Magn. **30**(2), 232 (1994)
74. K. Takanashi, H. Fujimori, J. Watanabe, M. Shoji, A. Nagai, IEEE Translation J. Magn. Jpn. **5**(5), 348 (1990)
75. M. Yazawa, S. Iwata, S. Uchiyama, IEEE Transl. J. Magn. Jpn. **4**(7), 437 (1989)
76. C.S. Kuroda, T. Taniyama, Y. Kitamoto, Y. Yamazaki, T. Hirano, IEEE Trans. Magn. **37**(4), 2432 (2001)
77. Y. Iwasaki, T. Fukumura, H. Kimura, A. Ohkubo, T. Hasegawa, Y. Hirose, T. Makino, K. Ueno, M. Kawasaki, Appl. Phys. Express **3**, 103001 (2010)
78. E.F. Vila, X.M.B. Sueiro, J.B. Gonzalez-Diaz, A. Garcia-Martin, J.M. Garcia-Martin, A.C. Navarro, G.A. Reig, D.M. Rodriguez, E.M. Sandoval, IEEE Trans. Magn. **44**(11), 3303 (2008)
79. Y. Tanaka, M. Kurebayashi, T. Maeda, K. Torazawa, A. Takahashi, N. Ohta, S. Yonezawa, IEEE Trans. Consum. Electron. **43** (3), 476 (1997)
80. J. Hong, E. Kay, S.X. Wang, IEEE Trans. Magn. **32**(5), 4475 (1996)
81. K-U Barholz, R. Mattheis, IEEE Trans. Magn. **38**(5), 2767 (2002)
82. C.C. Faulkner, D.A. Allwood, M.D. Cooke, G. Xiong, R.P. Cowburn, IEEE Trans. Magn. **39** (5), 2860 (2003)
83. S. Kumar, D.E. Laughlin, IEEE Trans. Magn. **41**(3), 1200 (2005)

Chapter 5
Thermoelectrics, Thermomagnetics, Magnetoelectrics, and Multiferroics

Abstract Several approaches to harvesting of waste energy by incorporating thermoelectric, thermomagnetic, magnetoelectric, and even multiferroic generators have been proposed and fabricated over the last few decades. So far, a significant drawback has been the *efficient* transformation of otherwise lost energy into electricity. Furthermore, depending on the application, some devices require that they function independently. Unfortunately for many devices that combine more than one phenomenon, their somehow traditional design with dependence on external, nonportable parts has been an additional inconvenience. Nevertheless, more moveable and scaled-down versions have experienced a variety of challenges of their own, one being that of optimizing the energy collecting source parameters as a function of the provided heat power. And there is also the problem of matching the generator to the load. Despite all shortcomings, new generations of micro-integrated thermoelectric, thermomagnetic, magnetoelectric, and to some extent, multiferroic devices have continued to emerge showing a lot of promise, while having something in common: Whether in the laboratory or in commercial applications as miniaturized power generators, these devices are expected to make effective use of waste energy.

This chapter deals with some of the physical principles underlying today's thermoelectric, thermomagnetic, magnetoelectric, and multiferroic devices, placing an emphasis on materials properties. By no means is this meant to review the historical development of these devices, or any concepts of thermodynamics. A textbook on thermodynamics is highly recommended for this purpose. Nor does the discussion in this chapter evolve into actual device design with an equivalent circuit analysis. This is left to the ingenious mind who wants to take these concepts into practical applications. However, this chapter does highlight the difference between thermoelectric, thermomagnetic, magnetoelectric, and multiferroic devices that rely on more than one physical phenomenon to function, and therefore reveals what is fundamentally distinct about each device category from a thermodynamic, electric, or magnetic point of view.

Most thermoelectric devices are not magnetic, nor based on magnetic principles. So, why are they discussed here? They warrant a place in this book because power generation has been traditionally regarded mainly as a thermodynamic

C.-G. Stefanita, *Magnetism*, DOI 10.1007/978-3-642-22977-0_5,
© Springer-Verlag Berlin Heidelberg 2012

subject, while there is no reason why modern magnetics should not be part of it. Additionally, power generation is not the only application that requires further improvements in the twenty-first century. These days, it is desired for many devices to exploit a variety of combinations of materials with unusual thermal, electric, and even magnetic properties. For instance, lightweight moveable heat pumps or embedded solid-state refrigerators are expected to become part of tomorrow's buildings or outfits. So, why not build these devices with a special class of magnetic materials while considering the laws of heat transfer? It is hoped that more and more engineers will take the thermodynamic principles exploited in, for example, thermoelectric devices and combine them with the cross-coupling of electric and magnetic properties of, for instance, multiferroic materials. It may be that a temperature change such as that necessary in thermoelectric or thermomagnetic devices could be created by a change in magnetostriction, replacing the traditional thermocouple with a magnetostrictive material. Exciting results obtained from such revolutionary combinations of properties and materials would lead in the near future to a new generation of embedded and self-sustainable magnetic mobile devices.

5.1 Thermoelectric Devices

5.1.1 Thermoelectric Voltage

Thermoelectric devices are energy conversion systems that rely at least upon the *Peltier* or *Thomson effects*, but more often also on a third related phenomenon: the *Seebeck effect*. Thermoelectric devices in the most general form are valuable information sources, whether used simply for measurements (thermometers, pyrometers, pyrheliometers, and calorimeters), or in power/cooling equipment (e.g., refrigerators, heat pumps, and power supplies) [1]. In many applications not all parts of the device are active measurement or heat/energy transfer agents. For instance, the basic thermocouple part plays only a passive role. In spite of this, analyzing the constituent parts of a thermoelectric device allows optimization of overall device performance. Even if improvements are made of only one aspect, e.g., new materials, these developments open wide possibilities, such as of more efficient energy conversion.

We shall begin by making the distinction between two commonly encountered thermoelectric effects. The first one occurs if an electric current flows through two junctions of two dissimilar materials, traditionally metals. In this case, heat is produced at one junction, and absorbed at the other. This is the *Peltier effect*, and is characterized by the fact that a thermal flux is induced in the system composed of the two conducting media. On the other hand, a second effect can take place in only one material through which a current is passing, provided the material has one end at a higher potential and a temperature T_1, while the other end is at a lower potential and thereby a different temperature T_2. This is the *Thomson effect*. Just like for the Peltier effect, in the Thomson effect there is also a heat flow associated

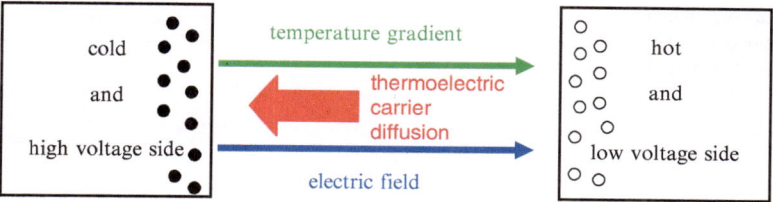

POSITIVE SEEBECK COEFFICIENT

Fig. 5.1 Schematic illustration of a positive Seebeck coefficient situation

with the electric current passing through the material. A body in contact with the low-temperature end will have its temperature lowered, while the converse is true for the other end. Both effects are commonly encountered in thermoelectric devices.

The *Seebeck effect* takes place when a thermoelectric voltage appears at junctions of dissimilar materials that experience a temperature-induced transport of electrons or holes. In this case, heat is produced or the materials are cooled, or even power is generated. The Seebeck effect occurs because the temperature difference between the two junctions causes charged carriers to diffuse from the hot side to the cold side of the material, as shown schematically in Fig. 5.1. When this happens, they leave behind oppositely charged carriers so that an electric field is created between the two sides. Eventually, the buildup process on the cold side ceases, and equilibrium is reached. As the electric field enables carriers to drift back to the hot side, fewer carriers accumulate at the cold side. A *thermoelectric voltage* is created, also known as the *Seebeck coefficient* because it is actually a change in voltage per unit of temperature. It depends significantly on the crystal structure, as this determines what carriers participate in the thermoelectric process.

The Seebeck coefficient is quite small in metals, while being higher in semiconductors. The latter offer the possibility of doping with excess carriers, allowing more control over the Seebeck coefficient. While negative coefficient values can be obtained, they depend on the charge of the excess carriers. In principle, a positive Seebeck coefficient means that the lower voltage side is at higher temperature, and the electric field and temperature gradient point in the same direction. The field will point from high to low voltage, whereas the temperature gradient points from low to high temperature. In a p-type material the Seebeck coefficient is positive, as the material has carrier *holes* so that at equilibrium both temperature gradient and electric field point in the same direction (Fig. 5.1). It is more likely that materials used in practice for thermoelectric devices would have both carrier *electrons* and *holes*; therefore the sign of their Seebeck coefficient is determined by the carriers that dominate.

These days, thermoelectric devices are mainly semiconductor-based, in contrast to those built earlier with metal components. Semiconductor thermoelectric devices have the advantage that they can be miniaturized and incorporated as cooling or power supply systems in electronic equipment. Acoustically silent, lightweight,

Fig. 5.2 A p–n pair forming
a thermocouple. The heat sink
connects to external electrical
leads

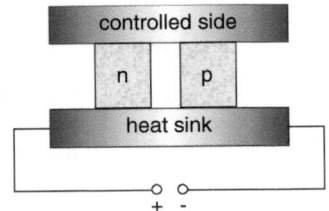

and free of moving parts, they seem ideal for a large variety of small-scale
applications. However, there is one significant problem: they are not efficient enough
in comparison to traditional vapor-compression or other fluid-based systems. Partly
to blame are the contact resistances, as external leads are necessary to connect to
an instrument that will either measure or use the thermoelectric voltage (Seebeck
coefficient). Quite often, a thermoelectric voltage is induced over the length of
the measurement electrode. Nevertheless, this fact can be used as an advantage in
some applications because the two parts (thermoelectric material of interest and
measurement electrode) form a thermocouple, and the measured thermoelectric
voltage is due to the contribution of both.

To illustrate how a semiconductor thermoelectric structure is built, consider
the system depicted in Fig. 5.2 containing a p-type and an n-type semiconductor
material. Quite often, the individual n-type and p-type structures are obtained
through dry-etching techniques. The Seebeck coefficient in the n-type section is
negative α_n, while in the p-type section it is positive α_p. Then heat is reversibly
absorbed at the controlled side, and reversibly released at the uncontrolled side when
the current is positive. The thermoelectric system operates in refrigeration mode
when current flows clockwise through it. On the other hand, the system is a heat
pump for a counterclockwise current flow. The structure depicted in Fig. 5.2 is only
one of the many built into a thermoelectric device. Devices with up to 8,000 p–n
pairs (also called *thermoelectric legs* or *thermolegs*) per cm^2 have been built. This
was accomplished by using the Micropelt platform technology [2] with an achieved
open circuit total voltage of ∼2.3 V at a temperature difference of 10 K. The
maximum reported total power output for such a system is 2.8 mW. High packing
densities are common for these devices, as they are employed in microsystems for
wireless data transfer. The thermolegs are thin films of usually ∼20μm thickness.

5.1.2 Figure-of-Merit

Of course, of interest is how efficient these devices are. The quantity used to
measure efficiency of a thermoelectric system is the *figure-of-merit*. Aside from
the thermoelectric voltage α which is a material property, the figure-of-merit ZT
depends on other factors, too

$$ZT = \frac{\alpha^2 \sigma}{k} T \tag{5.1}$$

such as the electrical conductivity σ, and the thermal conductivity k of the thermoelectric material at temperature T. Larger induced thermoelectric voltages result in higher efficiencies. Therefore, the quantity $\alpha^2 \sigma$ needs to be as large as possible to ensure maximum conversion of electrical power to heat and to minimize Joule losses in the material. On the other hand, the thermal conductivity k has to be as small as possible. A high efficiency is obtained when these parameters are optimized, which means a figure-of-merit larger than unity. Refrigeration systems based on Bi_2Te_3 come close to a figure-of-merit of unity at room temperature, whereas a Si–Ge system has a ZT of only half at temperatures of 600°C. Many research groups aim at developing thermoelectric materials capable of delivering a large figure-of-merit (e.g., at least 3). One problem is that in bulk semiconductors σ and k are interdependent, and that limits what values can be obtained for the figure-of-merit. Therefore, at the moment, most research work is focused on developing new and more thermoelectrically efficient material combinations, in particular arrangements of small-scale structural geometries that could be used as replacements for conventional thermoelectric cooling geometries.

5.1.3   Heat Transfer in Thermoelectric Structures

A few more words should be said about the carriers of thermal energy (electrons or holes vs. phonons), and about the efficiency of thermal transport in materials of various geometrical size, meaning bulk, micro-, or nanostructures. Heat conduction by charge carriers such as electrons occurs mainly in metals, but also in bulk semiconductors where holes can participate in thermal conduction, as well. In *low-dimensional structures* of semiconductor materials, thermal transport occurs primarily through *phonons* that are, roughly speaking, associated with vibrations of the crystalline lattice. Low-dimensional structures do not have the same properties as their bulk counterpart, even if fabricated from the same type of parent material. Therefore, the favorable influence of size effects on phonon transport can lead to an improved figure-of-merit ZT than otherwise obtainable in bulk materials, as many current studies show. Size effects will be discussed again in a subsequent section.

It should be pointed out that in thermoelectric structures, *electrical contact resistivities* ρ at the interface between two materials can pose serious problems in maintaining a high figure-of-merit. For example, in semiconductor structures this is due to the dependence of the resistivity ρ on potential barrier height E_0 at the interface, barrier thickness d, and tunneling probability P_t through the barrier

$$\rho = \frac{e^2 \sqrt{2m^-}}{\hbar^2} \frac{\sqrt{E_0}}{d} P_t. \tag{5.2}$$

The other quantities in (5.2) are electron charge e, effective mass m^-, and Planck's constant h. A higher tunneling probability through a thinner barrier results in a higher electrical contact resistivity. Nevertheless, we shall not concern ourselves with tunneling probabilities and barrier effects in this chapter, but merely highlight their importance. A textbook on semiconductor physics is highly recommended for further study.

Phonon transport and the scattering of phonons at interfaces and boundaries play a major role in enhancing the figure-of-merit of semiconductors. For example, the low value of the cross-plane thermal conductivity k_y of the very popular Bi_2Te_3/Sb_2Te_3 superlattices has led to a figure-of-merit of [3] $\sim$2.4. Similarly, the figure-of-merit of the newly developed two-dimensional (thin film) or quasi-one-dimensional (nanowire) structures is shown in many studies to be better than in bulk materials. Because of their efficiency, thermoelectric materials in thin film form show potential of use in module arrays fabricated on common 2-inch wafers used in the semiconductor industry. It has been noticed that a considerable enhancement in the device figure-of-merit ZT is obtained at room temperature for thin film modules with heating/cooling power densities in excess of $100\,W/cm^2$ and ultrafast response times [4]. Furthermore, microdevices that contain an integrated heat distributor of the type described below [3] show a much reduced thermal heating of the IC package, and are therefore regularly included in the IC design.

Consider Fig. 5.3 which shows an array of p-type and n-type thermoelectric structures of the typed depicted earlier in Fig. 5.2. Each pair forms thermocouples connected electrically in series, so that the thermoelectric system releases heat at the uncontrolled side. As the current is increased, the thermoelectric device

Fig. 5.3 Partial sketch of a thermoelectric device consisting of p-type and n-type pair sections electrically connected in series. The thermal coupling is in parallel, and each p–n pair forms a thermocouple. Ceramic layers cover the top and bottom of the p–n pairs, but only one layer connects to external electrical leads, and is referred to as the uncontrolled side. This side constitutes the heat sink or energy exchanger

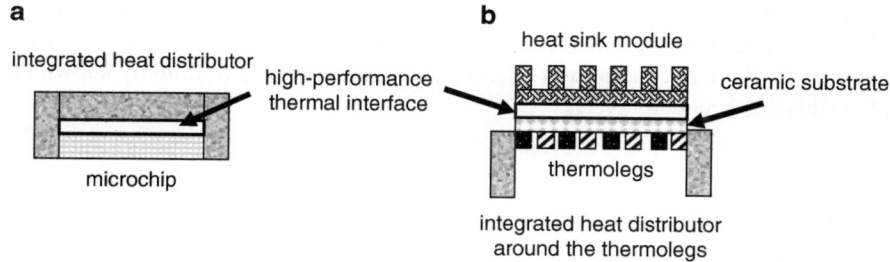

Fig. 5.4 (**a**) An integrated heat distributor attached to a microchip through a high-performance thermal interface reduces the on-chip thermal gradient by dispersing it over a wider area; (**b**) Schematic of a package containing an integrated heat distributor around the thermolegs which in turn are connected to a heat sink module through a high-performance thermal interface attached to the ceramic substrate

starts to cool down, and a temperature decrease is obtained at the cold side. However, in some electronic components it is necessary to attach an integrated heat distributor. This is accomplished by adding the heat distributor through a high-performance thermal interface to a microchip containing thermoelectric p–n structures. Figure 5.4a shows more details. The dispersion of heat is then facilitated from the chip across a wider area, especially if the heat distributor is a material that has a higher thermal conductivity. This heat dispersion reduces the on-chip thermal gradient and improves the heat sink cooling efficiency, in particular given the nonuniform heat generation on the microchip. Additionally, high-performance thermal interface materials between the heat sink module and the ceramic substrate containing the thermolegs ensure dissipation of heat to the ambient environment. This is shown schematically in Fig. 5.4b where the system is depicted without the microchip. The heat distributor and the extra thermal layers are particularly useful [5] when considering the continued increase in not only packaging density, but also frequency and power density of high-performance microprocessors.

In terms of thermal performance of integrated circuit packages, the performance of an IC package is evaluated by determining the junction-to-ambient thermal resistance R_{ja}

$$R_{\mathrm{ja}} = \frac{(T_{\mathrm{j}} - T_{\mathrm{a}})}{P}, \tag{5.3}$$

where T_{j} and T_{a} are junction and ambient temperature, respectively, and P is the power dissipation. The junction temperature is determined by using a thermal test die, such as the one described by Liu et al. [3]. A standard electrical–thermal conversion method is used to measure the junction-to-ambient thermal resistance in natural and forced convection conditions [3]. The contact pressure is one of the factors that influence most the junction-to-ambient thermal resistance of devices built with bulk materials. This is because the thermal resistance is basically the ratio between the temperature variation at the interface of two materials at different temperatures, and the average heat flow across the interface. One of the materials can be air. Because heat travels vertically through the structures, another significant physical quantity in thermoelectric devices of the type depicted in Fig. 5.3 is the *cross-plane thermal conductivity* k_y, as opposed to the more commonly known in-plane thermal conductivity.

5.1.4 Thermal Conductivity

We divert our attention toward the important material property known as *thermal conductivity* which appears in the mathematical expression for the figure-of-merit. It should be noted in passing that thermal conductivity is influenced not only by electrical contact resistivities ρ of the type mentioned above, but also by the *thermal contact resistances* $\gamma_{c,th}$ due to electron and phonon energy carriers, as we shall see in a subsequent section. For now, we will look at a few simple principles governing thermal conductivity.

The thermal conductivity indicates the material's ability to conduct heat, and is given in its basic form by the following relationship:

$$k = -\frac{\Delta Q}{\Delta t} \cdot \frac{1}{A} \cdot \frac{x}{\Delta T}. \tag{5.4}$$

Specifically, the thermal conductivity k is the quantity of heat ΔQ transmitted during a known time interval Δt through a certain thickness x, normal to a surface area A, and due to a certain temperature difference ΔT, as shown schematically in Fig. 5.5. In this drawing, it is cooler toward the right end of the material, as heat flows in the opposite direction to the temperature gradient, and thus the temperature difference ΔT is negative.

The following example will take us back to fundamental thermodynamics and will illustrate the importance of thermal conductivity as a material property.

Example 5.1. A box made of three different but thermally conductive materials has a heat source within it. The temperature within the box is T_1. If the temperature outside the box is T_2, what is the rate of heat loss from the box? Materials have wall

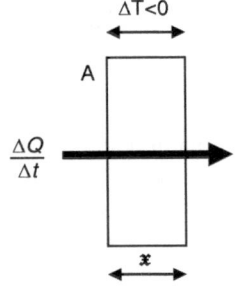

Fig. 5.5 Thermal conductivity defined as the property of a material to transmit heat ΔQ at a flow rate $\Delta Q / \Delta t$ normal to surface area A, through thickness x due to temperature difference ΔT.

Fig. E5.1 Box made of three different but thermally conductive materials. Materials 1, 2, 3 have wall areas A_1, A_2, A_3, wall thicknesses X_1, X_2, X_3, and thermal conductivities k_1, k_2, k_3

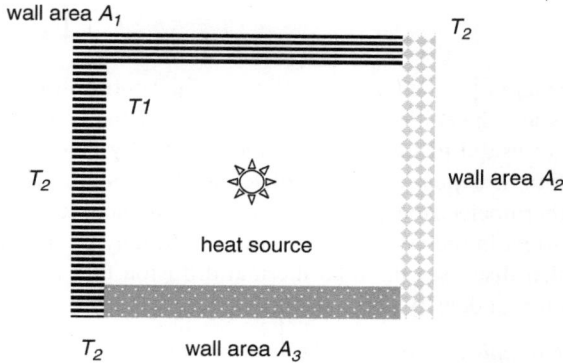

areas A_1, A_2, A_3, wall thicknesses X_1, X_2, X_3, and thermal conductivities k_1, k_2, k_3. Refer to Fig. E5.1 for details.

Answer: By analogy with terms in electricity, the expression $\frac{kA}{x}$ is known as *thermal conductance*. This follows the similar relationship of electrical conductivity and electrical conductance. Conductances add in parallel, so that we have

$$\frac{k_1 A_1}{x_1} + \frac{k_2 A_2}{x_2} + \frac{k_3 A_3}{x_3} \tag{E5.1.1}$$

contributing to the rate of heat loss (or heat flow rate) as follows:

$$\frac{\Delta Q}{\Delta t} = \left(\frac{k_1 A_1}{x_1} + \frac{k_2 A_2}{x_2} + \frac{k_3 A_3}{x_3} \right) (T_2 - T_1). \tag{E5.1.2}$$

This is the rate at which heat escapes from the box through the walls made of three different materials. Of course, the problem can be formulated in different ways, such as for instance knowing the heat flow rate, calculate the volume of the box or the inside temperature.

Let us look again at the interface between two different materials. We have from relationship (5.4) the quantity $\Delta \vec{q}_{\text{td}} = (\Delta q / \Delta t x / A)$ as the induced thermal diffusion flux. This flux can also be written as

$$\Delta \vec{q}_{\text{td}} = -k \Delta T. \tag{5.5}$$

Nonetheless, expression (5.5) is actually incomplete, as an additional term is required to take into account the Peltier effect which occurs at the interface. Specifically, this term is $(\Pi_1 - \Pi_2) j$ where Π_1 and Π_2 are the Peltier coefficients of material 1 and 2, respectively, and j is the electric current density (i.e., in a metal). The Peltier coefficients represent how much heat current is carried per unit charge through the material (i.e., metal), and therefore cannot be neglected. Taking that into account gives the following expression for the total thermal diffusion flux:

$$\vec{q} = -k\nabla T + \prod j, \tag{5.6}$$

where $\prod$ is the Peltier coefficient of the entire structure. By the way, such a structure where the Peltier effect is significant is actually a thermocouple. These relationships are useful to illustrate in principle that by determining the total thermal diffusion flux, it is possible to calculate the temperature distribution in a thermocouple. If a thermoelectric device contains a thermocouple or the Peltier effect is a dominant effect in the device, relationship (5.6) may be important. Here is another example that discusses the total thermal diffusion flux in the simple case when the carrier current density is constant.

Example 5.2. Derive the heat balance equation in a thermoelectric structure with a constant carrier current density j so that $\text{div}\,\vec{j} = 0$. Also use the vector identity $\text{div}(\varphi\vec{j}) = \varphi\text{div}\,\vec{j} + \nabla\varphi.\vec{j}$.

Answer: See also [6], relationships (E5.2.1)–(E5.2.3).

$$\text{div}(\vec{q} + \varphi\vec{j}) = 0, \tag{E5.2.1}$$

$$\text{div}(\tilde{\varphi}\vec{j}) = \tilde{\varphi}\text{div}\,\vec{j} + \nabla\tilde{\varphi}\cdot\vec{j}, \tag{E5.2.2}$$

$$\text{div}\,\vec{j} = 0, \tag{E5.2.3}$$

$$\text{div}\vec{q} = -\vec{j}\cdot\nabla\tilde{\varphi}, \tag{E5.2.4}$$

$$\vec{j} = -\sigma(\nabla\tilde{\varphi} + \alpha\nabla T), \tag{E5.2.5}$$

$$\nabla\tilde{\varphi} = -\frac{\vec{j}}{\sigma} - \alpha\nabla T, \tag{E5.2.6}$$

$$\text{div}\vec{q} = \frac{j^2}{\sigma} + \sigma\vec{j}\nabla T. \tag{E5.2.7}$$

Relationship (E5.2.7) is the heat balance equation which shows two sources of heat: Joule and Thomson. Namely, the first term on the right side is the Joule heat, and the second term is the Thomson heat. It is significant to mention that the equation was derived under the assumption that only an equilibrium carrier current exists with a carrier current density $\vec{j}$ constant; thereby $\text{div}\,\vec{j} = 0$.

Let us also observe what happens if we assume that both electric and thermal conductivity, as well as the Seebeck coefficient, are constant. Taking the divergence of $\vec{q}$ in (5.6), and equating it to the heat balance (E5.2.7) in the previous example, we obtain for each material in the thermoelectric structure

$$k_{1,2}\frac{d^2 T_{1,2}(x)}{dx^2} + \rho_{1,2}j^2 = 0, \tag{5.7}$$

where $k_{1,2}$ are the thermal conductivities, $T_{1,2}$ are the temperatures, and $\rho_{1,2}$ are the resistivities of the two materials. To keep things simple we consider the surface Joule heating at the boundary between the two materials as negligible. Furthermore, we make the statement that the thermal diffusion flux is continuous at the interface. This represents one of the boundary conditions for (5.7). The second boundary condition is given by

$$T_1(\text{at boundary}) = T_2(\text{at boundary}). \tag{5.8}$$

With the two boundary conditions in place, (5.8) allows determination of the temperature distribution in the thermoelectric structure formed by the two materials. However, the minimum temperature that can be obtained in a cooling mode of functioning still depends significantly on the figure-of-merit of this thermoelectric structure [6]. The best figure-of-merit materials are found among heavily doped semiconductors, as metals have relatively low Seebeck coefficients. Furthermore, the thermal conductivity of metals is dominated by electrons, and proportional to the electrical conductivity. Therefore, it is quite difficult to attain a high figure-of-merit in metals that have a high electrical and thermal conductivity. Remember, a low thermal conductivity is needed for a high figure-of-merit. In contrast to metals, the thermal conductivity of semiconductors is mostly due to phonons, with a lesser contribution from electrons. This means that the thermal conductivity of semiconductors can be reduced without altering the electrical conductivity. Reduction of thermal conductivity in semiconductors can be obtained through alloying. Commercial thermoelectric cooling materials are for instance alloys of Bi_2Te_3 with Sb_2Te_3, giving rise to the p-type $Bi_{0.5}Sb_{1.5}Te_3$. Alternatively, alloying Bi_2Te_3 with Bi_2Se_3 results in the n-type [5] $Bi_2Te_{2.7}Se_{0.3}$. These materials have a figure-of-merit $ZT \sim 1$ at room temperature and are very frequently employed in thermoelectric devices.

5.1.5 Nanoscale Effects on Heat Transfer

As already mentioned, electrons are the main thermal energy carriers in metals, while phonons are believed to be primarily responsible for heat transfer in semiconductors. This fact is of dominant significance for semiconductor thermoelectric devices, as their figure-of-merit can be manipulated in ways not possible with all-metal materials. Since phonons are actually lattice vibrations in a nonmetal, there is no motion of media as heat propagates, and therefore thermal transport obeys different rules than in metals where electrons are the carriers. Nevertheless, just like for electrons, in semiconductors there is the concept of *phonon mean free path* Λ which enters the mathematical expression for the thermal conductivity

$$k = \frac{1}{3}Cv\Lambda, \tag{5.9}$$

where C is the phonon specific heat per unit volume, and v is the speed of sound. An entire mathematical apparatus is constructed around concepts familiar for electron behavior, but in this case phonons are the dominant carriers of thermal energy. Heat conduction through phonons is modeled using various methods, depending on the length scale of the device and on the size of the entities where thermal aspects matter. Whether thermal modeling occurs on the scale of the lattice spacing (a few tens of nm), or the phonon mean free path (a few hundred nm), certain compromises have to be made regarding which physical quantities are dominant or more relevant. For instance, the length scale of the problem determines if phonons should be modeled as particles or whether their wave nature is more important. Typical semiconductor device dimensions lie in between the phonon mean free path Λ and the dominant phonon wavelength at room temperature. Quite often, semiclassical or even classical models are more than adequate for this regime. On the other hand, this approach may not be feasible with the devices we encounter nowadays. More devices, in particular magnetic ones have shrunk so much that the dimensions of the problem are comparable to the phonon wavelength, and the temperature is small compared to the Debye temperature. In these cases, quantum mechanical models are recommended. When modeling devices electrothermally, the transport of charge carriers could be simulated first, in order to calculate the distribution of the heat source in the device. Only afterward should the transport of phonons be modeled to obtain the resulting temperature distribution in the lattice. It is often a good idea to consider heat generation and transport as two distinct issues and treat them separately. Heat generation is computed through electron–phonon scattering.

In view of the above, and through the physical quantities of (5.9), it becomes obvious that dimensions of the material have an effect on thermal conductivity, just like they would if electrons were the main carriers. If scattering of phonons by interfaces can be neglected in bulk materials, this is certainly not the case in small-scale or low-dimensional structures. For instance, it has to be taken into account in thin films where the characteristic length of materials such as film thickness can be less than the phonon mean free path Λ. This in fact happens in bulk crystals with thin layers of different materials (i.e., multilayer thin films) where the phonon mean free path is usually much longer than the film thickness. Not surprisingly, long mean free paths are expected for phonons, as the films are so thin that they range in thickness from only monoatomic layers to thousands of angstroms. In these structures, heat conduction is likely to be dominated by phonon scattering at interfaces, resulting in a reduced thermal conductivity. Nevertheless, this is beneficial for the figure-of-merit of thermoelectric devices. It should also be appropriately mentioned that theoretical work from the mid-60s indicates that acoustic phonons can significantly enlarge the value of the thermoelectric voltage, whereas optical phonons can reduce it [7]. Subsequent experimental work spanning more than two decades confirmed these early predictions, and therefore these concepts are most likely incorporated nowadays into new designs of thermoelectric devices.

The thermal conductivity of commonly employed semiconductors is known to vary with temperature as [8] $k \sim 1/T^n, n = 1$–1.5. If this news is not good enough on its own, consider this: Thermal conductivity values of their binary or ternary alloys are even lower due to the increased phonon scattering. The

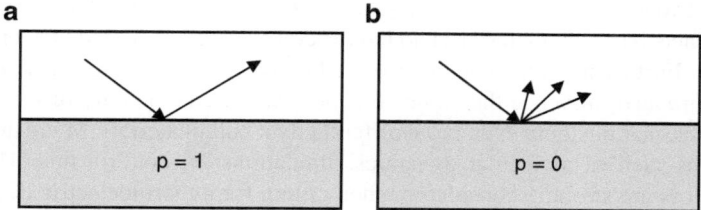

Fig. 5.6 Schematic representation of (**a**) elastic or specular ($p = 1$) and (**b**) inelastic or diffuse ($p = 0$) scattering at an interface between two materials

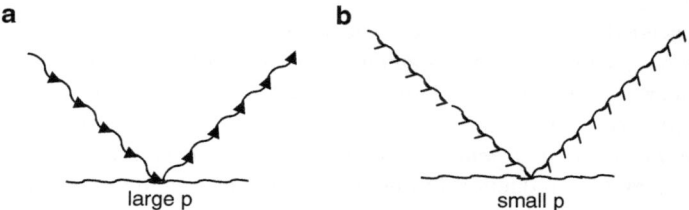

Fig. 5.7 The scattering parameter p depends on surface roughness and wavelength. For the same surface roughness, the scattering parameter is (**a**) approaching a value of 1 for large wavelengths, and (**b**) closer to zero when the wavelength is small. This means, the surface appears as rough to the short waves, and smooth to the longer ones

conductivities may still be more reduced in multilayers of monoatomic thickness, even if compositionally the materials are the same as bulk. There is a simple way to quantify scattering through a parameter p, which measures the proportion of energy carriers (e.g., phonons) specularly reflected from the surface as the carriers lose momentum in the heat flow direction. The values for the scattering parameter p range from 0 to 1, and the two extremes are depicted in Fig. 5.6 which shows schematically that (a) when $p = 1$ the scattering is fully elastic (or specular), and (b) for $p = 0$ the scattering is fully diffuse. In most situations, p has a value in between.

It should be noted as well that the scattering parameter depends on surface roughness and wavelength (Fig. 5.7). When the wavelength is small, the scattering parameter is very small (Fig. 5.7b) because the waves gain or lose phase and are widely scattered by the rough asperities of the surface. For large wavelengths, the scattering parameter approaches a value of 1 (Fig. 5.7a), as the surface appears smooth to the incoming wave. Size effects can thus be utilized to alter the transport properties like manipulating boundaries for phonons to scatter more efficiently. Interestingly, the same effects mostly attributed to scattering and highly undesirable for other microelectronic or optoelectronic devices; such as reduced heat conduction capability can benefit many semiconductor thermoelectric devices, especially those routinely used in thermal management [5].

There is a growing consensus among materials scientists that low-dimensional structures are new materials, in spite of their similar atomic structure as the parent material. Monoatomic multilayers display by far more phonon scattering and loss of coherence than bulk materials, and have therefore a unique range of properties of their own. Furthermore, parameters involved in material fabrication such as growth temperature also influence the values for the thermal conductivity of monoatomic multilayers, making them even more different than bulk materials. Many theoretical predictions such as molecular dynamics simulations [8] confirm this. These and other factors are strongly considered when opting for a thermoelectric design with low-dimensional structures where reduced thermal conductivities can be obtained. Thereby, size effects can result in significantly higher device temperature increases than in bulk materials which, as mentioned previously, may be detrimental to optoelectronic devices, transistors, lasers, or high power detectors, and nevertheless are much needed for many thermoelectric devices.

Following up on the above statements, it was confirmed by various studies [9, 10] that an increased phonon-boundary scattering and a reduced phonon thermal conductivity lead to a much higher thermoelectric figure-of-merit ZT in so-called *quasi-one-dimensional structures*. For instance, silicon nanowires register a two orders of magnitude reduction in phonon thermal conductivity as opposed to the bulk silicon value of [11] 148 W/m K. Furthermore, nanodots may offer additional improvements in device performance due to their strict electron confinement [12]. It is believed that if the position of each electron is fixed by confining it in a restricted region such as a "dot," the energy distribution in that region will not experience large temperature variations. Each region could be a nano- or a quantum dot that effectively prevents the electrons from being thermally excited to higher energy states. Pb–Te-based quantum dot superlattices have a reported figure-of-merit between 1.6 and 2 at room temperature [4].

As seen, effects that play a significant role at very small length scales but not necessarily in bulk can have an influence on the figure-of-merit of thermoelectric devices. In addition to scattering, various losses, as well as other electrical and thermal resistances, may also come into play. For this reason, consider [13, 14] a slightly different figure-of-merit such as the *effective figure-of-merit*

$$ZT_{\text{eff}} = \frac{\alpha^2 \sigma_{\text{eff}}}{k_{\text{eff}}} T, \qquad (5.10)$$

where the *effective electrical conductivity* σ_{eff} and the *effective thermal conductivity* k_{eff} are influenced by the electrical contact resistivity ρ and thermal contact resistance $r_{\text{c,th}}$

$$\sigma_{\text{eff}} = \frac{H}{H\rho_{\text{bulk}} + 2\rho}, \qquad (5.11)$$

$$k_{\text{eff}} = \frac{H}{Hk_y^{-1} + 2r_{\text{c,th}}}, \qquad (5.12)$$

while k_y is the cross-plane thermal conductivity instead of the commonly used bulk or in-plane thermal conductivity. The change in the figure-of-merit is determined by comparison

$$\frac{Z T_{\text{eff}}}{Z T} = \frac{\sigma_{\text{eff}}/\sigma}{k_{\text{eff}}/k_y} = \frac{H + 2k_y r_{\text{c,th}}}{H + 2\sigma\rho}. \tag{5.13}$$

It is interesting to note that ZT_{eff}/ZT will be greater than or equal to unity when k_y/σ (also called a *reduction function*) is greater than or equal to $\rho/r_{\text{c,th}}$. This happens when the p-type and n-type structures, and hence the whole thermoelectric device is built on a nanoscale. In contrast, micron-sized structures would lead to relationship (5.13) being less than unity. These concepts are illustrated in the following example.

Example 5.3. Find the change in the figure-of-merit ZT of a nanoscale bismuth telluride thermoelectric device due to electrical contact resistivity and thermal contact resistance at the interface of the p-type and n-type structures with the substrate. The height of the p-type or n-type structure is $H = 50$ nm. The Seebeck coefficient for the n-type or p-type structure is $\alpha_{\text{p,n}} = 4\times10^{-4}$ V/K, and the bismuth telluride resistivity is $\rho_{\text{bulk}} = 10^{-5}\Omega$m. *Note:* All relevant relationships and further values for the bismuth telluride material properties can be found in [13, 14].

Answer: Consider the effective figure-of-merit

$$ZT_{\text{eff}} = \frac{\alpha^2 \sigma_{\text{eff}}}{k_{\text{eff}}} T, \tag{E5.3.1}$$

where the effective electrical σ_{eff} and thermal conductivities k_{eff} are influenced by the electrical contact resistivity ρ and thermal contact resistance $r_{\text{c,th}}$

$$\sigma_{\text{eff}} = \frac{H}{H\rho_{\text{bulk}} + 2\rho}, \tag{E5.3.2}$$

$$k_{\text{eff}} = \frac{H}{Hk_y^{-1} + 2r_{\text{c,th}}}, \tag{E5.3.3}$$

while k_y is the cross-plane thermal conductivity considered instead of the bulk or in-plane. The change in the figure-of-merit is determined by comparison

$$\frac{Z T_{\text{eff}}}{Z T} = \frac{\sigma_{\text{eff}}/\sigma}{k_{\text{eff}}/k_y} = \frac{H + 2k_y r_{\text{c,th}}}{H + 2\sigma\rho}. \tag{E5.3.4}$$

The reduced film thickness is $\delta = H/\vartheta$, where $H = 50$ nm is the height of the n-type or p-type structure. In our case (see mentioned references) $\vartheta = 12.5$ Å, which gives $\delta = 40$ and

Table 5.1 Some piezoelectric materials showing promise in magnetoelectric composites

Piezoelectric material
$Bi_4Ti_3O_{12}$
$PbMg_{1/3}V_{2/3}O_3$
$PbMg_{1/3}Nb_{2/3}O_3–PbTiO_3$
$PbZn_{1/3}Nb_{2/3}O_3–PbTiO_3$
$BaTiO_3$
$BiFeO_3$
$BiFe_{1-x}Ti_xO_3$
PZT

$$k_y = k_{bulk}\frac{1}{3}\left(3 - \frac{1}{\delta}\right).$$

(E5.3.5)

Then $k_y = 1.375\,\mathrm{W/(m\,K)}$ since $k_{bulk} = 1.5\,\mathrm{W/(m\,K)}$. From Table 5.1 in [13], $\rho = 2.625 \times 10^{-12}\,\Omega\mathrm{m}^2$ and $r_{c,th} = 9.316 \times 10^{-9}\,\mathrm{m}^2\mathrm{K/W}$. Then $ZT_{eff}/ZT \approx 1.5$, which is obviously higher than unity.

5.1.6 A Few Thermoelectric Material Considerations

Many components in electronic devices require protection from overheating, which means it is necessary to have an efficient redistribution of heat. For this purpose, thermoelectric devices are being used. An approach to improving thermal transport in thermoelectric devices is to use as a substrate for the thermolegs a good heat conductor. Aluminum has a thermal conductivity of $0.235\,\mathrm{W/mm°C}$. However, using metal substrates such as aluminum alone is not an option due to their high electrical conductivity. Coating metal plates with dense alumina ceramics (5–30 μm) significantly increases the thermal conductivity, as opposed to using just alumina ceramic substrates which have a thermal conductivity of only [15] $0.026\,\mathrm{W/mm°C}$. On the other hand, polymer coatings may not be an improvement, as they reduce thermal conductivity and also become detached from the metal substrate due to thermal degradation. A technique that has been in existence since the 1930s, called *plasma electrolytic oxidation* (PEO), can produce durable oxide coatings on metals such as Al, Mg, Ti, Zn, Zr, or Nb. Anodic oxidation in an aqueous solution creates an oxide coating that has been reported to resist over 7,000 h of salt water exposure, and have a dielectric breakdown voltage much higher than [15] 270 V ac or 160 V dc. They show promise of use in thermoelectric cooling systems, provided further research can improve their figure-of-merit.

Bismuth telluride alloys are some of the most popular thermoelectric materials employed in commercial applications due to their high figure-of-merit. Regrettably, traditional fabrication methods render materials with poor mechanical properties, prone to cleavage and brittleness. Alternative preparation methods such as *melt spinning* combined with *spark plasma sintering* have produced materials of better mechanical quality. For instance, a nanostructured p-type $Bi_{0.5}Sb_{1.5}Te_3$ compound prepared by these techniques, and displaying a fine grain structure of only 5–15 nm, showed intense phonon scattering, and thereby a decreased lattice thermal conductivity. The high phonon scattering is attributed to defects and local lattice distortion resulting from residual heat stress during the melt spinning process. High-resolution transmission electron investigations show that these occur within the nanometer grains. The influence of the fabrication method is reinforced when a figure-of-merit of 1.2 is obtained at 300 K for specimens prepared under certain, well-controlled melt spinning conditions [16].

Skutterudites are minerals of cobalt arsenide containing nickel and iron substituting for cobalt. They are known for their hydrothermal properties, i.e., the ability to circulate hot water. In similar skutterudite compounds where As is replaced by Sb such as $Co_{1-x}Fe_xSb_3$, Yang [17] et al. noticed that the presence of iron reduces the thermal conductivity. Although these compounds are unlikely to have a high figure-of-merit, they give a good idea of the role of iron, a ferromagnetic element, in thermoelectric materials. Without iron, $CoSb_3$ is diamagnetic; however doping with iron renders the $Co_{1-x}Fe_xSb_3$ compound paramagnetic. For this reason, the compound can also be regarded as thermomagnetic, as it is likely it will exhibit a Nernst or Ettingshausen effect under the influence of a transverse magnetic field, as discussed in a subsequent section. Hall measurements of $Co_{1-x}Fe_xSb_3$ compounds indicate positive, i.e., p-type carriers dominating charge transport, with as much as a ninefold increase in the number of carriers for higher (up to $x = 10\%$) Fe doping concentrations. These latter results are somehow surprising considering that magnetization studies reveal an electronic structure of $Co_{1-x}Fe_xSb_3$ the same as that of $CoSb_3$. Nevertheless, further analysis suggests that the carrier concentration is not due to Fe^{2+} replacing Co^{3+}, but rather that the concentration of vacancies increases on the Co site. Hence, as the Fe doping increases, the hole concentration increases as well, as more vacancies become available on the Co site. With the increase in hole concentration, the Seebeck coefficient decreases, which is in contrast to the significant increase in thermal conductivity with Fe doping. This enhancement in thermal conductivity can be explained by the increase in concentration of point defects in $Co_{1-x}Fe_xSb_3$ most likely due to excess Sb. There is also the possibility that the Hall effect diminishes the influence of thermomagnetic effects; however this likelihood requires further investigation.

Since the 1950s, thermoelectric phenomena in anisotropic media have occupied the interests of A.G. Samoilovich [18] and his collaborators in the Russian Institute of Thermoelectricity. They generalized the theory of thermoelectric phenomena for isotropic materials, extending it to anisotropic media. This fact changes the way thermoelectric devices are designed, as the p–n thermolegs can be replaced with an anisotropic material alone. In this design, there are no pairs of p and n elements, but rather one thermoelectrically anisotropic crystal is used. In this case, the appearance of a transverse thermoelectric voltage V_S (Fig. 5.8) is attributed to the anisotropic nature of the Seebeck coefficient, while the anisotropy of the electric and thermal conductivity leads to a temperature gradient. The Seebeck coefficient

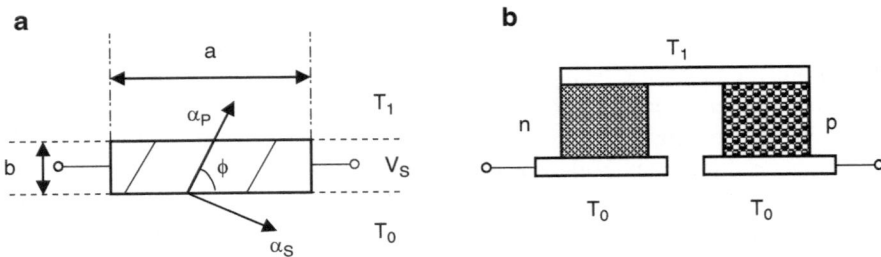

Fig. 5.8 (**a**) A sketch of an anisotropic thermoelectric element illustrating Samoilovich's idea of using only one material. A transverse thermoelectric voltage V_S appears due to the anisotropic Seebeck coefficient. Directional variations in the Seebeck coefficient appear along the parallel and the perpendicular directions with respect to the principal thermoelectric axis. This design is in contrast to the traditional thermoelectric element with p and n materials (**b**)

has a perpendicular component α_S and a parallel component α_P as related to the principal thermoelectric axis. The principal thermoelectric axis makes an angle with respect to the length of the crystal. A difference in temperature $\Delta T = T_1 - T_0$ exists between the upper and lower side of the crystal, while a transverse thermoelectric voltage V_S is developed perpendicular to the height of the crystal (i.e., end faces). The latter is given by the expression

$$V_S = \frac{1}{2}\Delta\alpha \cdot \Delta T \cdot \frac{a}{b}, \qquad (5.14)$$

when the length a exceeds by far the height b of the crystal. The thermoelectric anisotropy is found in the directional variation of the Seebeck coefficient $\Delta\alpha = \alpha_p - \alpha_S$. These variations are due to the fact that α has a different value along the direction parallel to the principal thermoelectric axis than along the perpendicular direction. Relationship (5.14) shows that the larger the thermoelectric anisotropy, the larger the obtained transverse voltage. Thus, in principle, larger voltages could be obtained in thermoelectrically anisotropic materials than in traditional p–n thermoelectric elements. The thermoelectric anisotropy has been found to be sensitive to impurity concentration, a fact that could be exploited in practical applications.

5.1.7 Heusler Alloys with Thermoelectric Potential

Ferromagnetic metal alloys based on a Heusler phase are an interesting class of materials that are discussed in several other chapters in this book. This is because their magnetic properties are not due to expected reasons, such as the ferromagnetism of individual constituents, but rather attributed to the double-exchange mechanism between neighboring spin-uncompensated ions of a different species. Quite often, these compounds exhibit localized magnetism of the rare-earth ions. Some show antiferromagnetic ordering at very low temperatures (2–3 K), whereas others (e.g., compounds containing Er and Pd) remain paramagnetic down to the lowest temperatures studied. Some yttrium-based compounds alloyed with Pd exhibit a weak form of diamagnetism [19]. What happens in these materials is a consequence of phase transitions and possibly spin reorientations that occur with temperature variations.

Many Heusler alloys are considered narrow-gap semiconductors. A positive Seebeck coefficient hints at holes being the dominant carrier, holes arising from some acceptor centers. The overall temperature dependencies of the Seebeck coefficient are similar to those of ternary compounds containing Ni, Sn, and another metal or semimetal. Particularly, it is assumed that low carrier densities are responsible for the variations observed in the Seebeck coefficient. A figure-of-merit ZT of only 0.15 was evaluated at room temperature for a particular compound, ErPdSb. However, many ternary compounds containing Pd, Bi, or Sb, together with a variety of rare-earth elements, have indicated a metallic-like conductivity at particular temperatures. Data [19] obtained at low temperatures suggest an overlap of an impurity band with a conduction band, giving rise to a metallic-like character for the resistivity. Conversely, higher temperatures show a resistivity behavior best approximated by [19]

$$\frac{1}{\rho(T)} = \sigma_0 + \sigma_\alpha \exp\left(\frac{-Ea}{2k_B T}\right), \qquad (5.15)$$

where σ_0 is the *residual conductivity* due to defect scattering, and σ_α is a constant with units of conductivity that depends on several factors, among them the velocity of the dominant carriers at the Fermi surface, the density of the dominant carriers, and the Debye temperature. E_a is the energy gap, k_B is the Boltzmann constant, and T is absolute temperature. In compounds that order magnetically, anomalies in resistivity are attributed to magnetic phase transitions occurring at those particular temperatures. Interestingly, the Seebeck coefficient shows no temperature variation from 300 K down to those transition temperatures. Also, rare-earth compounds containing Pd/Bi or Pd/Sb show rather large Seebeck coefficients, up to $200\mu V/K$ at room temperature for the latter type of compounds [15]. It is suspected that Pd compounds have multiple bands of electrons and holes, indicative of a semimetallic character. Unlike in metals, the Hall coefficient is strongly temperature-dependent, reflecting the complex electronic structure with different temperature variations of the mobilities.

In spite of the promising combination of somewhat large Seebeck coefficients and moderate electrical resistivity, the figure-of-merit obtained for ternary compounds containing rare earth, Bi or Sb, and Pd is still too modest. It is not competitive enough with respect to compounds such as bismuth tellurides used regularly in commercial devices displaying a figure-of-merit close to unity. Nevertheless, 3d-transition metals are studied in many areas of magnetism for their yet unexploited, but reassuring potential. Maybe thermoelectric devices will benefit from compounds containing these metals, too.

5.2 Thermomagnetic Phenomena

5.2.1 The Nernst and Ettingshausen Effects

Just like their thermoelectric counterpart, thermomagnetic effects have been studied in the hope of developing efficient energy conversion devices. Nevertheless, the boundary between thermoelectric and thermomagnetic materials is sometimes indistinct, as many magnetic compounds exploited for thermoelectric devices are likely to exhibit thermomagnetic effects under the influence of a transverse magnetic field. In this case, thermoelectric and thermomagnetic effects are actually combined, and may quite well either interfere with, or enhance each other. A thermomagnetic effect is observed, for instance, when a potential difference is induced in the y-direction in the presence of a magnetic field B_z applied perpendicular to a conductor in which a thermal variation exists in the x-direction. This is known as the *Nernst effect*. A different situation arises when a perpendicular magnetic field B_z and an existing electric current flowing through the sample in the y-direction create a thermal variation in the x-direction. This is known as the *Ettingshausen effect*. The two effects are shown schematically in Fig. 5.9. Both effects are described by an experimentally measurable coefficient. The isothermal Nernst coefficient N is given by

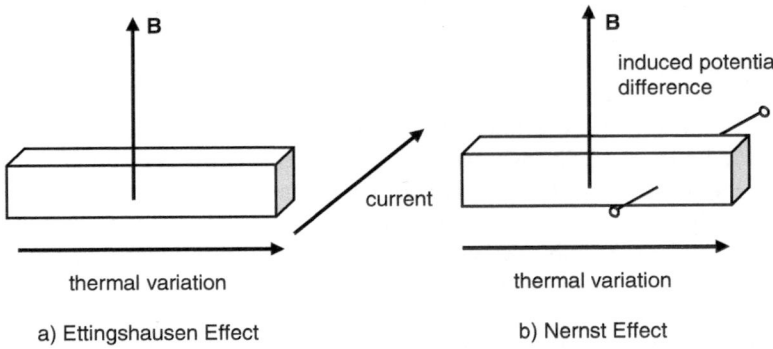

a) Ettingshausen Effect   b) Nernst Effect

Fig. 5.9 (**a**) The Ettingshausen effect is measured when a thermal variation is created along the x-axis in the presence of a magnetic field applied along the z-axis, and an existing current along the y-axis. (**b**) Conversely, when an existing thermal variation along the x-direction makes an angle of $90°$ with a z-axis applied magnetic field, a potential difference is created in the y-direction, and the Nernst effect is determined

$$N = \frac{\partial \tilde{\varphi}/\partial y}{B_z(\partial T/\partial x)} \quad \text{with} \quad \frac{\partial T}{\partial y} = 0, \tag{5.16}$$

where $\tilde{\varphi}$ is the electrochemical potential, and T is the temperature. Similarly, the isothermal Ettingshausen coefficient E_{tt} per unit volume can be determined from the relationship

$$E_{tt} = \frac{\partial T/\partial x}{B_z j_y} \quad \text{with} \quad \frac{\partial T}{\partial y} = 0, \tag{5.17}$$

where j_y is the electric current density in the y-direction. If the thermal conductivity k in the magnetic field is known, the Ettingshausen coefficient can be determined from the Nernst coefficient by using the Bridgman relationship

$$NT = E_{tt} k. \tag{5.18}$$

For the two coefficients, N and E_{tt}, interesting quantum oscillations that take place with magnetic field changes have long been observed in some materials, effects attributable to *magnetic breakdown* [20]. In an applied magnetic field, boundary conditions on the side of the specimen are important. Specifically, the limiting conditions are usually of two types: isothermal and adiabatic.

The isothermal condition, discussed above, involves a zero temperature gradient in the direction perpendicular to the direction of current or thermal variation. On the other hand, the adiabatic condition is defined as zero heat gradient in the direction perpendicular to current flow or thermal variation. The isothermal and adiabatic conditions are connected through the Heurlinger relation dating back to 1915. Specimen geometry, shape, and configuration of electrodes, as well as orientation of magnetic field, are all factors that determine the transport properties of the specimen. Two situations are usually considered, longitudinal and transverse, depending on the orientation of the magnetic field with respect to the thermal flow.

5.2.2 Influence of Specimen Geometry

Size and geometry considerations have long influenced what kind of power generation systems or energy harvesting devices can be obtained with magnetic materials. This is because the scalability of magnetic devices is limited by several factors, one being for instance the currents that are necessary to establish the magnetic fields. As such, researchers have normally stayed away from electrically excited magnetic systems. However, the market demand for power supplies under 100 W requiring more exotic energy conversion mechanisms have sparked an interest in novel magnetic devices of reduced dimensions. Because magnetic flux density obtainable from a permanent magnet is independent of its size, permanent magnets are often incorporated into miniaturized magnetic devices. With a magnetic field source assured, the problem remains of the optimum size and geometry of the rest of the magnetic device. Geometry is likely to play a role in what thermal gradients can be observed due to the presence of magnetic fields. It is therefore not surprising that there are differences in results obtained between the longitudinal and lateral sides of a device. For instance, thermal gradients of 23 K were measured along the 10 mm width of a $Bi_{88}Sb_{12}$ sample kept at an ambient temperature of 100 K, while a 15.4 K gradient was observed longitudinally in a perpendicular magnetic field of 1 T, with a current density of [21] 1 A/mm^2. This particular $Bi_{88}Sb_{12}$ compound in these element proportions has proven to be the best n-type Seebeck material ($\alpha \sim 150\,\mu V/K$) at low temperatures (~ 100 K), showing that thermoelectric and thermomagnetic effects can couple in the same material [22]. Interestingly, the thermal conductivity of this compound decreases with an increasing magnetic field which is beneficial for the figure-of-merit. Just like in the case of magnetoresistance experiments, the geometry of the specimens is expected to affect some of the values of the quantities being measured, as was noted earlier. At least, this was shown to be the case when [23] a larger Seebeck coefficient was recorded in longer Bi/Sb samples. Furthermore, the Nernst coefficient measured in some experiments involving magnetic fields up to 4 T was 12% smaller in the wider specimens [24]. However, not all experimental results agree with these findings that the geometry affects thermomagnetic quantities, and as such the subject requires further investigation. Some theoretical considerations are discussed in the next section.

5.2.3 Summarized Theoretical Model

In order to explain the experimental variations in the results obtained from thermomagnetic measurements, several theoretical models have been proposed. Of these, we summarize the approach followed by Okumura et al. [25], which could potentially be implemented into commercial software.

Consider a specimen in a thermomagnetic experiment depicted in Fig. 5.10 and described by

$$-\nabla\tilde{\varphi} = \rho\vec{j} + a\nabla T + R_H\vec{B} \times \vec{j} + N\vec{B} \times \nabla T, \qquad (5.19)$$

$$\vec{Q}_{tr} = \tilde{\varphi}\vec{j} - k\nabla T + aT\vec{j} + NT\vec{B} \times \vec{j} + kR_L\vec{B} \times \nabla T. \qquad (5.20)$$

The quantities in these relationships are electrochemical potential $\tilde{\varphi}$, electrical resistivity ρ, electric current density j, Seebeck coefficient α, temperature T, Hall coefficient R_H, magnetic flux density B, Nernst coefficient N, energy flux density Q_{tr}, thermal conductivity k, and R_L the Righi–Leduc coefficient. The latter is the thermal analog of the Hall coefficient R_H and describes a deviation from thermal equilibrium for electrons when the sample experiences a temperature gradient. Specifically, in the Righi–Leduc effect a temperature gradient develops perpendicular to an applied temperature gradient. The Righi–Leduc coefficient R_L is related to the Hall coefficient R_H and electrical resistivity ρ through the simple relationship $R_L = R_H/\rho$. Because a lot of quantities are contained in the description of the specimen under discussion, many effects can influence the results we are hoping to obtain. In fact, the last three terms of relationship (5.19) represent the Seebeck, Hall, and Nernst effects, respectively, while the last term in (5.20) stands for the Righi–Leduc effect. The Righi–Leduc effect may provide information on the electronic rather than phononic heat transfer. Nevertheless, some simplifications can be made if the assumption is made that the material is homogeneous, and quantities such as ρ, α, R_H, N, R_L, and k involved in (5.19) and (5.20) are *isothermal*. Note that the temperature is independent of the z coordinate. Also, the magnetic field is along the z-axis and does not depend on position, whereas the current density has no z-component. The system is in steady state; therefore the current density and energy flux density do not diverge.

The quantities that are expected to be determined from (5.19) and (5.20) are the electrochemical potential $\tilde{\varphi}$, the electric current density j, and the temperature T. In fact, it is desired to find the *distribution* of these quantities within an irregularly shaped sample of a semiconductor material. Finite element analysis software such as Ansys or Cosmos could be used to solve for the desired quantities. For instance, a mesh formed of squares is constructed within the sample. A sketch of a meshed specimen is shown in Fig. 5.10. T and $\tilde{\varphi}$ are considered on each corner (i, j) of a square, whereas current density j is considered on each side of a square, between corners. In this case, the temperature T is determined by taking the divergence of (5.20), equating it to zero (the energy flux density Q_{tr} does not diverge), and using (5.19). A second-order partial differential equation is obtained for T, which can be written in abbreviated form as

$$\nabla^2 T(x, y) = F(x, y). \tag{5.21}$$

Equation (5.21) can be discretized by taking into account the location of the point (i, j) with respect to the boundaries of the problem. On the boundary, appropriate boundary conditions (Dirichlet or Neumann) should be given.

The electric current density j can be determined from (5.18) by rewriting the latter in x and y component form, discretizing it, and applying boundary conditions $j_y = 0$ and $Q_{tr}^y = 0$. Finally, the electrochemical potential $\tilde{\varphi}$ can be determined

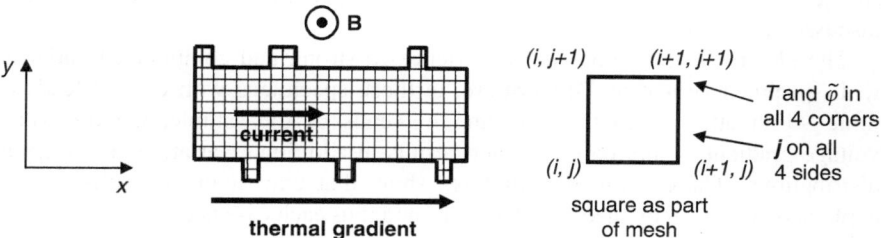

Fig. 5.10 Meshed specimen in a thermomagnetic experiment. A square as part of the mesh is shown on the right

from the expressions for the just calculated j, and by taking into account the continuity equation $\nabla \cdot j = 0$. In this case, an equation for j_{ij}^x with the following $\tilde{\varphi}_{ij}$ dependence is obtained

$$j_{ij}^x = \frac{\rho \tilde{\varphi}_{ij}}{h(\rho^2 + R^2 B^2)} + \text{terms independent of } \tilde{\varphi}_{ij}. \qquad (5.22)$$

The discretized value j_{ij}^x is taken along the line segment connecting two points with potentials $\tilde{\varphi}_{ij}$ and $\tilde{\varphi}_{i+1,j}$. The quantity j_{ij}^x is the current outgoing from point (i, j) in the positive x-direction, with three more such expressions for j_{ij}^y, $-j_{i-1,j}^x$, and $-j_{i,j-1}^y$. Adding these four expressions, the outgoing current density is

$$j_{ij}^{\text{out}} = \frac{4 \rho \tilde{\varphi}_{ij}}{h(r^2 + R^2 B^2)} + \text{terms independent of } \tilde{\varphi}_{ij}. \qquad (5.23)$$

Updating $\tilde{\varphi}_{ij}$ with a new value gives

$$\tilde{\varphi}_{ij}^{\text{new}} = \tilde{\varphi}_{ij} - j_{ij}^{\text{out}} \frac{h(\rho^2 + R^2 B^2)}{4 \rho}, \qquad (5.24)$$

where the right-hand side of (5.23) vanishes. Relationship (5.24) is used at different mesh points to update $\tilde{\varphi}_{ij}$.

In summary, after setting suitable initial conditions, T, $\tilde{\varphi}$, and j are continuously updated as explained above. Further details as well as complete equations can be found in Okumura et al.'s work [25]. Their results indicate that the Seebeck effect does not depend on specimen geometry if the current leads that measure longitudinal voltage difference are narrow enough. This does not mean that specimen geometry plays no role; it only shows that in some limiting case it would not influence the Seebeck effect.

On the other hand, the Nernst effect is sensitive to sample geometry even with narrow leads. This fact is attributed to a Righi–Leduc induced effect, a transverse temperature gradient (i.e., y-direction) that is proportional to the magnetic field B. Consequently, a transverse voltage gradient is produced as per the Seebeck effect giving a nearly constant bias to the Nernst coefficient. This would explain the smaller Nernst coefficient measured in wider specimens by Ikeda et al. [24], discussed in the previous section.

The electrochemical potential experiences a strong and complicated variation with specimen geometry. Furthermore, in those specimens with current leads of finite width attached to the cold and hot ends of the sample, the transverse voltage gradient is shorted out, which results in additional variations in potential distributions. These simulation findings show that experimental results can be explained quite well by theoretical models, and thus each case needs to be evaluated individually before reaching a conclusion.

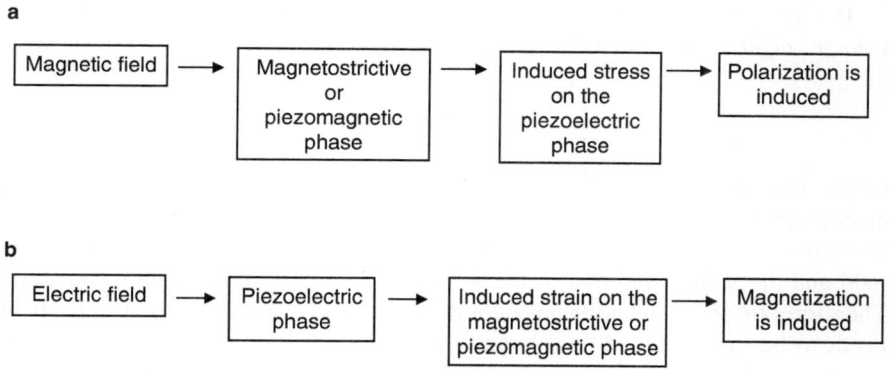

Fig. 5.11 (a) An electric polarization is induced by stress coupling of the piezoelectric phase to the magnetostrictive or piezomagnetic phase under the influence of a magnetic field. (b) A magnetization is induced by strain coupling of the magnetostrictive or piezomagnetic phase to the piezoelectric phase when an electric field is present

5.3 The Magnetoelectric Effect

5.3.1 Magnetoelectric Voltage Coefficient

The magnetoelectric effect occurs when a coupling between magnetic and electric fields takes place so that a magnetization is induced by an electric field or conversely, an electric polarization is induced by a magnetic field. The effect takes place in composites where a magnetostrictive or piezomagnetic phase exists, while there is also a piezoelectric phase present. In this case, an electric polarization is produced by stress coupling of the piezoelectric phase to the magnetostrictive or piezomagnetic phase under the influence of a magnetic field (Fig. 5.11a). Conversely, a magnetization is produced by strain coupling of the magnetostrictive or piezomagnetic phase to the piezoelectric phase when an electric field is present (Fig. 5.11b). Notice that in contrast to previous sections in this chapter, no thermal effects are involved.

Historically, the magnetoelectric effect gained popularity in the 1960s after its commercial potential became of interest. However, it had been predicted by Curie as early as 1894, but only identified experimentally by Debye in 1926. Since then, traditional magnetoelectric materials have included Cr_2O_3, Ti_2O_3, $GaFeO_3$, garnet films, and solid solutions such as $PbFe_{0.5}Nb_{0.5}O_3$, boracites, and phosphates. Less than a hundred single-phase materials display the effect. Unfortunately, the magnetoelectric effect is too weak to be exploited at room temperature. At present, no single-phase material is known that has a high enough coupling between magnetization and polarization for realistic practical applications. Therefore, these days more than a single-phase material is used.

To characterize a magnetoelectric effect in a material, the best way is to use a *voltage coefficient* α similar to the Seebeck coefficient

$$\alpha = \frac{\partial E}{\partial H}, \qquad (5.25)$$

where E is the electric field produced by an applied ac magnetic field H. The electric field is measured via the voltage V detected across the sample thickness t. In contrast to single-phase materials, the relationship between the applied magnetic field and the voltage induced in the detection circuit is not linear in composites. Therefore, to obtain linearity in a composite material, a weak ac magnetic field needs to be superimposed on a large dc bias field [26]. In this case, over a short range around this bias, the magnetoelectric effect can be approximated as linear, so that the voltage coefficient α varies linearly with the dc bias field.

It is not surprising that for an enhanced coefficient, the modulation frequency of the applied field needs to coincide with the magnetic, electrical, or mechanical eigenmodes of the system. As expected in this case, a significant increase in the voltage coefficient is obtained if the ac magnetic field is tuned to electromechanical resonance. Several observations show that α displays a strong frequency dependence in the microwave range [27, 28]. However, besides frequency, the orientation of the dc magnetic field and ac probe field also determine the value of the voltage coefficient. A longitudinal coefficient of a certain magnitude is measured when the ac and dc fields, as well as the induced signal, are all parallel to each other. On the other hand, when the induced electric field is perpendicular to the applied dc and ac magnetic fields a transverse coefficient of a different value is determined. A smaller voltage coefficient is expected in a direction with reduced magnetostriction because the latter is known to be anisotropic. Conversely, if the magnetostriction is increased in some direction by applying an increased magnetic field, a stronger magnetoelectric coupling is obtained and therefore a larger voltage coefficient.

5.3.2 *Fabrication Conditions*

Among the many factors that can lead to a ME response, crystal structure is highly probable to be significant in determining whether or not a magnetoelectric effect is observed. Fabrication conditions and mechanical processing are also expected to play an important role. As a matter of fact, these influences were confirmed experimentally [29, 30] in $LiCoPO_4$, $TbPO_4$, or yttrium iron garnet films, as well as predicted theoretically through mathematical considerations [31]. Many reports have been published detailing the consequences of microscopic mechanisms taking place in the material and obstructing the induction of larger magnetizations or polarizations. Hence, how you prepare a compound and of what materials remain important issues.

A classification of magnetoelectric compounds is sometimes used to differentiate between the types of composites that can usually be obtained. For example, it was observed that higher voltage coefficients are achieved in composites formed by two or three constituents than in single-phase materials. In these cases where the coefficient exceeds the value measured in single-phase materials it is said that the compound displays *scaling or combination properties*. However, if the single-phase materials are not magnetoelectric at all, whereas their composite is, a *product property* is attained. The voltage coefficient is then a product tensor property. Lastly, the composite may display a *sum property* if it is a result of the weighted contribution of each material. The effect is then proportional to the volume of weight fractions of each phase. Overall, this classification is just a guideline, and therefore not always enforceable, especially since sometimes magnetoelectric compounds are obtained through techniques and with materials that do not fall clearly into a well-defined category.

Over time, a series of methods have been developed through which magneto-electric compounds can be fabricated, some with better ME response than others. For instance, it was observed that assembling magnetoelectric composites from particulate materials or by hot molding leads to larger voltage coefficients. A widespread technique is setting mixtures of PZT/PVDF or Terfenol-D/PVDF in molds and hot pressing. While some fabrication recipes have failed, others are still in use. See the example below.

Example 5.4.   What are the effects of sintering on the magnetoelectric effect?

Answer: One of the many advantages of sintering is that sintered composites are more cost effective to prepare than unidirectionally solidified in situ materials. It is easy to control the molar ratio of phases in sintered composites, the grain size of each phase, and the sintering temperature. Nevertheless, with sintered composites a variety of undesired complexities arise, such as the difficulty in transferring the mechanical stress between the magnetostrictive and piezoelectric materials without losses, or the reduction in the magnetoelectric response due to leakage currents in low resistivity magnetostrictive phases. Microcracks would appear as a result of sintering, due to the thermal expansion mismatch between constituents. This would diminish ME behavior, together with impurities or undesired phases that are also likely to reduce the magnetoelectric effect.

More recently, it has become common to employ layering (or lamination) techniques for better magnetoelectric results. Instead of sintering the constituents which has its drawbacks, lamination can be used with much improved voltage coefficients. At least, this is the case when placing a PZT (lead–zirconate–titanate) layer between two Terfenol-D plates and bonding using a conductive epoxy resin. Some curing at 80°C for 3 to 4 h under load is also necessary [32]. Many compounds are obtained these days through lamination, among which we can count the popular combinations of ferrite/piezoelectric ceramics and ferrite/PZT composites. The large voltage coefficients attained in laminated composites can be attributed to the lack of chemical reactions between independently prepared constituents. This is in contrast to sintered granular composites or unidirectionally solidified composites where the temperature of processing is too high to avoid chemical reactions of the phases [33]. In any case, voltage coefficients measured in composites exceed the largest values observed in single-phase compounds.

It should be noted that when fabricating magnetoelectric materials, it is required to have a strict control over the composition of the constituents to ensure the effect is observed, as any small change in element concentration can result in losing the effect altogether. We will illustrate in the example below a few ideas related to compositional effects.

Example 5.5. For instance, a way to improve or even obtain a magnetoelectric effect in a material that would normally not be magnetoelectric, or alternatively would display a very weak effect, is to substitute some atoms of the host compound with a guest element. As such, it has been observed that substitution of the element Ti in the popular compound $BiFeO_3$ has led to an observable magnetoelectric effect [34]. The fact that the effect is noticeable at all is attributed to an increase in volume fraction of grain boundaries believed to lead to an enhanced interfacial polarization. Additionally, the Ti substitution also refined the microstructure and reduced the grain size. However, at the moment, despite all improvements in the fabrication of $BiFeO_3$, they are not significant enough to meet industrial expectations.

It is sometimes possible to further enhance the magnetoelectric effect, as studies on the much used compound $BaTiO_3$–$CoFe_2O_4$ show where an excess of TiO_2 has resulted in a larger voltage coefficient. These values improve every year due to the large variety of techniques employed, and therefore the research literature should be consulted for the most updated data.

Embedding nanostructures in composites can also have a beneficial effect on the magnitude of the magnetoelectric coefficient. Tb–Fe nanoclusters deposited on a PZT film by using a mask and a magnetron-sputtering–gas-aggregation technique (Fig. 5.12) have proven to give a larger mechanical stress-mediated magnetoelectric coefficient than otherwise obtainable through other fabrication methods [35]. The nanosize of the ferromagnetic clusters renders the overall composite characteristics not generally accessible in bulk compounds, such as strong coupling between ferromagnetic and ferroelectric phases. This is attributed to the unavoidable diffusion of some portion of the magnetostrictive clusters into the piezoelectric film. However, the coupling efficiency is also influenced and restricted by the bias magnetic field.

5.3.3 Incorporating Piezomagnetic Materials

Most ferromagnetic materials display the *magnetostrictive effect*. This is characterized by a field-induced strain proportional to the *square of the field strength*. On the other hand, in *piezomagnetic materials* the strain induced by a magnetic field is *linearly proportional to the field strength* [36]. Consider additionally that piezoelectricity is also a linear effect. Thus, if a piezoelectric phase is coupled to a magnetostrictive phase, the magnetic field-induced magnetostriction will reach saturation before the piezoelectric response does. In this case, the magnetoelectric coefficient of magnetostrictive–piezoelectric composites will increase with magnetic field for a while, but will decrease after the magnetic field reaches the saturation value for the magnetostriction. For this reason, piezomagnetic/piezoelectric composites tend to have better magnetoelectric coefficients than magnetostrictive/piezoelectric ones, as the response in the two phases is linear in the magnetic field. However, some ingenious fabrication techniques have been recently developed to incorporate highly magnetostrictive materials such as Terfenol-D into magnetoelectric compounds. Additionally, appropriate electronic circuitry has been employed to compensate for some of the undesired effects.

A typical composite material is made of layers of piezomagnetic and piezo-electric constituents such that a strain produced by a magnetic field applied to the piezomagnetic material is passed along to the piezoelectric part where it induces a polarization. Although ferrites such as Cr_2O_3 (or a compound containing Cr_2O_3) are not piezomagnetic, a magnetostriction induced in an ac magnetic field gives rise to *pseudopiezomagnetic effects*, and a magnetoelectric effect is still observed [37, 38]. Hence, the piezomagnetic properties can be extended to include pseudo effects, as well, and we shall use quite often (but not always) the word piezomagnetic instead of magnetostrictive when referring to magnetic field-induced strains in a magnetic material.

Before we proceed further on the subject of piezomagnetic materials, a word of caution should be said regarding incorporating popular or well-known magnetic constituents, whether piezomagnetic or magnetostrictive, as some of their electrical properties may not be beneficial for obtaining magnetoelectric behavior.

Example 5.6. Popular magnetic materials are ferrites. But ferrites may have an undesirable influence on the magnetoelectric effect observed in composites, as the low resistivity of ferrites produces leakage currents in the sample, reducing the magnetoelectric effect. There are some measures that can be taken for avoiding the leakage currents. One way is to increase the resistivity of the ferrite constituent. This means that the sintering temperature needs to be decreased during the fabrication process [39]. However, this results in losing other properties in the composite.

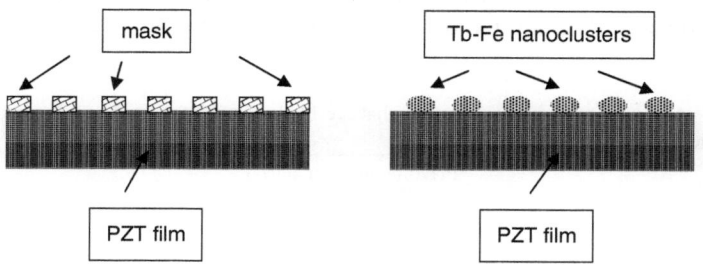

Fig. 5.12 Sketch of a fabricated Tb–Fe/PZT magnetoelectric composite: (**a**) mask on the PZT film; (**b**) Tb–Fe nanoclusters sputtered onto the PZT film

Therefore, if this is not an option, the ferrite phase should be replaced entirely by a different material. Safe alternative are compounds that contain one rare-earth element. In fact, the magnetoelectric properties of composites containing a rare-earth-iron alloy such as Terfenol-D combined with piezoelectric ceramics like polyvinylidine-fluoride (PVDF) have been calculated using the Green's function technique and found likely to exhibit a giant magnetoelectric effect [40].

In general, the piezomagnetic behavior observed in [41] $La_{0.7}Sr_{0.3}MnO_3$, $La_{0.7}Ca_{0.3}MnO_3$, or $Tb_{1-x}Dy_xFe_2$ alloys (Terfenol-D) makes them popular substitutes for ferrites as constituents in magnetoelectric composites. Nevertheless, Terfenol-D has its own problems, as it exhibits a high eddy current loss and is quite brittle.

Piezoelectric materials that show promise in magnetoelectric composites are $Bi_4Ti_3O_{12}$, $PbMg_{1/3}V_{2/3}O_3$, $PbMg_{1/3}Nb_{2/3}O_3$–$PbTiO_3$, or $PbZn_{1/3}Nb_{2/3}O_3$–$PbTiO_3$(Table 5.1) [42]. Piezocrystals with large piezoelectric voltage coefficients and strong elastic compliances are better suited for obtaining enhanced magnetoelectric signals. Furthermore, piezoelectric coupling has to be good and the volume fraction of the piezoelectric phase needs to be sufficiently high. By combining piezoelectric materials such as $BaTiO_3$, $BiFeO_3$, or $BiFe_{1-x}Ti_xO_3$ (Table 5.1) and piezomagnetic $CoFe_2O_4$, researchers are able to obtain a new composite material with an enhanced magnetoelectric effect strongly dependent on the fabrication conditions [43]. For instance, solid-state reactions take place among the constituent oxides in compounds such as $BaCO_3$, Co_3O_4, Fe_2O_3, or TiO_2 so that particles adhere to each other when they are heated below their melting point. As a result, various levels of mechanical coupling are obtained during different fabrication methods.

The extent of mechanical coupling between the piezomagnetic and piezoelectric phases is likely to influence the magnetoelectric effect. As such, PZT is only weakly coupled to $La_{0.7}Sr_{0.3}MnO_3$, $La_{0.7}Ca_{0.3}MnO_3$, or $CoFe_2O_4$, while displaying a stronger coupling to $NiFe_2O_4$and an enhanced magnetoelectric effect (Table 5.2). Mechanical defects such as pores or microscopic cracks can limit the mechanical bond between particles. On the other hand, doping Ni-ferrites with Co, Cu, and Mn results in a stronger coupling. Further experimentation and bringing fabrication down to a nanoscale is likely to lead to significant improvements in mechanically coupling diverse materials in order to obtain more reliable practical applications.

Table 5.2 A few piezomagnetic materials that are likely candidates for being incorporated into magnetoelectric composites

Piezomagnetic material
$La_{0.7}Sr_{0.3}MnO_3$
$La_{0.7}Ca_{0.3}MnO_3$
$CoFe_2O_4$
$NiFe_2O_4$
Terfenol-D

To get a qualitative feel for the importance of mechanical coupling between the piezomagnetic and piezoelectric phases, even if these are individually strong, consider the effect of an applied magnetic field on a magnetoelectric composite material. The piezomagnetic phase experiences in this case a linear change in strain ε with magnetic field intensity H

$$\frac{\partial \varepsilon}{\partial H} = \epsilon_m . \tag{5.26}$$

The piezomagnetic coefficient ϵ_m measures such field-induced infinitesimal changes. The effect of the field-induced strain is transmitted to the piezoelectric phase which responds in turn by a change in polarization P

$$\frac{\partial P}{\partial \varepsilon} = \varepsilon_e. \tag{5.27}$$

The piezoelectric coefficient ε_e measures the extent of this infinitesimal change. As a result, we can write that the (infinitesimal) changes in polarization P with applied field H are given by

$$\frac{\partial P}{\partial H} = k_c \, \epsilon_m \, \varepsilon_e, \tag{5.28}$$

where the mechanical coupling factor k_c had to be introduced for practical reasons. Values in the range $0 \leq |k_c| \leq 1$ are obtained, depending on the fabrication method. Hence, a strong mechanical coupling favors a large ME response, provided the piezomagnetic and piezoelectric coefficients are high, too.

5.4 Multiferroics

5.4.1 Combining Ferroic Phases

Ferroic phases, such as ferromagnetic or ferroelectric (or even ferroelastic), are characterized by the fact that ferromagnetic or ferroelectric long-range ordering can be induced within that phase. This ordering can be the result of an externally applied or an internally existing, intrinsic magnetic or electric field. The effect takes place naturally only in some material systems at low temperatures where ferromagnetic and ferroelectric phases coexist within the same material. However, in most cases, materials need to be engineered to display the effect. Furthermore, there is a distinction between what type of effect is observed, depending on the sequence in which the ferroic ordering is induced: It is said that the magnetoelectric effect is *direct* if an electric polarization appears upon applying a magnetic field. On the other hand, we are talking about a *converse* magnetoelectric effect if a magnetization is induced as a result of an applied electric field.

One would say that these are characteristics of magnetoelectric compounds, so why use the term *ferroic* or *multiferroic*, as the title of this section implies? While magnetoelectric compounds can be a coarse combination of ferromagnetic and ferroelectric materials with various degrees of mixing, multiferroic composites enjoy a more subtle coupling between the ferroic phases. Quite often, the multiferroic term is reserved for composites where the constituent ferroic phases cannot be easily distinguished by SEM or AFM imaging. However, it is not always the case, and many researchers are less cautious in using one term over the other in published scientific literature. It usually becomes apparent from the context of a specific work if the composite is a true multiferroic, or whether it fits rather with a more traditional definition of a magnetoelectric.

Example 5.7. Comment on the magnetoelectric effect on a nanoscale.

Answer: Theoretical predictions involving Green's function or finite element methods have been employed only for bulk composites. However, fewer theoretical approaches exist for nanostructured compounds. Aside from quantum mechanical phenomena revealed on a nanoscale, the magnetic, dielectric, magnetostrictive, and piezoelectric properties are also strongly dependent on mechanical boundary conditions, or mechanical constraints caused by lattice misfit between film and substrate. Choosing appropriate ferroelectric and magnetic materials for an epitaxial nanocomposite requires consideration of lattice match and elastic properties, solid solubility, as well as chemical compatibility. Several scientific papers contain appropriate information on these theoretical models and their conditions. For experimental work consult for instance [44, 45].

Experimentation by the latter research group [45] determined that the coupling interaction in a nanostructured multiferroic $BaTiO_3$–$CoFe_2O_4$ film is different from that in bulk composites due to large longitudinal residual stresses and strains created by the $CoFe_2O_4$ nanopillars embedded in the $BaTiO_3$ matrix. These results were confirmed by theoretical models. The magnetoelectric parameters calculated from theoretical considerations for nanostructured systems are well agreeable with the experimentally measured values, thereby verifying the residual strain-induced variations in the magnetoelectric films. Therefore, nanostructured magnetoelectric compounds can act as a transducer between magnetic and electric field signals, proving that magnetoelectric materials can be obtained by nanofabrication, while intensifying the anticipation of revolutionary changes in miniaturization, integration, and length-scale reduction in instrumentation and electronics. Novel nanomaterials may at last bring the magnetoelectric effect to practical levels.

In a composite with ferromagnetic and ferroelectric properties, manipulating the ferromagnetic phase implies that setting or reading of a magnetic state can be obtained by means of a coexisting ferroelectric state. When a magnetic field is applied to the composite, the ferromagnetic particles change their shape magnetostrictively, and the strain is transferred to the piezoelectric phase, resulting in an electric polarization. The reverse is true for the ferroelectric phase when an electric field is applied. Because two or more ferroic phases are present, each with its own ferroic properties, these materials are known as *multiferroics*. So far, it has not been

Table 5.3 Summary of phases involved in the iron-rich $BiFeO_3$-based composite films fabricated using chemical solution deposition [46]

$BiFeO_3$	Primary ferroelectric phase
$Bi_2Fe_4O_9$	Nonferroelectric secondary phase needs restraining
Fe_3O_4	Reliable ferromagnetic phase
α-Fe_2O_3	Antiferromagnetic secondary phase needs restraining

feasible to combine magnetic and electric ordering in the same phase due to the mutually exclusive nature of atomic-level mechanisms responsible for ferromagnetism or ferroelectricity. Sometimes the suppression of secondary phase formation is necessary to increase an induced magnetization or polarization. This is the case for example, for $BiFeO_3$-based composite films fabricated with a high excess iron composition where a spontaneous magnetization coexists with a remanent polarization [46]. By suppressing the antiferromagnetic α-Fe_2O_3 secondary phase, and the nonferroelectric $Bi_2Fe_4O_9$, the remanent polarization is likely to increase. This is because the antiferromagnetic phase works against obtaining a high spontaneous magnetization or magnetic coercive field, and therefore needs to be restrained. This leaves Fe_3O_4 as the reliable ferromagnetic phase. The nonferroelectric $Bi_2Fe_4O_9$, which also forms when high Fe concentrations are present, does not show the inversion symmetry structure that would render it ferroelectric. Nevertheless, the primary $BiFeO_3$ phase displays the desired ferroelectric properties. The four phases are shown in Table 5.3. A small amount of Fe_3O_4 embedded in the $BiFeO_3$ matrix will result in the coexistence of ferromagnetism and ferroelectricity in one composite.

Multiferroicity depends on nanoscale effects within the phases present in the material so that formation of switchable domains becomes possible, and ultimately long-range ferromagnetic or ferroelectric ordering. As such, there is a cross correlation between coexisting forms of ordering. Certain symmetry operations can take place in the crystal when either the magnetization or polarization, or both, rotate through well-defined angles. It is possible for an electric field-induced reversal of polarization to lead to a rotation of the magnetization [47]. Some rotations are not necessarily symmetry operations. Nevertheless, coupling between polarization and magnetization conserves the symmetry and makes it possible to obtain a large magnetoelectric voltage coefficient [48]. Therefore, combining ferroic phases into a multiferroic composite with multiple types of domains can lead to interesting practical applications.

In time it has been noticed that for instance, a ferroelectric restructuring can be produced by an applied magnetic field in orthorhombic manganites, or an electric field-induced ferromagnetic ordering can be obtained in hexagonal manganites. These compounds are not the only ones displaying these properties, as such, materials like $Ni_3B_7O_{13}I$ join the multiferroics classification [49]. A three-phase ferroic is for example, the Terfenol-D/PZT/PVDF composite obtained by molding and pressing at 10 MPa and 190°C, containing distinct particles of the blended parts. Terfenol-D is the ferromagnetic constituent, while the PZT/PVDF particles constitute the ferroelectric part. Terfenol-D particles display a low resistance, and

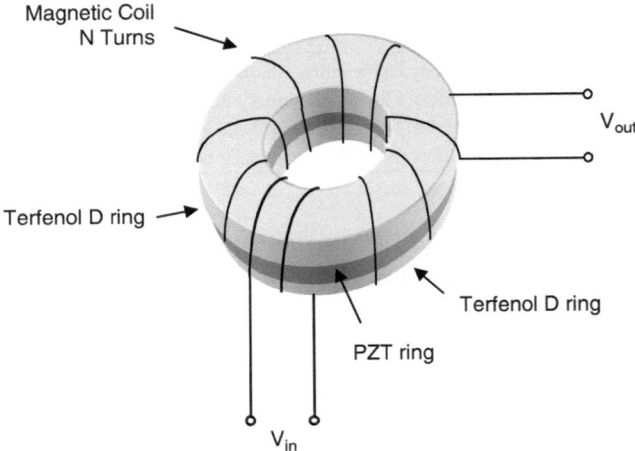

Fig. 5.13 Schematic illustration of a three ring layer multiferroic PZT/Terfenol-D transformer

hence it is required that they are well dispersed in the PZT/PVDF matrix in order to maintain the electrically insulating property of the composite. Otherwise, the low resistance path created by the Terfenol-D particles will cause the polarization charges to leak via this path. A magnetostrictive change in the Terfenol-D particles is transmitted to the PZT/PVDF matrix inducing polarization charges, giving rise to a magnetoelectric response [50].

5.4.2 Multiferoics in Applications

Experiments in miniature transformers have been performed, involving a combination of magnetostrictive Terfenol-D and piezoelectric PZT [51]. Specifically, the transformer is built of a ring-type laminated composite containing Terfenol-D that has been circumferentially magnetized, and a circumferentially polarized PZT. By making the conductive magnetostrictive layers very thin, unwanted eddy currents are minimized or even eliminated. Circumferentially applied fields give rise to strains in the circumferential direction. A rough sketch of the transformer is shown in Fig. 5.13.

A toroidal coil wrapped around the three ring layers carries a current that induces an ac magnetic excitation field. The magnetostrictive rings shrink or expand radially due the vortex magnetic field contained circumferentially within the rings, and an output voltage is produced. There is a strong magnetoelectric coupling that is enhanced in resonance drive. This occurs when the frequency of the excitation field is equal to the resonance frequency of the ring. At that moment, the output voltage is much higher than the input, and a large voltage gain is obtained of maximum value of $\sim$25, for a resonance frequency of 53.3 kHz [51].

Ring-shaped laminates of Terfenol-D and PZT have also been used as sensors for ac rotating or vortex magnetic fields [51] based on similar construction and design as the transformer in Fig. 5.13. Specifically, a vortex magnetic field applied to the Terfenol-D rings gives rise to a strain in the circumferential direction. This sets the piezoelectric PZT ring in a radial vibration mode. A high sensitivity of 10^{-8}T allows detection of very small magnetic fields. This ring-shape geometry has the advantage that it allows detection of vortex magnetic fields, as opposed to designs where the magnetic field is parallel, and the electric field is perpendicular to the principal strain direction [52] so that only magnetic fields of constant direction can be detected.

The right material design can enhance multiferroicity or even acquire it. For instance, transition-metal ions can be partly replaced by paramagnetic ions rendering magnetoelectric properties to a compound such as $PbFe_{1/2}Nb_{1/2}O_{1/3}$. Similarly, doping a perovskite manganite results in an induced displacement of Mn^{3+} ions along the hexagonal axis when a magnetic field is applied. This gives rise to a geometric charge ordering that generates in turn a ferroelectric polarization, and therefore a magnetoelectric effect. Many examples can be given involving manganites. A static magnetic field applied to $Pr_{1-x}Ca_xMnO_3(0.3 < x < 0.7)$ results in an insulator–metal transition. Consequently, a transformation from an insulating antiferromagnetic to a ferromagnetic metallic state takes place that is accompanied by changes in dielectric properties [53]. It is not surprising that rare-earth manganites $RMnO_3$ (where R can be Ho, Er, Tm, or Yb) can induce phase transitions that generate in turn high magnetoelectric signals. This is because an applied magnetic field induces an antiferromagnetic reorientation of the Mn^{3+} moments and a ferromagnetic ordering of the rare-earth sublattices occurs [54]. Even lowering the porosity and increasing the resistivity in a ferroic compound can strengthen magnetoelectric interactions. Hot pressing can also contribute to controlling key magnetoelectric parameters that depend on grain size and density, by increasing the initial permeability and strengthening the electromechanical coupling [55].

The group of multiferroics is highly interesting because of the coexistence of spontaneous long-range magnetic and dipolar ordering. These materials represent an attractive class of compounds combining fundamental physics with a wide range of technologically appealing applications, maybe even showing some potential in the new field of spintronics discussed in later chapters. Theoretical calculations have significantly contributed to recent progress in this area, elucidating different mechanisms for multiferroicity and providing essential information on various compounds where these effects manifest themselves. Density-functional theory investigations show two main classes of materials: (1) multiferroics where ferroelectricity is due to hybridization or just structural effects (e.g., the material $BiFeO_3$), and (2) multiferroics where ferroelectricity is driven by correlation effects and is strongly linked to electronic degrees of freedom such as spin-, charge-, or orbital-ordering (e.g., rare-earth manganites). For this latter type of multiferroics, calculations show that spin-ordering breaks inversion symmetry (e.g., in anti-ferromagnetic E-type $HoMnO_3$) leading to a magnetically induced ferroelectric polarization as large as a few $\mu C\,cm^{-2}$. These material examples point to several possible avenues for future technological approaches where new mechanisms for multiferroicity are explored. For instance, electronic correlations are observed that are at play in transition metal oxides, enabling ways to maximize the strength of these effects as well as the corresponding ordering temperatures.

These days, there are many research groups in industry and academia working on improving materials properties by fine tuning fabrication conditions, and as time passes, new developments allow for device functionalities not available in previous designs. Not so long ago, thermoelectricity was used only for measuring and controlling temperature. Now, the focus has shifted to power generation and refrigeration. Electrical and thermal transport properties of electrons and phonons are connected in any modern treatment of thermoelectricity, an idea that not many had embraced in the past. There is a wealth of detailed theory and experimentation related to energy converters or cooling devices. The dual nature of devices based on electric and magnetic phenomena together with modern nanofabrication technologies provide additional means for a new class of energy generators, heat pumps, or solid-state refrigerators also capable of incorporating principles from other disciplines of physics and materials science.

Tests, Exercises, and Further Study

P5.1 Draw a schematic of a Micropelt device with p–n thermolegs. What are the power factors for the n- and p-materials? *Note: See* [2]. What is the highest achievable packing density for leg pairs/mm² for industrial devices, as quoted in the reference?

P5.2 Describe the method for measuring the junction temperature of a thermo-electric device, by using a thermal test die. *Note: See* [5].

P5.3 (a) Give a few examples of Hall coefficient values obtained for different Heusler alloys with thermoelectric potential. (b) What are the Hall carrier densities for those compounds? *Note: See* [19].

P5.4 What is the thermal conductivity of commonly used (e.g., in microelectronics and microphotonics) IV and III–V semiconductors at room temperature? What happens when these materials are alloyed, e.g., instead of individual Si or Ge you have Si_xGe_{1-x} or you have a ternary compound such as $Ga_xAl_{1-x}As$?

P5.5 Give two examples of reported thermoelectric anisotropies as measured at lower temperatures ($<100\,K$) through directional variations in the Seebeck coefficient of single semiconductor crystals used as a one material thermo-electric device. *Note: See* [56].

P5.6 Give three examples of piezomagnetic materials that are usually combined with the piezoelectric material known as PZT ($PbZr_{1-x}Ti_xO_3$). *Note: See* [57,58].

P5.7 Name two or three factors that prevent attaining higher experimental magnetoelectric coefficients in particulate composite ceramics, in spite of many theoretically predicted possibilities.

P5.8 Magnetoelectric compounds can encompass various geometrical combinations between the piezomagnetic and piezoelectric phases. For instance: (a) particles from one phase can be dispersed uniformly throughout the other phase, (b) layers of the two phases can alternate horizontally within a composite (i.e., laminated composite), or (c) fibers of one phase can be embedded within the matrix of the other phase. Depending on the coupling magnetic–mechanical–electric response and other factors, these geometrical combinations result in various magnetoelectrical performances. How is this performance assessed, and how do the various combinations compare with each other? Be quantitative in performance evaluation, and refer to [59] for guidance.

P5.9 Consider commercially available heat exchange solid-state devices. What would be a very large heat exchange coefficient? What is a realistically obtainable heat exchange coefficient? What is an obtainable output power? Because of the limited heat exchange, what is a realistic temperature variation?

P5.10 Why is there a decrease in the temperature gradient on the outer surface of a thermoelectric device? What is the ideal temperature gradient? What is the real temperature gradient and what can be done to increase it?

P5.11 How do you measure performance of a thermoelectric device?

P5.12 With what can you replace a good thermoelectric material so that the overall output power is about ten times larger?

P5.13 How do you optimize a thermoelectric device as opposed to a pyroelectric one?

P5.14 What is the thermoelectric conversion efficiency η for $ZT \approx 1$ and 20°C peak-to-peak temperature variation compared to the Carnot efficiency η_{Carnot}? For what ZT can we obtain $= 50\% \times \eta_{Carnot}$?

P5.15 Comment on the practical implementation challenges for miniaturizing magnetic machines.

P5.16 Give examples of magnetic microgenerators and areas of use.

P5.17 From the expressions of the Hall coefficient and electrical resistivity, find an expression for the Righi–Leduc coefficient.

P5.18 Suggest two routes to effectively improve the figure-of-merit ZT of established thermoelectric materials. (*Hint:* Think low-dimensional structures for both cases).

P5.19 Give an example of a commonly used magnetostrictive material and a method of fabrication that improves its properties.

Hints and Partial Answers to Problems in Chap. 5

The answers to these problems are in many instances just a hint as to where further information is to be found. They contain the main idea, but the reader needs to find the references that are sometimes recommended for each exercise and complete the answer through further bibliographic reading and study.

P5.1 See [2]. In 2007, the power factors were $\sim 30\mu\text{W}/\text{cmK}^2$ for n-materials, and $\sim 40\mu\text{W}/\text{cmK}^2$ for p-materials. The reference quotes 100 thermoleg pairs/mm^2.

P5.2 Follow the method described in [5].

P5.3 (a) From [19] positive Hall coefficients are obtained as follows: $ErPd_2Bi$ has $R_H = 10^{-9}\text{m}^3/\text{C}$, and $ErPd_2Sb$ has $R_H = 10^{-10}\text{m}^3/\text{C}$ in a magnetic field of 12 T, and measured at room temperature. These values are similar to those for semimetals. (b) Hall carrier densities of 10^{21}cm^{-3} for $ErPd_2Bi$ and 10^{22}cm^{-3} for $ErPd_2Sb$ have been obtained.

P5.4 After [3]: Si $k = 145\,\text{W}/\text{m K}$; Ge $k = 60\,\text{W}/\text{m K}$; GaAs $k = 45\,\text{W}/\text{mK}$; AlAs $k = 91\,\text{W}/\text{mK}$; InAs $k = 27\,\text{W}/\text{mK}$; AlSb $k = 57\,\text{W}/\text{mK}$; InP $i = 68\,\text{W}/\text{m K}$. The thermal conductivity is lower in compounds such as $Si_{0.5}Ge_{0.5} k = 6\,\text{W}/\text{mK}$, and $Ga_{0.5}Al_{0.5}$ As $k \approx 10\,\text{W}/\text{mK}$ at room temperature.

P5.5 See, for example, a single crystal of ZnSb studied in [56]. Directional variations in thermoelectric anisotropy of single crystal ZnSb with a hole concentration of $p \sim 6 \times 10^{16}\text{cm}^{-3}$ were reported for $T = 90\,\text{K}$ as $\Delta\alpha = \alpha_{22} - \alpha_{33} = 136\mu\text{V}/\text{K}$ and $\Delta\alpha = 236\mu V/K$ at $T = 80\,\text{K}$. The directional Seebeck coefficients α_{22} and α_{33} are the diagonal elements of the α tensor and are due to thermoelectric anisotropy in the ZnSb single crystal.

P5.6 $PbZr_{1-x}Ti_xO_3$ (PZT) is often combined with constituents such as manganites, $LiFe_5O_8$, Permendur, and YIG [57]. A more familiar choice is the ferrite $NiFe_2O_4$ or the doped version $Ni_{0.8}Zn_{0.2}Fe_2O_4$, as well as compounds where Ni is substituted with Co such as in [58] $CoFe_2O_4$.

P5.7 Particulate ceramic composites are fabricated by high-temperature sintering. During the fabrication process, an interdiffusion takes place, often combined with chemical reactions between the piezoelectric and piezomagnetic phases. Additionally, the high sintering temperature leads to a thermal expansion mismatch between the two phases. These factors contribute to the reasons why lower experimental magnetoelectric coefficients are obtained, in spite of more optimistic theoretical predictions.

P5.8 See [59].

P5.9 A very large heat exchange coefficient would be $1,000\,\text{W}/(\text{m}^2\text{K})$. A realistically obtainable heat exchange coefficient is $1\text{--}10\,\text{W}/(\text{m}^2\text{K})$. An obtainable output power is $P = 13.4\mu\text{W}/\text{cm}^3$. Because of the limited heat exchange, a realistic temperature variation is 5–6 K.

P5.10 There is usually a decrease in the temperature gradient because of the limited heat exchange on the outer surface. The ideal temperature gradient is

10–20°C. But because of the limited heat exchange, it is only a few degrees Celsius, and therefore the power output is very limited. It is impossible to ensure a temperature gradient as large as the external one. To ensure a higher temperature gradient, forced air convection or a fluid is usually employed, as well as an increased exchange area.

P5.11 Measuring performance means determining the efficiency of the thermoelectric device. Efficiency is measured using the figure-of-merit ZT, where T is the temperature of the hot junction.

P5.12 Using a linear pyroelectric material is superior to using a good thermoelectric material. The overall output power is about ten times larger.

P5.13 In a thermoelectric device you optimize heat flow, whereas in a pyroelectric device you optimize the periodic heat transfer.

P5.14 The Carnot efficiency for 20°C peak-to-peak temperature variation or a gradient around the room temperature is 6.7%. The thermoelectric conversion efficiency η is

$$\eta = \frac{\theta_{\text{hot}} - \theta_{\text{cold}}}{\theta_{\text{hot}}} \frac{\sqrt{ZT+1} - 1}{\sqrt{ZT+1} + (\theta_{\text{hot}}/\theta_{\text{cold}})}.$$

These temperatures need to be measured on the element, and not outside the device. For $ZT \approx 1$, you get $\eta = 17\% \times \eta_{\text{Carnot}}$. To get $\eta = 50\% \times \eta_{\text{Carnot}}$ you need $ZT \approx 9$.

P5.15 When miniaturizing magnetic machines, certain methods become necessary for fabricating components at the appropriate sizes and even maintaining strict dimensional tolerances in some cases. Therefore, it is not uncommon to integrate conventional machining with micro- or nanofabrication because the devices possess both bulk and micro- or nanofeatures. A particularly difficult miniaturization challenge is obtaining strong magnetic properties in physically small devices. The materials produced by microfabrication methods (sputtering/electroplating) do not approach the levels possible by bulk manufacture. This is because it is difficult to produce high energy density micromagnets considering that output power scales quadratically with magnetic remanence. Most of these materials also require high-temperature annealing. In recent decades, great technological progress has been made in the design and fabrication capabilities of MEMS and NEMS, resulting in a wide variety of compact sensors and actuators. Some of these advances are based on research results obtained in the magnetic recording industry, making use for instance of magnetization patterns in thin films or nanoscale magnetic laminations.

P5.16 There is a type of devices relying on SmCo or ferrite permanent magnet rotors with multiturn Cu windings designed for consumer applications such as pedal-powered bicycle lights, wristwatch power sources, and flow-powered turbines for water meters and irrigation systems. The smallest device can generate 10 mW of power for a power density of 0.36 W/cm^3.

P5.17 With $R_H = 1/ne$, and $\rho = m^*/ne^2\tau$, one finds $R_L = e\tau/m^*$.

P5.18 It was long believed that the enhanced Seebeck coefficients obtained when creating low-dimensional structures with a feature size smaller than the spatial extent of the electron wave function were due to an increase in the density of states of the electrons near the Fermi level. However, it was subsequently determined that the observed improvements were a result of increased phonon scattering, and hence acknowledging the important role phonons play in heat transport. As such, semiconductor device sizes are comparable these days to the phonon mean free path Λ, especially the dominant one at room temperature. Thus, heat transport through phonons has changed the way we view thermoelectric devices fabricated on small scales. The second route that can be followed for improving the figure-of-merit of already established thermoelectric materials is to fabricate them in nanocomposite form, so that nanocrystals become embedded in a host matrix. Superlattices of GaN, PbTe, $CoSb_3$, and $BiTe_3$ have revealed the beneficial effects of such nanostructuring, albeit some material limitations encountered regarding the optimum size of these structures.

P5.19 Terfenol-D has been recognized as a key magnetostrictive material because of its giant magnetostriction and low magnetic anisotropy at room temperature. To improve its properties, a common fabrication technique is to epoxy-bind Terfenol-D particulates, creating a composite. The advantages of obtaining such a composite are reduced eddy current losses, lower brittleness, and better cost. A volume percent of roughly 50 has proven to be optimal for magnetoelectric devices. A substitute material for Tb or Dy could be the lower cost, light rare-earths such as Pr.

References

1. L.I. Anatychuk, *Proc. 17th Int. Conf. Thermoelectrics* (1998), p. 9
2. H. Böttner, J. Nurnus, A. Schubert, F. Volkert, *Proc. 26th Int. Conf. Thermoelectrics* (2007), p. 306
3. G. Chen, *IEEE Trans. Comp. Packaging Technol.* **29**(2), 238 (2006)
4. R. Alley, K. Coonley, P. Addepalli, E. Siivola, M. Mantini, R. Venkatasubramanian, *Proc. 21st Int. Conf. Thermoelectrics* (2002), p. 528
5. C.-K. Liu, C.-K. Yu, C.-Y. Hsu, S.-L. Kuo, M.-J. Dai, *Microsystems, Packaging, Assembly and Circuits Technology, IMPACT International* (2007), p. 88
6. Y.G. Gurevich, G.N. Logvinov, O.A. Fragoso, J.L. del Rio V., *4th Int. Conf. Elec. Electron. Eng.* (2007), p. 369
7. C.N. Plavitu, Phys. Stat. Sol. **12**, 265 (1965)
8. R.C. Chu, R.E. Simons, *18th Int. Conf. Thermoelectrics* (1999), p. 270 – *and references therein*
9. E.B. Ramayya, D. Vasileska, S.M. Goodnick, I. Knezevic, *8th IEEE Conf. Nanotech.* (2008), p. 339
10. L. Hicks, M. Dresselhaus, Phys. Rev. B **47**, 16631 (1993) – *and references therein*
11. A. Hochbaum, R. Chen, R. Delgado, W. Liang, E. Garnett, M. Najarian, A. Majumdar, P. Yang, Nature **451**, 169 (2008)
12. B.C. Sales, D. Mandrus, R.K. Williams, Science **272**(5266), 1325 (1996)

13. A.M. Pettes, R. Melamud, S. Higuchi, K.E. Goodson, *Proc. 26th Int. Conf. Thermoelectrics* (2007), p. 283
14. M.I. Flik, C.L. Tien, *J. Heat Transfer* **112**, 872 (1990)
15. B.-Y. Kim, W.S. Seo, M.H. Lee, H.-S. Choi, J.-W. Moon, D.Y. Lee, K.Y. Kim, *Proc. 26th Int. Conf. Thermoelectrics* (2007), p. 336
16. W.J. Xie, X.F. Tang, G. Chen, Q. Jin, Q.J. Zhang, *Proc. 26th Int. Conf. Thermoelectrics* (2007), p. 23
17. J. Yang, G.P. Meisner, C. Usher, *18th Int. Conf. Thermoelectrics* (1999), p. 458
18. G.N. Logvinov, L.P. Bulat, V.V. Chernysh, I.Y. Korenblit, A.A. Snarskii, *Proc. 25th Int. Conf. Thermoelectrics* (2006), p. 1
19. K. Gofryk, D. Kaczorowski, A. Leithe-Jasper, Y. Grin, *24th Int. Conf. Thermoelectrics* (2005), p. 383
20. R. Fletcher, Phys. Rev. B **28**(12), 6670 (1983)
21. M. Sakurai, N. Satoh, S. Tanuma, I. Yoshida, *Proc. 16th Int. Conf. Thermoelectrics*, 180 (1997), p. 180
22. H. Okumura, Y. Hasegawa, H. Nakamura, S. Yamaguchi, *Proc. 18th Int. Conf. Thermoelectrics* (1999), p. 209
23. M.E. Ertl, G.R. Pfister, H.J. Goldsmid, *Brit. J. Appl. Phys.* **14**, 161 (1963)
24. K. Ikeda, H. Nakamura, S. Yamaguchi, *Proc. 16th Int. Conf. Thermoelectrics* (1997)
25. H. Okumura, S. Yamaguchi, H. Nakamura, K. Ikeda, K. Sawada, *Proc. 17th Int. Conf. Thermoelectrics* (1998), p. 89
26. L.H. Ong, J. Osman, D.R. Tilley, Phys. Rev. B **65**, 134108 (2002)
27. H. Schmid et al. (ed.), Proc. 2nd Int. Conf. Magnetoelectric Interaction Phen. Cryst. (MEIPIC-2). Ferroelectrics **161–162**, 748 (1994)
28. J.F. Scott, JETP Lett. **49**(4), 233 (1989)
29. J.P. Rivera, Ferroelectrics **161**, 147 (1993)
30. G.T. Rado, J.M. Ferrari, W.G. Maisch, Phys. Rev. B **29**, 4041 (1984)
31. W.F. Brown Jr., R.M. Hornreich, S. Shtrikman, Phys. Rev. **168**, 574 (1968)
32. S. Dong, J. Li, D. Viehland, J. Appl. Phys. **95**(5), 2625 (2004)
33. S.W. Or, N. Nersessian, G.P. Carman, IEEE Trans. Magn. **40**, 1 (2004)
34. M. Kumar, K.L. Yadav, J Phys.: Condens. Matter **18**, L503 (2006)
35. S-F Zhao, C-H Yao, M-L Yao, Y-W Mu, J-G Wan, G-H Wang, *Symp. Piezoelectricity, Acoustic Waves, and Device Applic.* SPAWDA (2008), p. 51
36. J. Wang, R. Guo, A.S. Bhalla, *17th IEEE International Symposium on the Applications of Ferroelectrics (ISAF)* **2** (2008), p. 1
37. K.K. Patankar, S.A. Patil, K.V. Sivakumar, R.P. Mahajan, Y.D. Kolekar, M.B. Kothale, Mat. Chem. Phys. **65**, 97 (2000)
38. S. Shastry, G. Srinivasan, M.I. Bichurin, V.M. Petrov, A.S. Tatarenko, Phys. Rev. B **70**, 064416 (2004)
39. K.K. Patankar, R.P. Nipankar, V.L. Mathe, R.P. Mahajan, S.A. Patil, Ceramics Int. **27**, 853 (2001)
40. C.W. Nan, J. Appl. Phys. **76**, 1155 (1994)
41. A.P. Ramirez, J. Phys.: Condens. Matter **9**, 8171 (1997)
42. M.I. Bichurin, V.M. Petrov, Ferroelectrics **162**, 33 (1994)
43. J. van den Boomgard, R.A.J. Born, J. Mat. Sci. **13**, 1538 (1978)
44. F. Zavaliche, H. Zheng, L. Mohaddes-Ardabili, S.Y. Yang, Q. Zhan, P. Shafer, E. Reilly, R. Chopdekar, Y. Jia, P. Wright, D.G. Schlom, Y. Suzuki, R. Ramesh, Nano Lett. **5**(9), 1793 (2005)
45. G. Liu, C.W. Nan, J. Sun, Acta Mat. **54**, 917 (2006)
46. H. Naganuma, T. Okubo, K. Kamishima, K. Kakizaki, N. Hiratsuka, S. Okamura, IEEE Trans. Ultrasonics, Ferroelectrics, Freq. Control **55**(5), 1051 (2008)
47. H. Schmid, Bull. Mater. Sci. **17**, 1411 (1994)
48. S.M. Skinner, IEEE Trans. Parts, Mater. Packaging **6**, 68 (1970)
49. H Schmid, *Magnetoelectric Interaction Phenomena in Crystals* (Kluwer, New York, 2004)

50. C.W. Nan, N. Cai, L. Liu, J. Zhai, Y. Ye, Y.H. Lin, J. Appl. Phys. **94**(9), 5930 (2003)
51. S. Dong, J.F. Li, D. Viehland, Appl. Phys. Lett. **84**(21), 4188 (2004)
52. S. Dong, F. Bai, J.F. Li, D. Viehland, IEEE Trans. Ultrason. Ferroelectr. Freq. Control **50**(10) (2003)
53. Y. Tomioka, A. Asamitsu, H. Kuwahara, Y. Morimoto, Y. Tokura, Phys. Rev. B **53**, R1689 (1996)
54. M. Fiebig, Th. Lottermoser, M.K. Kneip, M. Bayer, J. Appl. Phys. **99**, 08E302 (2006)
55. V.K. Babbar, R.K. Puri, J. Appl. Phys. **79**, 6515 (1996)
56. E.V. Osipov, A. Aulas, *Proc. 16th Int. Conf. Thermoelectrics* (1997), p. 757
57. Proc. 4th Int. Conf. Magnetoelectric Interaction Phen. Cryst. (MEIPIC-4), Ferroelectrics **282**, 1 (2002)
58. M.I. Bichurin, I.A. Kornev, V.M. Petrov, A.S. Tatarenko, Y.V. Kiliba, G. Srinivasan, Phys. Rev. B **64**, 094409 (2001)
59. C.-W. Nan, M.I. Bichurin, S. Dong, D. Viehland, G. Srinivasan, J. Appl. Phys. **103**, 031101 (2008)

Chapter 6
Giant Magnetoresistance. Spin Valves

Abstract With different areas of magnetism being incorporated these days into conventional electronics, there is no surprise that magnetoresistance (MR) has found its way into common everyday usage, like the read head in personal computers. But read heads are not the only devices employing the magnetoresistive effect. In fact, MR is one of those fields that result in many other technologically exploitable applications, proving that the magnetoresistive effect cannot be underestimated and has yet to be used at its full potential. Whether rotational speed control devices, high current monitoring devices for power lines, or positioning control devices in robotic systems, MR has been incorporated into systems that require a high sensitivity to magnetic fields. Unfortunately, for many readers it is not always obvious where the consequences of MR are demonstrated within a device, especially when overshadowed by other more dominant physical phenomena. Therefore, this chapter discusses the physical basics of the magnetoresistive effect and emphasizes a few material aspects of what characterizes magnetoresistive configurations. In the end, it is hoped that the existence of the magnetoresistive effect in devices other than read heads is unveiled to the reader who will more readily recognize MR configurations in other applications.

In principle, magnetoresistive structures are usually composed of several layers of different materials that can be combinations of ferromagnetic, antiferromagnetic, or nonmagnetic materials, metals or semiconductors, or even organic compounds. Predicted by calculations based on spin-dependent energy bands, and confirmed through a variety of experiments, GMR can be explained by a spin-dependent conductivity in these multilayer stacks, highly influenced by scattering at interfaces between ferromagnetic and nonmagnetic layers. A common feature of MR structures is that their layers are coupled in a specific way, rendering them certain magnetic and electric properties while allowing selective passage for one spin component of the electronic current density. Changes in electrical resistance occur when a varying external magnetic field overcomes the coupling between the layers of the compound. Only those variations in magnetoresistance that are significant enough are of technological interest, and these large MR variations are commonly known as *giant magnetoresistance* (GMR). In view of the spin-dependent conductivity

C.-G. Stefanita, *Magnetism*, DOI 10.1007/978-3-642-22977-0_6,
© Springer-Verlag Berlin Heidelberg 2012

variation, the GMR mechanism is different from the better known MR phenomena studied in the past. GMR is also different from colossal magnetoresistance (CMR) for which a variety of mechanisms have been proposed such as electron–phonon coupling, double exchange, electron–magnon interactions, or phase- and charge-segregation.

Historically, GMR was initially reported in the late 1980s when the results of now celebrated experiments were first published [1, 2]. However, it was not until November 1997 that they made their appearance on the market, when IBM introduced commercial multilayer GMR sensors as magnetic recording read heads. These were incorporated into disk drive products Deskstar 16 GP where extremely small magnetic bits at an areal density of $2.69\,Gb/in^2$ were read. Deskstar 16 GP contained 95-mm-diameter disks, each with a storage capacity of more than 3.2 GB, resulting in a total data storage capacity of 16.8 GB. The current flowed parallel to the layers in the device in a so-called *current-in-plane* (CIP) geometry, requiring the sensor to be electrically insulated from the conducting magnetic shields. Since then, many advances have occurred in the world of GMR sensors, and it may be easy to forget how it started and where technological progress has taken us in the meantime.

6.1 Basic Considerations about Giant Magnetoresistance

6.1.1 Changes in Magnetoresistance

Many magnetoresistive structures have a ferromagnetic/nonmagnetic/ferromagnetic trilayer configuration, nevertheless, different arrangements are used with many more layers or other geometric combinations. For these different layer geometries, changes in the ferromagnet resistance with an applied magnetic field occur as follows: when the increasing applied magnetic field reaches the coercivity of one of the ferromagnetic layers, the magnetization of that layer aligns parallel to the applied field (Fig. 6.1a). The result is that the magnetoresistance of the system increases or decreases depending on the direction of magnetization in the next ferromagnetic layer. If the magnetizations in the two ferromagnetic layers are parallel to each other, the resistance decreases, while the opposite happens for antiparallel layers. With further increase in the applied magnetic field and upon reaching the coercivity of the second ferromagnetic layer, the magnetization in that layer aligns in the direction of the applied field as well (Fig. 6.1b), and the magnetoresistance of the system changes again. In order to calculate the magnetoresistance (MR), the expression

$$MR = \frac{R(0) - R(H)}{R(0)} \tag{6.1}$$

is used, where $R(0)$ is the resistance of the system at zero magnetic field, and $R(H)$ is the resistance of the system at a magnetic field value H. Depending on the shape of the magnetoresistance curve, the system is said to have *positive*

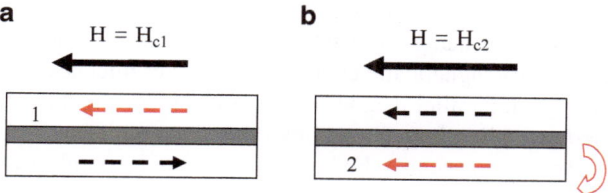

Fig. 6.1 The magnetizations in the two ferromagnetic layers respond to an external magnetic field: (**a**) when the field reaches the coercivity of one of the ferromagnetic layers (i.e., 1), the magnetization of that layer aligns parallel to the applied field and (**b**) upon further increasing the field until it reaches the coercivity of the second ferromagnetic layer (i.e., 2), the magnetization in that layer aligns with the applied field as well. The gray layer in between the ferromagnets is the "spacer"

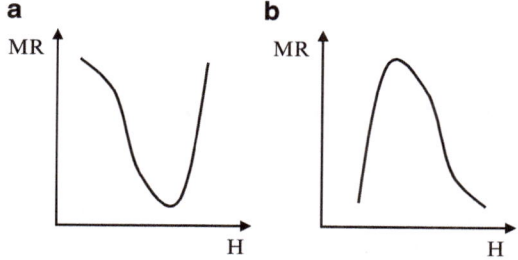

Fig. 6.2 The shape of the magnetoresistance curve is indicative of: (**a**) positive and (**b**) negative magnetoresistance

magnetoresistance (i.e., "holds water" as in Fig. 6.2a) or in the converse case, *negative magnetoresistance* (Fig. 6.2b).

Usually, and by design, the relative orientation of magnetization differs in consecutive ferromagnetic layers. Also, the magnetization of one ferromagnetic layer is fixed, while the magnetization of the other ferromagnetic layer is free to rotate in the plane of the layer in response to an external magnetic field. The most common geometry is the *current-perpendicular-to-plane* (CPP) where the current flows perpendicular to the plane, as the name literally suggests. These days, the CPP geometry replaces the earlier version of current-in-plane CIP; nevertheless, more about these geometries will be said in the next chapter which deals with other aspects of spin transport.

Baibich et al. used in the 1988 experiment multilayer structures of ∼30 layers or more that were formed from (001)Fe/(001)Cr superlattices on substrates of (001) GaAs. The Cr layer constituted a spacer, whereas the Fe layers served as the ferromagnetic layers. The highest magnetoresistance was observed at low temperatures (4.2 K) in structures with Cr layers only 9 Å thin, and was measured as a two times resistivity *decrease*. Interestingly, in this experiment the GMR effect was still observable at room temperature. It has been observed since then that the antiferromagnetic coupling is particularly important for obtaining a GMR response. Actually, a useful detail had been established in an earlier study: (100)Fe/Cr/Fe structures are antiferromagnetically coupled for Cr layer thicknesses of 9 Å, consequently displaying antiparallel magnetizations even in zero field [3]. This type of antiferromagnetic coupling is facilitated by the reduced thickness of the Cr layers (below 30 Å) between consecutive Fe layers, but it is not the only way of obtaining antiferromagnetic coupling, as we shall see in a subsequent section. Nevertheless, if a spacer layer facilitates the antiferromagnetic coupling, the spacer layer cannot be too thin, as an effect termed *oscillatory interlayer exchange coupling* is known to set in [4] and reduce the GMR response. Subsequent sections in this chapter will discuss further fine points related to the configuration of the individual layers comprising a GMR structure. For now, it suffices to say that the orientation of magnetization in the layers, the thickness of the layers, and the type of magnetism displayed by the layer itself determine the GMR response.

6.1.2 Magnetoresistance and Electron Spin

The term "magnetoresistance" is used in research papers and some books rather loosely, although a distinction should be made between the commonly known *magnetoresistance* of the past, and the much talked about *giant magnetoresistance* of the present and future applications. The former is usually attributed to the Lorentz force acting on the charge of the electrons in a paramagnetic material. In contrast, giant magnetoresistance is encountered in combinations of ferromagnets with other materials, and the spin rather than the charge of the electron is held responsible for its appearance. This is the effect we are discussing in this chapter, and it will be revealed more and more why it is so important that it has become an independent area of magnetics. However, before we go deeper into our examination of GMR, it should be mentioned that there is a phenomenon called *anisotropic magnetoresistance* manifested in the dependence of resistance on the angle between the magnetization in a ferromagnet and the direction of the current. This dependence is actually believed to be due to the interaction of the spin with the orbit. Thereby, differentiating between the distinct types of magnetoresistance will help us understand what exactly we are after, and how to recognize GMR when other effects are simultaneously present. Since we have mentioned "spin" which is a quantum mechanical concept, it is unavoidable not to think about the relationship between this field of physics and GMR.

Quantum mechanics tells us that the spin of the electron can be divided into two categories, a classification that depends on the relative spin direction with respect to the multilayer spin axis: "spin-up" denoted $\uparrow$ and "spin-down" represented by $\downarrow$. It is also known that linear combinations of the two are possible, but we assume for now that the $\uparrow$ and $\downarrow$ spins are independent. Remember that the discussion in this chapter revolves around the uncompensated spins of electrons in a ferromagnet or rather a combination of ferromagnets, specifically those spins

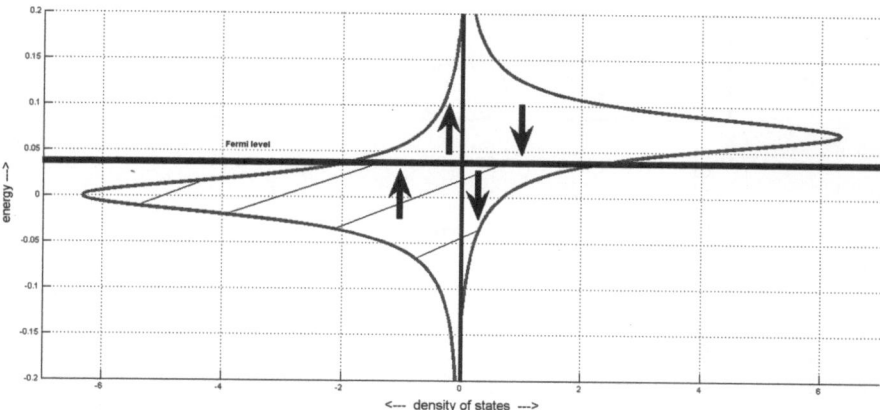

Fig. 6.3 For the uncompensated spins of electrons in a ferromagnet, the electron density states for the two spin directions are often displaced in energy relative to one another so that the Fermi energy level goes through a larger density of states of spin electrons of one type, at the expense of the other type. It should be noted that these graphs were obtained in Matlab by implementing semiconductor physics concepts. These concepts are reviewed briefly through examples in this chapter

that are responsible for ferromagnetism. For these spins, the electron density states for the two spin directions can be displaced in energy relative to one another under certain circumstances. Such a situation can be schematically represented as shown in Fig. 6.3, and it has important consequences for spin transport. Throughout this chapter, we shall come across concepts that may already be familiar to some readers; nevertheless some re-examinations should be welcome. As such, Example 6.1 serves as a short review of an electronic property known as the *density of states*. It is important for GMR because the configuration of the density of states of spin-carrying electrons determines whether giant magnetoresistance effects are observed at all, as well as capabilities of obtaining other spin-based devices.

Example 6.1. When does an energy level become broadened into a continuous density of states?

Answer: Before the channel in a device gets coupled to a source and a drain, the density of states is a straight "line" in the channel, represented by the energy level ε (see Fig. E6.1). Once coupled, some of the states in the channel spread into the source and drain; thus the channel loses them. However, the channel also gains some of the states from the source and drain, states which spread into the channel. The overall effect is that the density of states broadens from the initial sharp "line" into a distribution that can be described by a variety of functions. One of the simplest functions to describe the broadened density of states in the channel is the Lorentzian function $D_\varepsilon(E)$

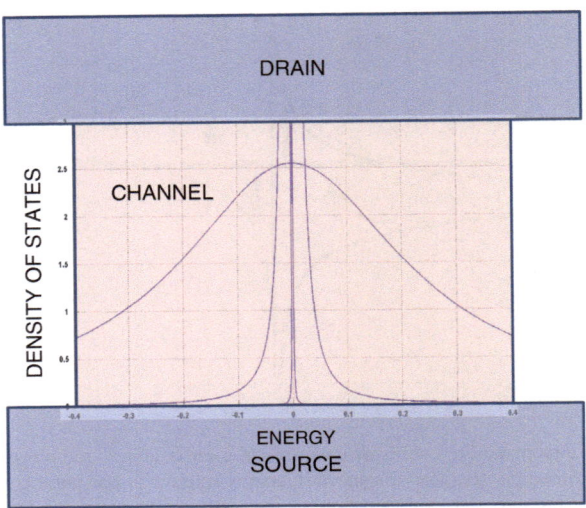

Fig. E6.1 A density of states of electrons for $\gamma = 0.00001$ (sharp "line" or delta function), and broadened for $\gamma = 0.05$ and $\gamma = 0.5$. The density of states is represented as a Lorentzian function, and values were chosen for illustrative purposes only. Curves were obtained in Matlab. The source, drain, and channel sketches place the concept of density of states into context, although this should not be taken as an accurate representation of actual processes in the channel

$$D_3(E) = \frac{\frac{\gamma}{2\pi}}{(E - \varepsilon)^2 + \left(\frac{\gamma}{2}\right)^2}, \qquad (E6.1.1)$$

where E is the energy, ε is that particular energy level, and γ is the coupling to the source or drain. In the limiting case, we obtain the delta function for the density of states when the coupling γ tends to zero. The sharp "line" mentioned above is actually the delta function obtained for $\gamma \to 0$, and it is not exactly a "line." All calculations in Fig. E6.1 were performed in Matlab, and the reader is encouraged to verify them.

When an energy displacement occurs for the "spin-up" and "spin-down" densities of states, the Fermi energy level will likely go through a larger density of states of spin electrons of one type, at the expense of the other type. As a reminder of the importance of the Fermi level, Example 6.2 describes the notion of the Fermi level in a few words. When the energy displacement for the "spin-up" and "spin-down" electrons occurs, the details will depend on the actual configuration of the density of states. In this case, it means that electrons with different spin projections parallel or antiparallel to the magnetization of the ferromagnet contribute in unequal parts to the conduction process. The inequality in spin contribution, also known as the *two current model*, provides a reasonably good explanation for many magnetoresistive phenomena, and is frequently encountered in published research papers.

Example 6.2. What is the Fermi level?

Answer: When a source and a drain region are coupled to a channel held at zero potential, electrons flow in and out of this device. The consequence is that all three regions, source, drain, and channel, will reach equilibrium at an electrochemical potential μ. Once this equilibrium has been reached, the average number of electrons over time in any energy level is given by a distribution known as the Fermi function

$$f_0 (E - \mu) = \frac{1}{1 + \exp\left[\frac{E - \mu}{k_B T}\right]}. \qquad \text{(E6.2.1)}$$

The Fermi level is the energy μ. The Fermi function is the probability that there is an electron in a particular single-particle state with energy E. This means that the energy levels far below the electrochemical potential μ are always full. Consequently, the probability is $f_0 = 1$. In contrast, the energy levels above μ are always empty with $f_0 = 0$. Those energy levels with energy values within a few $k_B T$ of μ are sometimes empty and other times full. Then, the average number of electrons is between 0 and 1.

Since transport properties are determined by the states at the Fermi energy, the difference between the two spin species contributes to a spin-asymmetry in the transport properties of a ferromagnet. A smaller density of states at the Fermi level implies a larger Fermi velocity and a smaller relaxation rate. The smaller available density of states into which electrons can scatter also reduces the relaxation rate; hence a large conductivity for that particular spin species of electrons is expected. The majority of the current will be carried by the electrons with the larger conductivity, and the current is spin polarized to some extent. Transport phenomena for electron spin are usually conveniently described by the nonequilibrium Green's function formalism which requires advanced knowledge of semiconductor physics and quantum mechanics. It is strongly recommended that a textbook on quantum transport or mesoscopic physics is consulted for a more comprehensive description [5, 6]. Some basic ideas related to the nonequilibrium Green's function approach will be reviewed in this chapter, in a later section.

6.1.3   Quantifying Magnetoresistance

Johnson and Silsbee did a series of experiments in the late 1980s and confirmed some of the findings of giant magnetoresistance studies. In particular, they focused on the topic of spin injection from a ferromagnet [7]. The spin injection concept would prove particularly useful in the field of spin electronics or *spintronics* that has emerged since. However, to be able to discuss spin injection and transport more comprehensively, a simplified quantitative relationship needs to be taken into account involving the resistivity due to the two types of spin. The notation $\rho_\uparrow$ and $\rho_\downarrow$ is used to distinguish between the resistivity due to $\uparrow$ and $\downarrow$ spins. Furthermore, the notion of *minority* ρ_- or *majority spin* ρ_+ is introduced, nevertheless it will be discussed in more detail in a later section. That being said, the $\uparrow$ and $\downarrow$ spins are related, in their turn, to the resistivities of minority ρ_- and majority spins ρ_+ as follows:

(a.) For a ferromagnetic alignment, we have

$$\rho_\uparrow = N\rho_+ \quad \text{and} \quad \rho_\downarrow = N\rho_-, \tag{6.2a}$$

(b.) Whereas for an antiferromagnetic alignment the resistivities are the same

$$\rho_\uparrow = \rho_\downarrow = \frac{N\,(\rho_+ + \rho_-)}{2}. \tag{6.2b}$$

N is the total number of interfaces. It is assumed that the resistivity due to scattering accompanying spin reversal (also known as spin flipping) is negligible, in particular at low temperatures where most experiments are performed. However, in practice this is not the case, especially since most devices need to function at room temperature. Nevertheless, within the limits of our assumption, we can write the overall resistivity as:

(a.) for a ferromagnetic alignment

$$\rho_F = \left(\frac{1}{\rho_\uparrow} + \frac{1}{\rho_\downarrow}\right)^{-1} = \frac{N\rho_+\rho_-}{\rho_+ + \rho_-} \tag{6.3a}$$

(b.) and for an antiferromagnetic alignment

$$\rho_{AF} = \frac{N\,(\rho_+ + \rho_-)}{4}. \tag{6.3b}$$

Remember the definition of magnetoresistance in relationship (6.1). The magnetoresistance can then be written in terms of resistivity rather than resistance as follows:

$$MR = \frac{\Delta\rho}{\rho} = \frac{\rho_{AF} - \rho_F}{\rho_{AF}} = \frac{(\rho_+ - \rho_-)^2}{(\rho_+ + \rho_-)^2}. \tag{6.4}$$

A magnetic material for which the resistivity depends on spin orientation is likely to contribute to magnetoresistance effects, due to the lower resistivity experienced by one of the spin components. Therefore, if $\rho_+ \ll \rho$ or $\rho_+ \gg \rho_-$, we have $\Delta\rho\rho \cong 1$ and a giant magnetoresistance effect is measured. In GMR structures, the magnetization vectors in the layers have to be ferromagnetically aligned at some point, so that the spin component with the preferred spin polarization can pass through the layers. Many researchers use a parameter $\alpha = \rho_\downarrow/\rho_\uparrow \neq 1$ that measures the spin transport asymmetry through an interface. It was noticed experimentally that a correlation exists between α and the magnitude of the GMR response, as well as whether it is positive or negative. A material dependence is evident from relationship (6.4). Some of these ideas will become more understandable as the discussion in this chapter and the next progresses.

6.1.4 Using the Nonequilibrium Green's Function Formalism

The purpose of the following summary is to introduce the reader succinctly to the usefulness of the nonequilibrium Green's function approach for describing electron spin transport. The whole section is structured in an intuitive way, leaving out any complicated mathematical issues, while focusing on transparency and compactness. For a proper understanding of the examples, as well as for solving the problems at the end of the chapter, a review of matrix algebra is recommended. The Greens' function formalism is used because spin valves are devices in which electrons flow, even though it is the spin we are actually interested in. Nevertheless, the spin is carried by electrons, so we end up studying electrical currents. These spin valve devices have contacts and a central portion that we refer to as "channel," although it may not be a "channel" in the traditional sense, but rather a group of multilayers. Nonetheless, a real-life spin valve device has a channel with n energy levels, so many quantities that would describe spin transport for only one energy level are replaced in the case of n energy levels by a corresponding matrix of size $(n \times n)$. As such, we no longer speak of the energy level ε, but rather a Hamiltonian matrix H. Note that the energy level is not the only quantity which becomes a matrix, and the matrix dimensions differ for different quantities.

Consider Fig. 6.4. The Hamiltonian of our system, the left and right contacts (attached to the source and drain), and the central portion (channel) can be partitioned into three subsystems: H_L, H_R, and H_c. The coupling of the channel to the leads forms an "extended channel," and it is particularly important for the theoretical modeling of the entire system that the "extended channel" includes enough lead atoms from the source and drain contacts. The Hamiltonian of the system to be analyzed becomes in this case

$$H = \begin{pmatrix} H_L & \tau_L & 0 \\ \tau_L^+ & H_c & \tau_R^+ \\ 0 & \tau_R & H_R \end{pmatrix}, \tag{6.5}$$

where $\tau_{L,R}$ and $\tau_{L,R}^+$ describe the coupling interaction matrices between the left and right leads, respectively, and the channel. The "+" sign represents the conjugate and transposed (thereby Hermitian) matrix, and it appears because of the left, respectively, right-moving waves (i.e., moving towards the right lead). We assume there is no direct coupling between the two leads. The leads are taken to be ideal Fermi metals. Electrons entering the contact regions from the left or right obey the Fermi distribution, and their electrochemical potentials are time-independent. The leads are perfect ballistic conductors and reflectionless, the latter meaning that

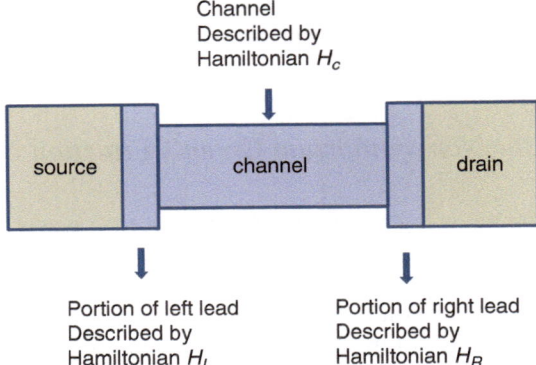

Fig. 6.4 Sketch of a device (system) described by a partitioned Hamiltonian in the nonequilibrium Green's function approach. The partitioned Hamiltonian comprises: left and right contact Hamiltonians H_L and H_R, and the Hamiltonian for the central portion (channel) H_c. The channel has a length comparable to other important length-scales of the system, such as the phase-breaking length or impurity mean free path

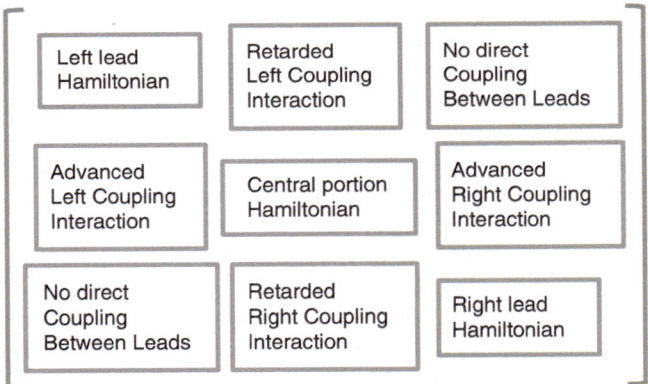

Fig. 6.5 Schematic explanation of the different terms in the Hamiltonian matrix of a system composed of two leads and a channel with n energy levels

electrons entering the leads will not reflect. In view of the above, the Hamiltonian of the system has the structure schematically explained in Fig. 6.5. We assume for now that transport in the channel is ballistic. This implies that no inelastic scattering takes place during the flow of spin-carrying electrons. That assumption will change later in this chapter when we discuss incoherent scattering, and at that time we have to introduce so-called Büttiker probes. However, at the moment, we shall not concern ourselves with diffusive transport.

The Green's function $G(E)$ is defined as

$$(E - H)G(E) = I, \qquad (6.6)$$

where I is the identity matrix. The Green's function gives the response of the system to a constant perturbation in the Schrödinger's equation. Knowing the Green's function facilitates calculation of different properties of the system. This approach is very often adopted because it is easier than solving the whole eigenvalue problem, allowing a simpler treatment of many complicated device issues. In this formalism, two types of Green's functions occur, retarded (outgoing) G and advanced (incoming) G^+, corresponding to outgoing and incoming waves in the contact. The nonequilibrium nature of the Green's function approach becomes apparent when considering that the symmetry of the distant past and the distant future has been broken; therefore calculations take place on a two-side time contour. A "full" Green's function refers to the entire system and is denoted by capital G, whereas g identifies a subsystem, a matrix of reduced dimensions. For instance, if we analyze only certain states in the contact to the right, then g_R describes these states. A Green's function of the central portion G_c can be calculated separately, without calculating the whole Green's function G. The same is true for the other parts of the system. Actually, only a small number of matrix elements describing the contact are needed, specifically those next to the channel. The fact that we can do this simplifies the problem to be solved enormously, as we do not need to deal with huge matrices. In this simplified case, we can write

$$\begin{pmatrix} E - H_L & -\tau_L & 0 \\ -\tau_L^+ & E - H_d & -\tau_R^+ \\ 0 & -\tau_R & E - H_R \end{pmatrix} \begin{pmatrix} G_L & G_{Lc} & G_{LR} \\ G_{cL} & G_c & G_{cR} \\ G_{RL} & G_{Rc} & G_R \end{pmatrix} = \begin{pmatrix} 1 & 0 & 0 \\ 0 & 1 & 0 \\ 0 & 0 & 1 \end{pmatrix} \qquad (6.7)$$

from which, after some calculations, the Green's function of the channel G_c can be found to be

$$G_c = \left(E - H_c - \sum_L - \sum_R \right)^{-1}. \qquad (6.8)$$

The quantities $\Sigma_L = \tau_L^+ g_L \tau_L$ and $\Sigma_R = \tau_R^+ g_R \tau_R$ are the so-called *self-energies*. It seems that the contacts have added extra terms to the Hamiltonian of the channel, namely the self-energies. These account for the coupling of the channel to the leads. Thereby, all information on the transport properties of electrons in the channel will be contained in the Green's function of the channel, and numerical calculations can be performed within this finite set. That being said, the Green's function also helps in calculating the density of states, which in this more general case is better described by the spectral function $A(E)$, instead of $D_\varepsilon(E)$

$$A = i(G - G^+). \qquad (6.9)$$

The two are connected through $A(E)2\pi \equiv D(E)$. Remember relationship (6.9), as we will use it shortly.

At this point, let us see how the nonequilibrium Green's function approach can model densities of states and electron transport in the channel under the influence of a magnetic field. We know that when a magnetic field is acting on the channel, its expected impact on the spin of the electrons is the Zeeman effect. Furthermore, the electron itself will experience a Lorentz force, if the magnetic field is perpendicular to the direction of electron flow. However, if the transport is *coherent*, i.e., no phase-breaking scattering occurs that would change the state of the electron, then the *transmission (and hence conductance or resistance)will not be affected for either "spin-up" or "spin-down" electrons.* This is illustrated explicitly in Example 6.3. On the other hand, unlike transmission, the *spectral function changes when a magnetic field is acting on the channel.* We show in Example 6.4 how the spectral function A for ↑ spin and ↓ spin electrons is affected by the presence of a magnetic field applied parallel to the device channel. See also Problem P6.4 for a perpendicular magnetic field.

Example 6.3. What is the influence of a magnetic field on the transmission of the two spin species (↑ spin and ↓ spin) inside the channel if the field is acting perpendicular to the channel? It is assumed there are no impurities present in the channel.

Answer: Consider the action of a magnetic field $\boldsymbol{B}$ with a component B_z along the z direction, i.e., perpendicular to the channel. We take $B_z(x)$ as a Lorentzian profile centered at x_0 and of half-width at half-maximum γ

$$B_z(x) = \frac{B_0}{1 + \frac{x-x_0}{\gamma}}. \tag{E6.3.1}$$

A broadening magnetic field is chosen to simulate in a more realistic way how a perpendicular field would spread over the channel. If the magnetic field is centered at x_0 along the channel, and applied perpendicularly in the z direction, the magnetic vector potential will increase linearly along the channel. The vector potential $\vec{A}$ can be chosen in the Landau gauge $(0, A(x), 0)$. In our case, for the perpendicular field, it has the expression

$$A(x) = \int B_z(x)dx = \frac{1}{2} + B_0\gamma \cdot arc\tan\left(\frac{x - x_0}{\gamma}\right). \tag{E6.3.2}$$

These extra terms due to the applied magnetic field need to be added to the channel Hamiltonian, which enters relationship (6.8) for the Green's function of the channel. Under the influence of the magnetic field and taking into account the magnetic vector potential as well, the channel can be described by the following Hamiltonian

$$H_c = \frac{\left[i\hbar\nabla + q\vec{A}\right]^2}{2m} + U_c + g\mu_B\sigma\vec{B}, \tag{E6.3.3}$$

where m^* is the electron effective mass, g is the effective Landé g-factor, μ_B is the Bohr magneton, U_c is the potential energy in the channel, while $\sigma = \pm 1/2$ denotes ↑ spin and ↓ spin. Using relationships (6.8) and (E6.3.1), and (E6.3.2), the transmissions for the ↑ spins and ↓ spins can be calculated as follows, where Tr stands for "trace"

$$T_{\uparrow,\downarrow}(E) = \text{Tr}\left[\Gamma_L G_{\uparrow,\downarrow} \Gamma_R G_{\uparrow,\downarrow}^+\right]. \tag{E6.3.4}$$

The quantity Γ with left L and right R subscripts is the broadening of the density of states due to the left and right contacts. Just like the density of states, the broadening $\Gamma(E)$ is also matrix defined from the self-energy matrix $\Sigma(E)$. It is equal to its anti-Hermitian component

$$\Gamma(E) = i\left[\Sigma(E) - [[\Sigma(E)]]^\uparrow +\right]. \tag{E6.3.5}$$

The matrix $[\Gamma]$ is of the same size as the Hamiltonian matrix $[H]$; however, unlike H, Γ can be energy-dependent, as already implied. The broadening Γ can be considered the matrix version of the level broadening γ discussed previously. It is important to emphasize once more that the influence of the coupling to the leads enters through the self-energy expression Σ only in the Green's function of the central part, which is the channel. This is an important concept in many-body physics where it is frequently employed to describe intricate connections; however, it is also quite helpful to us. It will be particularly handy in later sections when treating interactions with impurities. Relationship (E6.3.3) can be implemented into Matlab with the result shown in Fig. E6.2 for a magnetic field $B_z = 1\,\text{T}$ at its peak value. It needs to be mentioned that throughout this chapter subband energies are neglected, and a single transverse wavevector k is taken into account. Also, for simplicity, the self-energy is considered only for one-dimensional contacts (for instance, left designated by "1" and right designated by "N"). Therefore, it is a diagonal matrix with two nonzero entries

$$\Sigma(1,1) = -t_0 \exp(ika), \tag{E6.3.6a}$$

$$\Sigma(N,N) = -t_0 \exp(ika), \tag{E6.3.6b}$$

where $t_0 \equiv \hbar^2 2m^- a^2$ is a contact energy (in eV) between two discrete Hamiltonian points separated by a. The discretized Hamiltonian spans a range of Na points. The outcome shown in Fig. E6.2 may seem trivial; nevertheless, it should be remembered that no phase-breaking events have been taken into account so far in the discussion of electron transport in the channel. We considered that the transmission for both ↑ spin and ↓ spin electrons occurred without impediments. The reader is encouraged to repeat the calculations for a parallel magnetic field, to confirm that the transmission of electrons is not affected by the direction of the acting field when no impurities are present in the channel (see Problem P6.5).

Example 6.4. Show how an external magnetic field acting parallel to the channel affects the spectral function A for ↑ spin and ↓ spin electrons. See Problem P6.4 for

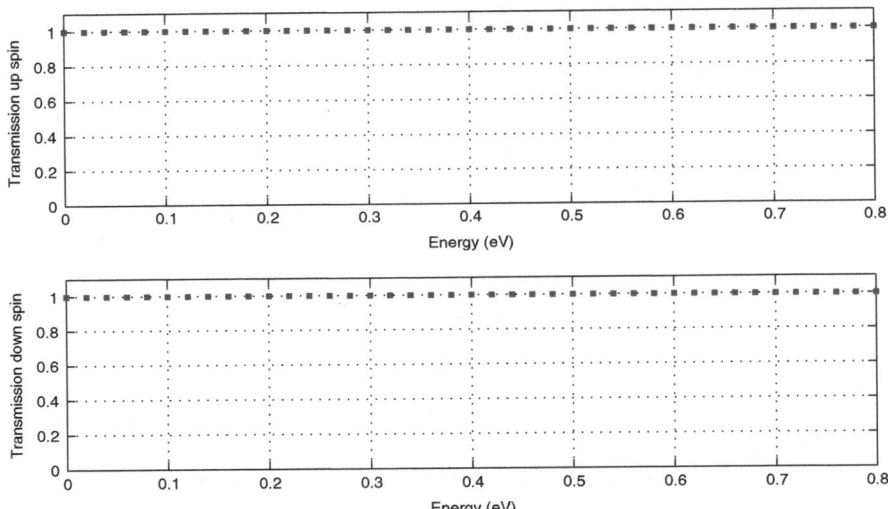

Fig. E6.2 Magnetic field effect on the channel. In the absence of incoherent scattering, both spin-up and spin-down electrons travel ballistically. The transmission of "spin-up" and "spin-down" electrons is, therefore, not affected, despite the magnetic field. This simple example serves as a comparison with Fig. 6.14 where an incoherent scatterer in the channel affects the transmission of electrons

how results change if the magnetic field is acting perpendicularly to the direction of current flow.

Answer: The solution is found in relationships (6.9) and (6.11), where the Green's function G in (6.9) is that of the channel, and the magnetic field is taken into account in the Hamiltonian for the channel. After implementing into Matlab, Fig. E6.3 is obtained. In the vicinity of the contacts at both ends of the channel, some states spill-over to the left contact, whereas other states spill-over to the right contact. For the case shown in this figure, a parallel magnetic field linearly varying along the channel is considered, giving rise to a constant magnetic vector potential. Although transmission is not affected by the direction in which the magnetic field is acting, the spectral function is (see Problems P6.4 and P6.5).

The subject of impurities that influence spin transport, and in particular the differences in spin transmission (or conductance) for up-spin and down-spin electrons will be presented in more detail in subsequent sections. Their effect on transport due to phase-breaking events is quite significant; therefore, more than one section in this chapter is dedicated to discussing them. For the time being, we shall focus our attention on design and material configurations for spin valves.

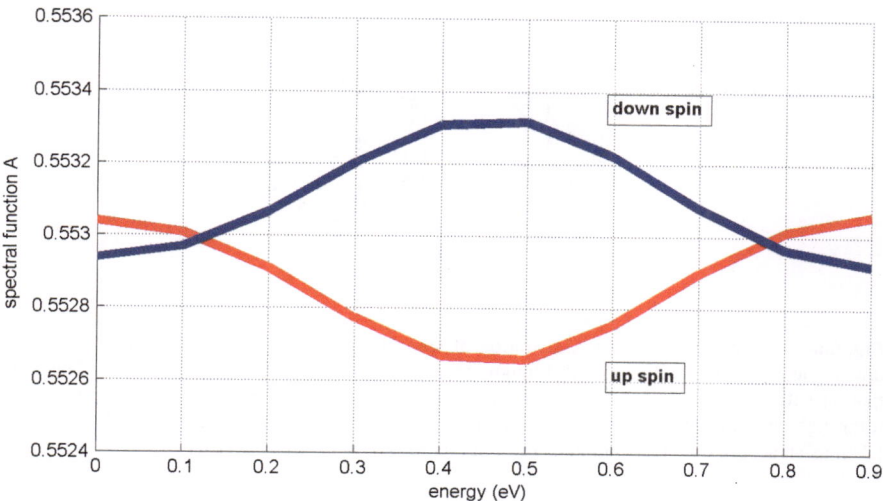

Fig. E6.3 An external magnetic field influences the spectral function $A(E)$ for ↑ spin and ↓ spin electrons. Depicted is only the effect of a linearly varying parallel magnetic field which gives rise to a constant magnetic vector potential. For a perpendicular magnetic field, see Problem P4. The magnetic field and the magnetic vector potential are included in the overall expression of the Hamiltonian for the channel. Results were obtained in Matlab

6.2 Configuration and Choice of Materials

6.2.1 The Term "Spin Valve"

An electron current flowing through the multilayers transports spin carried by electrons from one layer to another across interfaces. However, the two types of spin encounter scattering at the interfaces between the ferromagnets and the other materials, and the scattering is different for the ↑ spins and ↓ spins. Interfaces are where magnetic layers make contact with nonmagnetic layers. Remember that the actual variation in magnetoresistance occurs when the field changes the relative orientations of magnetizations in two adjacent magnetic regions. While these magnetizations are parallel to each other pointing in a common direction, electrons with spins oriented in the same direction (for instance, "spin-down") pass easily from one layer to another (Fig. 6.6). Electrons with antiparallel spins (for instance, "spin-up") are strongly scattered, resulting in a low resistivity for electrons with parallel spins. On the other hand, if adjacent regions have antiparallel magnetizations, both "spin-down" and "spin-up" electrons are significantly scattered, and the resistivity is high for all electrons. It can be said that the layered structure is *filtering spin* and is, therefore, often referred to as a *spin valve*.

The spins that pass easily through are referred to as *majority spins*, in contrast to the *minority spins* that are strongly scattered at the interface. Some minority spins

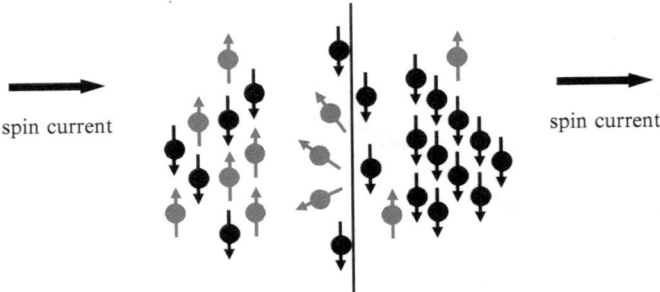

spin current   spin current

Fig. 6.6 A simplified schematic illustration of spin filtering at an interface. In this case, "spin-down" electrons are easily allowed through, whereas "spin-up" electrons are scattered. The spins that pass through with less difficulty are known as "majority spins." Some "minority spins" make it through, as well. This picture is only meant to illustrate some basic concepts, and should not be taken as an accurate representation of the complexities of interface scattering

pass through as well, but are not so numerous. The process repeats itself at the next ferromagnet/nonmagnet interface, with the role of "majority" and "minority" spins reversed, if the magnetizations are now in an opposite configuration. Note that the degree of interface roughness is known to have a significant influence on the amount of scattering for both majority and minority spins, as it results in more randomness in scattering. There are different types of disorder at interfaces, with various influences on scattering (see also Problem P6.1 and [8]). Therefore, the spin current will be unfortunately diminished, unless a better material and interface design is employed.

6.2.2   Hard and Soft Ferromagnetic Layers

Different types of spin valves have been developed over time; however, many spin valves consist of three types of layers shown schematically in Fig. 6.7 and described as follows. A *soft ferromagnet* is often used to allow a relatively low-field magnetization switching. It is also called a *free layer* due to the ease with which the direction of magnetization can be changed. Its hysteresis loop is centered and exhibits a low coercivity. On the other hand, a *hard ferromagnet* ensures that the magnetization is difficult to switch in that particular layer and stronger magnetic fields need to be used. Nevertheless, a hard ferromagnet does not consist of a single layer by itself, but is rather often formed by two, three, or even more layers. It is depicted schematically in its minimal form in Fig. 6.8. In many cases, one of these layers is a "pinned layer" exchange coupled to an "antiferromagnetic pinning layer." If the pinning layer is an insulator, it offers additional advantages. It can pin without shunting current, thereby allowing a higher GMR. But they can also act as part

Fig. 6.7 A typical spin valve structure contains three layers: (1) a soft ferromagnet, also known as a free layer, (2) a spacer, and (3) a hard ferromagnetic layer. The latter is comprised of two or several layers

Fig. 6.8 Schematic of a simple hard ferromagnetic layer comprising a "pinned" and a "pinning" layer. The latter is often antiferromagnetic. More layers can be incorporated into the design of a hard ferromagnet with the ultimate effect of making the magnetization in this layer hard to switch

of the insulating gap in a recording head, reducing the thickness requirements for conventional gaps.

The overall hard ferromagnet configuration is antiferromagnetic, and the hysteresis loop is shifted horizontally and is asymmetric. The term "exchange coupling" denotes an effect where an exchange interaction exists between magnetic moments in the two layers. Its origin is attributed to uncompensated interfacial spins locked in the antiferromagnet that cannot follow the external magnetic field in alignment. This interaction is much needed and its role is to ensure that the magnetization within the pinned layer cannot change directions easily. This is an important detail in the design of a spin valve, and has long been used in the magnetic recording industry for stabilizing the magnetization within the sensing layer. While the relative magnetization of the pinned and the free layer changes from parallel to antiparallel, the resistance of the spin valve increases. This is due to an increase of the spin-dependent scattering of the conduction electrons in those layers. Finally, the spin valve also has a *spacer* layer, usually a paramagnet. It ensures the separation between the soft and hard ferromagnets, as the name suggests. Overall, this trilayer design makes up a spin valve in its simplest form. However, demagnetizing fields associated with the magnetic moments of the ferromagnetic layers tend to increase as the volume of the whole structure is increased; therefore, the necessity arises of keeping the layers thin. It may be that progressively reducing the size of spin valve devices increases the magnetic field requirements for changing their magnetic state. In this eventuality, the reduced device size would constrain the smallest fields that can be detected with these devices, also limiting the packing density of neighboring devices.

The team of Baibich et al. acknowledged the spin filtering character of their multilayer structures and the reasons for achieving a giant magnetoresistance in them. They also observed that the orientation of the magnetization in the Fe layers with respect to the applied magnetic field plays a role in obtaining a giant magnetoresistance effect which was noticeably larger for ferromagnetic layers with antiparallel magnetization. It was since determined that a paramagnetic layer with

the thickness of a few lattice spacings such as the thin Cr spacer between two
ferromagnetic layers can result in antiferromagnetic behavior [9]. As a matter
of fact, IBM had introduced a design for GMR sensors in magnetic disk drives
consisting of an "artificial" antiferromagnet. This is a structure comprising two
ferromagnetic layers coupled antiferromagnetically by means of a Ru layer as
thin as $\sim 3\,\text{Å}$ such that magnetizations are antiparallel. In these structures, as the
external field is increased to higher values, the magnetization of the magnetic
layers switches and tends to orient parallel to the field. When coercivity is finally
reached, the parallel alignment is completed, and the magnetoresistance decreases
significantly. Artificial antiferromagnets are nowadays commonly used in spin valve
sensor designs because antiferromagnets allow for better GMR values.

6.2.3 Fabrication Dependence of Spin Valve Structures

Since the late 1980s, researchers have observed that aside from material selection,
fabrication methods also influence GMR values. For instance, high-quality Fe/Cu
granular films were obtained through a cluster beam deposition process [10].
This material combination was chosen because it had previously resulted in good
theoretical and experimental results. Cluster beam deposition involves the formation
of small metallic clusters as shown schematically in Fig. 6.9a. These are obtained by
adiabatic expansion of vaporized atoms through a crucible's nozzle in vacuum. An
electron beam ionizes the clusters, and a bias voltage is applied between the source
and the substrate. The clusters are propelled onto the substrate at the same time
as surface migration of adatoms is enhanced. In an alternative deposition method,
the clusters are preserved when neutral clusters are softly landed without ionization
or acceleration. Unfortunately, the cluster beam deposition technique proves to be
inadequate for GMR, as the granular nature of the clusters leads to a suppressed
effect. These results are in contrast to co-evaporation methods where chemically and
magnetically homogeneous films can be obtained (Fig. 6.9b). Hence, it is concluded
that giant magnetoresistance effects are highly sensitive to nanoscale heterogeneity,
as well as the magnetic state at the interface between the clusters and the host matrix.
This confirms previous research findings that determined that the GMR effect is
enhanced in multilayer structures rather than granular materials. Therefore, a lot
of time and effort have been dedicated to producing multilayers in various design
configurations. Several techniques were tried; nevertheless, a much used one is

a 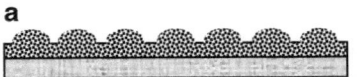   **b**

Fig. 6.9 Schematic representation of (**a**) cluster deposition of a film as compared to (**b**) a
homogeneous layer. The granular nature of the clusters leads to nanoscale heterogeneity

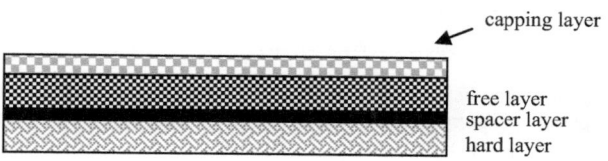

Fig. 6.10 Capping layer on top of the free layer may improve magnetic ordering and stabilize magnetic properties

magnetron sputtering due to its superior quality films. By employing a dc magnetron sputtering system, a group of researchers noticed that thin films of FePt/Fe display a more pronounced magnetocrystalline anisotropy, better chemical ordering, and a higher coercivity [11]. In particular, the presence of an additional Fe film renders the FePt layer more stability in its properties.

FePt films show promise in ultrahigh density recording media because the larger magnetocrystalline anisotropy is believed to delay the onset of superparamagnetism. Unfortunately, disordered FePt films are usually obtained at the commonly used room temperature where fabrication occurs. Post-fabrication annealing does not necessarily alleviate the film problems such as surface roughness and high noise. Therefore, a Fe cap layer is required to stabilize the FePt film because it lowers the kinetic ordering temperature. Figure 6.10 shows a sketch of where a Fe cap may fit in a spin valve structure. During fabrication, as the Fe-rich layer forms, it promotes a reorientation of magnetization in the FePt layer. A recrystallization of the film occurs during annealing as well, improving the film's magnetic anisotropy. The anisotropy is in proportion to the chemical ordering parameter of the FePt film. As the latter changes phases, it also changes its magnetic properties from soft ferromagnetism of a more disordered phase to those of a hard one. The Fe content plays a role in the degree of difficulty of ordering under the same conditions. The overall thickness of the Fe cap layer is under 2 nm; hence it does not add too much volume to the spin valve.

In general, GMR multilayer devices are not easy to design and fabricate at high yield. This is due to their relatively unstable response, as well as low repeatability and sensitivity. For this reason, aside from the layers discussed above, extra layers are usually added to a spin valve structure. These layers serve different purposes, depending on where they are located. For instance, a seed layer may be added to allow a magnetic texture to form in an adjacent layer, such as the one resulting from the crystallographic [111] orientation mentioned in the previous section. Also, it optimizes grain size while reducing grain boundary scattering. Figure 6.11 shows roughly where a seed layer may be added to the spin valve structure. On the other hand, a reference layer is usually added to the spin valve, as well, as shown schematically in Fig. 6.12. The reference layer is coupled to the pinned layer via a Ru layer which provides a powerful Rudeman–Kittel–Kasuya–Yosida (RKKY) interaction. This interaction is strongly antiferromagnetic, provided the Ru layer thickness is only a few Å. The magnetizations in the pinned layer and reference layer

Fig. 6.11 A seed layer may be necessary to allow the formation of a desired magnetic texture, which in turn is possible if a crystallographic texture is enforced first in the layer. The adjacent layer is then grown on top of the seed layer

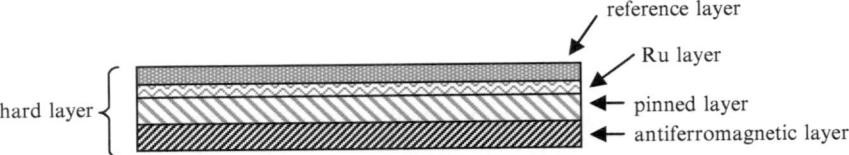

Fig. 6.12 Reference layer coupled to the pinned layer via a very thin Ru layer. The few Å thick layer provides a very strong RKKY coupling. On the other hand, the antiferromagnetic layer exerts an exchange bias on the pinned layer, keeping the magnetization of the latter in a fixed direction

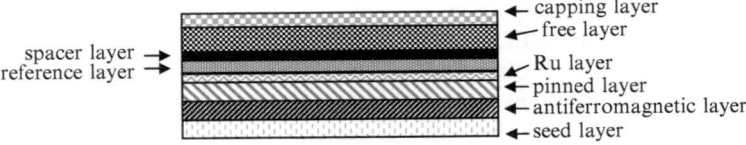

Fig. 6.13 Possible spin valve configuration illustrating various layers and their role

are thereby maintained in an antiparallel configuration, while their layer thicknesses are adjusted such that the magnetic moments are nearly equal. In this case, the net moment of the pinned layer/Ru/reference layer configuration is almost zero, and this configuration makes up the artificial antiferromagnet mentioned earlier in this chapter. The whole structure of layers constitutes the "hard layer" that does not respond to external fields, in addition to not creating stray fields that disturb the free layer (i.e., sensing layer).

Considering all of the above, the spin valve may have the layer configuration shown schematically in Fig. 6.13. Note that the signal is generated in the free layer/spacer/reference layer in response to the magnetization rotation of the free layer. This rotation occurs when an external field is acting, changing the resistance of the device and hence the GMR ratio. For upgrades in device performance, we need to focus on improving the signal amplitude. This is achieved by confining the current to the layer system where it is generated. For this purpose, it helps if the conductivity of the spacer layer is enhanced, the interfaces are sharpened, and interface scattering is minimized. The spin valve configuration illustrated in Fig. 6.13 is typical of current-in-plane (CIP) geometries where the current flows, as the name suggests, in the plane of the layers.

It is possible to use for pinning a very thin (~ 3 nm) $CoFe_2O_4$ layer instead of an antiferromagnet, as mentioned above. However, $CoFe_2O_4$ is ferrimagnetic and as such has a net magnetization that needs to be balanced to allow appropriate biasing of the spin valve structure. Nevertheless, IBM has figured out a way to employ $CoFe_2O_4$ for pinning while adjusting its ferrimagnetic properties. For this purpose, insulating antiparallel CoO and Co_3O_4 layers need to be deposited underneath the $CoFe_2O_4$ layer. To ensure the deposited layers have the desired characteristics, a reactive dc magnetron sputtering technique is employed. This is a popular fabrication method for layered spin valve structures that has proven to give good GMR results. It was noticed that it is important to grow epitaxially one layer on top of the other to maintain crystalline order. The additional insulating CoO and Co_3O_4 layers ensure that more $CoFe_2O_4$ grains order magnetically and the coercivity and magnetic moment are both higher [12]. (See also Problem P6.2 and reference [13]). One team of researchers went as far as eliminating altogether the need for a pinning layer [14]. They used nanopatterning to allow the spontaneous formation of a single domain that would fix the magnetization of the magnetic layer, thereby rendering the employment of an additional layer as redundant. Experimentation with various spin valve structures is subject to long-term verification to observe its validity. For the present, easier magnetization switching, and hence enhanced GMR values have been reported for magnetic layers that are magnetically textured displaying a preferred direction of magnetization. This is the case for layers containing nanowire arrays where the strong shape anisotropy of the latter has proven to be advantageous. It was observed [15] that multilayered films of $Fe/Ni_{80}Fe_{20}/Co/Cu/Co$ with a nanowire arrangement display larger GMR values than similar films without nanowires. The researchers attributed this improved response to the shape-induced easy axis of magnetization that allowed straightforward switching of magnetic moments without significant hindrance from the domain structure.

6.2.4   Dilute Magnetic Semiconductors as Spin Valve Materials

Until now, CoFe and NiFe alloys were commonly employed in the magnetic recording industry. They have a small, in-plane magnetocrystalline anisotropy and both spin species contribute to the current, each in its own conduction channel, provided spin-flip scattering does not occur. However, there has been a growing interest in a special class of semiconductors due to their intriguing magnetic properties that are observed when the materials are doped appropriately. These semiconductors offer not only an electron charge but also an electron spin for more diverse device applications. A material such as GaAs or a similar III–V semiconductor finds application in devices that use charge. If charge transport has been traditionally associated with processing of information and communication, the spin of the electron holds promise for the recording of information. However, how do you exploit both charge and spin in one material, and how are these materials obtained?

To answer these questions, we take a closer look at those semiconductors currently employed in microelectronics. III–V semiconductors have long been used in the semiconductor industry, in particular GaAs which is nonmagnetic. When doped with Mn impurities using molecular beam epitaxy under low-temperature equilibrium conditions, (Ga,Mn)As becomes a very different compound, not resembling its relative. This is because (Ga,Mn)As is a *diluted magnetic semiconductor* in which both ferromagnetism and semiconducting characteristics coexist. This material has been made magnetic, thereby exploiting electron spin rather than charge when appropriate concentrations of magnetic elements are introduced during fabrication. It is believed that in this compound, holes play a critical role in the magnetic coupling between ferromagnetic semiconductors and their impurities in such a way that exchange interactions between the electrons in the semiconducting band and the localized electrons at the impurity ions lead to interesting electronic, optical, and magnetic properties. Apparently, in (Ga,Mn)As, the hole-mediated exchange between nearby pairs of Mn ions is stronger than the antiferromagnetic superexchange. However, this is not the case in similar diluted magnetic semiconductor compounds such as for instance p-(Zn,Mn)Te. In the latter, a lack of electrical participation of Mn atoms is observed, typical of II–VI compounds. On the other hand, (Ga,Mn)As films have been noticed to display a considerable magnetic anisotropy attributed to the interaction between spin and orbital degrees of freedom of the magnetic electrons. In the case of (Ga,Mn)Sb with a few percent of Mn, a pronounced anomalous Hall effect and negative magnetoresistance below 50 K are obtained, indicating the formation of a ferromagnetic semiconductor where the Mn atoms are incorporated into the GaSb host [16]. Conversely, p-type (Ga,Mn)N is supposed to have a Curie temperature above room temperature [17], which would add more capabilities to GaN-based structures, already in use in photonics and high-power electronics. According to theoretical [18] and experimental [19, 20] investigations, the n-type (Ga,Mn)N does not display ferromagnetism above 1 K.

When doping ZnO with Mn, some interesting properties are observed, a fact that caught the attention of many research groups over the years. The diluted magnetic semiconductor $Zn_{1-x}Mn_xO$ has displayed a large magnetoresistance at low temperatures when in bulk form, although theoretical predictions show that it should be possible to obtain such results at room temperature [21]. This is because mathematical considerations indicate ferromagnetism at higher temperatures in spite of the fact that $Zn_{1-x}Mn_xO$ is regarded by many as paramagnetic. On the other hand, based on experimental observations of bulk samples, the related compound $Zn_{1-x}Co_xO$ is ferromagnetic at room temperature, at least when appropriately processed during fabrication. It should be noted that Zn vapor treatment is customarily used for both $Zn_{1-x}Co_xO$ and $Zn_{1-x}Mn_xO$ to overcome the naturally occurring insulating property of the as-prepared compounds [22]. This process renders the compounds semiconductor properties, but does not seem to influence their ferromagnetic or paramagnetic character. In any case, the fabrication of $Zn_{1-x}Mn_xO$ by a solid-state reaction process at relatively high temperatures results in a paramagnetic compound, at least when magnetic properties are measured at room temperature. Conversely, synthesizing $Zn_{1-x}Mn_xO$ by a low-temperature process apparently leads to ferromagnetic behavior which is attributed to an impurity metastable phase, and not to the actual $Zn_{1-x}Mn_xO$ itself. Hence, it is still believed that single phase $Zn_{1-x}Mn_xO$ is paramagnetic, but only when a single phase is achieved for

$x < 0.05$. Samples of $2\% < x < 10\%$ Mn concentrations show a positive magnetoresistance at low temperatures (< 35 K) with the greatest value registered at 5 K for $x = 5\%$. Above 50 K, the GMR becomes weak rapidly. The low-temperature GMR is attributed to the exchange interaction induced spin-splitting [20]. The latter signifies a splitting into two spin-subbands of the conduction band under the action of a magnetic field. Nevertheless, the formation of *bound magnetic polarons (BMPs)* is likely to occur in $Zn_{1-x}Mn_xO$. When magnetic carriers are trapped by any kind of attractive potential of a defect, they can form what can be described as a "cloud" of aligned spins, similar to a polarization of charge in trapped nonmagnetic carriers. These accumulated aligned spins are known as bound magnetic polarons. They lead to a delocalization of carriers when a magnetic field is applied, inducing an abrupt decrease in resistance and contributing to a negative GMR. As an added fact, an increase in spin alignment of both magnetic ions and carriers results in an increase in conductance. This suppression of spin-disorder scattering under the effect of a magnetic field may also be a reason for the observed negative GMR in some ZnO-based diluted magnetic semiconductors. Nevertheless, low magnetic fields only create a redistribution of spin carriers between the two subbands sufficient for a positive GMR to be observed. But this redistribution of carriers becomes weaker at higher magnetic fields. Most conduction electrons are in the low-energy subband, resistivity is decreased, and BMPs are delocalized, hence the observed peak in GMR. BMPs alone are not enough to explain the magnetic properties of Mn doped ZnO. Maybe ferromagnetism is exhibited when a large density of 3d states is pinned in the donor impurity band at the Fermi level [20]. In any case, $Zn_{1-x}Mn_xO$ is still regarded as a controversial compound and is being investigated further.

We leave for the time being the discussion about this interesting class of materials, namely dilute magnetic semiconductors, and return to the problem of incoherent scattering that we had placed earlier on hold. It is appropriate again to revisit the Green's function formalism that will aid in describing better the significance of incoherent scattering in transport considerations.

6.3   A Closer Look at Incoherent Scattering

6.3.1   Using Büttiker Probes

We know that by obtaining a larger conductance, transmission is enhanced, too, since the conductance is proportional to the transmission. However, it has been long acknowledged in practice that spin scattering due to impurities within the layers leads to a smaller spin transmission. As long as the spin is deflected without change (more exactly, flipping), spin propagation is considered coherent. However, if the spin flips, that constitutes a phase-breaking event and spin transmission is affected. To obtain enhanced GMR values, it is necessary to allow a higher transmission for the preferred spin species, at the expense of the other. A common method for increasing preferential transmission is by reducing the number of spin flipping events for the preferred spin, as these play a significant role in transmission not only at interfaces but also throughout the material. Measures are taken to ensure that spin-flipping impurities exist in certain carefully calibrated amounts, and only for the type of spin that needs a decreased transmission.

The nonequilibrium Green's function formalism can be helpful again in modeling electron transport through a channel, but this time by taking into account the presence of a phase-changing event. But how do we model this effect? Research literature indicates there are various mathematical techniques for doing so. In the present section, we use the Büttiker approach. A floating probe called a *Büttiker probe* is introduced in the expression of the Green's function, similar to the self-energies that are due to contact with the two ends of the channel. This probe is placed at the location in the channel where the scatterer is positioned. The role of this probe is to extract electrons from the device as they pass through, and re-inject them after (electron wavefunction) phase randomization. This actually models phenomenologically a phase-breaking event for an electron. The problem can be visualized as a simple circuit model to which we apply Kirchoff's law for current. Consider Fig. 6.14. Büttiker's approach treats the relationship between the left contact and the scatterer as conductance G_{13}, left and right contact, conductance G_{12}, and impurity and right contact, conductance G_{23}. Since conductance is proportional to transmission, the effective transmission through the three-piece system made of contact/channel/scatterer is the result of the addition of conductors in series and in parallel

$$T_{\text{eff}}(E) = T_{12}(E) + \frac{T_{13}(E)T_{23}(E)}{T_{13}(E) + T_{23}(E)}. \tag{6.10}$$

Figure 6.15 shows the effective transmission for such a system for a scatterer located midway through the channel. The Green's function for the channel has the form

$$G_{\text{C}} = (E - H_{\text{C}} - \Sigma_{\text{L}} - \Sigma_{\text{R}} - \Sigma_{\text{R}} - \Sigma_{\text{scatterer}})^{-1}, \tag{6.11}$$

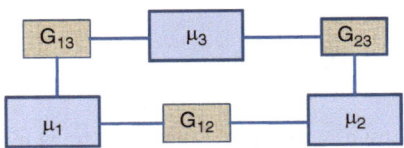

Fig. 6.14 In the Büttiker approach, the relationship between the contacts and scatterer can be represented as a conductor network provided the transmission between the individual elements is reciprocal. As such, the conductor between the left contact and the scatterer has a conductance G_{13}, the conductor between the left and right contact G_{12}, and scatterer and right contact G_{23}. Since conductance is proportional to transmission, the effective transmission through the network is the result of the addition of conductors in series and in parallel

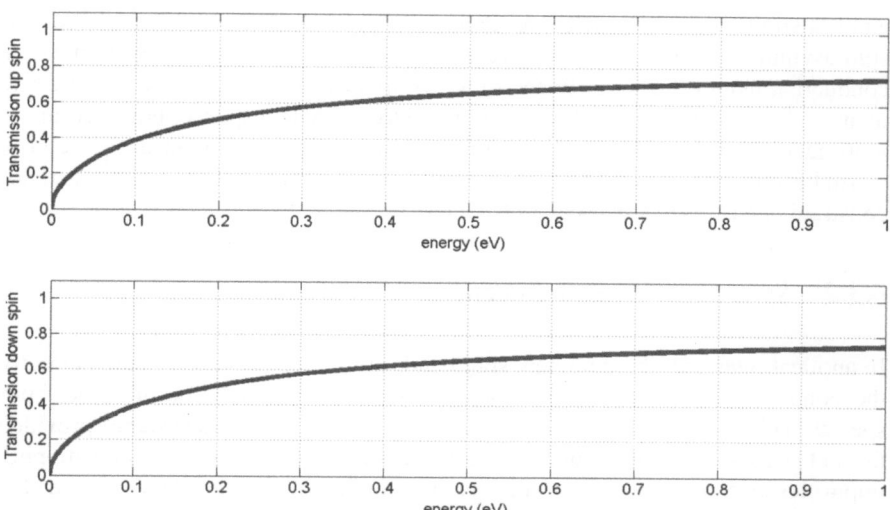

Fig. 6.15 A Büttiker probe can account for a scattering event in the channel. It is modeled as another self-energy Σ that appears in the expression of the Green's function of the channel. The scatterer has the effect that it lowers the transmission for both ↑ spin and ↓ spin electrons in an equal way. This figure was obtained by considering the energetic value of a scatterer about 1/3 of that of Σ_L or Σ_R. Both Σ_L and Σ_R were taken to have the same energetic value. A linearly increasing magnetic field applied parallel to the axis of the channel was added

where $\Sigma_{\text{scatterer}}$ is the added self-energy (*Büttiker probe*) due to the phase-breaking scattering event in the channel. It should be noted, however, that this approach is valid only if the transmissions T_{ij} are reciprocal

$$T_{ij} = T_{ji}, \tag{6.12}$$

where T_{ij} is the average transmission from contact j to i.

Remember that a way to improve the magnetoresistance ratio is to enhance spin conductance in the ferromagnetic and nonmagnetic layers, particularly for the dominant spin species. The trick is to find the "right Büttiker probe" for ↑ versus ↓ spins. In those experiments discussed in previous sections, it was determined that changes in magnetoresistance due to, for instance, Cr impurities present in the bulk Fe are approximately six times smaller when ↓ spins scatter on other ↓ spins, than for scattering of ↑ spins on similar ↑ spins. The Cr atoms in the layer of only three lattice constant thickness (the spacer layer in the spin valve structure) can be regarded as Cr impurities in bulk Fe, as the Fe layers are considerably thicker than the Cr layers. At zero field, both Cr and Fe atoms (of the layer with antiparallel magnetization) act as scattering centers for the mixed ↑ spins and ↓ spins, reducing transmission of both types of spin-carrying electrons through the layers. A variety of third elements (V, Mn, Ge, Ir, and Al) were also experimentally investigated as they were deposited between Fe and Cr layers. It was noticed that (Mn, V) impurities attain a similar scattering effect as Cr impurities in Fe. Ultimately, the larger the spin-asymmetry in transport, the larger the difference between the resistivities (or conductivities) for the majority and minority spin carriers is. This is considered by many to be the "top secret" for obtaining a larger GMR response; however, many more factors are considered when improving spin valve performance. Hence, it is worth keeping an open mind when analyzing magnetotransport phenomena and not constrain ourselves to a particular characteristic.

6.3.2   Spin Scattering at Interfaces

In an ideal interface, the dominant spin bands at the Fermi energy match perfectly the conduction bands of the next layer, with a poor match for the other spin species. This is, to some extent, the case for Co/Cu and CoFe/Cu systems, provided the [111] axis is perpendicular to the layers [23]. Furthermore, if spin flipping impurities exist at interfaces and the magnetization of a magnetic layer is reversed, the magnetic moment of impurity atoms near the interface is also reversed. As a result, the random exchange potential for ↑ and ↓ spin electrons switches too, and a change in resistivity occurs. When this happens, a spin-dependent interface resistance is felt by the electrons, resistance which depends on spin orientation and increases the GMR ratio. However, if an intermixing of atomic species takes place at the interface adding to the interface disorder, the electron states of the two spin orientations tend to blend. This results in a reduction of the transmission probability of the preferred spin orientation, and a reduced difference in conductivities for the two types of spin electrons. Consequently, the GMR ratio is reduced, as well. Preliminary calculations based on spin-dependent energy bands confirmed that spin-dependent conductivity is significantly influenced by scattering at the interfaces, especially between ferromagnetic and nonmagnetic layers. Thus, an important requirement is that the layer thickness is kept smaller than the mean free path for spin-flipping. In this case, a spin-carrying electron can make it through a layer without significant spin-flip scattering. Demagnetizing fields associated with the magnetic moments of the ferromagnetic layers are likely to have a higher influence for larger volumes. Thereby, it is even more necessary to keep the layers thin. There is the possibility that a progressive reduction of device size could change the magnetic field requirements for changing their magnetic state during recording and read-out. This would likely limit the smallest fields that can be detected with these devices, and their packing density. In view of all these considerations, tremendous effort is put into controlling interface impurity levels and type, as well as the sharpness of the interfaces during fabrication. That includes upgrading deposition equipment and developing new processing techniques.

6.3.3   Scattering in a Broader Context

We would think that scattering at interfaces is probably having the most effect on the quality of a spin valve, and this is in most part true. However, there are concurrent factors that play an equally important role. It may, therefore, help with a better understanding of the subject if we divide scattering into four categories. This is because the spin asymmetries of the transmission at interfaces and the mean free path within the ferromagnetic layers result in more than one type of spin scattering. For instance, we can speak of a "pure" interface scattering (type A) that occurs when only the transmission coefficients are spin-dependent, but not the mean free paths in different layers. For pure scattering, the mean free paths are assumed to be the same for both spin types. On the other hand, a different situation arises if the mean free paths are spin-independent but each layer has its own mean free path value. In this case, the scattering is impure (type B). There is also the scenario in which the mean free paths are the only ones that are spin-dependent while the transmission coefficients do not depend on spin type. Then, the scattering is said to occur in the bulk of the material (type C). Lastly, a more complex scattering circumstance develops when interface and bulk scattering take place at the same time. This happens when both transmission coefficients and mean free paths are spin-dependent (type D). All things considered, the type of scattering is thereby anticipated to play a major role in obtaining a certain GMR. Hence, we need to be careful about how we choose to model scattering, and what results to expect.

Scattering has not as much of an importance on its own unless geometrical and physical considerations such as layer thickness and type also come into play. We have already seen that the effects of layer thickness on spin flipping and spin scattering are significant, as well as the fact that the function of the layer determines which type of scattering is more consequential for the GMR response. It seems that in practice the most influential factor affecting the performance of the spin valves is not necessarily spin scattering but layer thickness which affects differently each layer depending on its function. Let us, therefore, always view scattering in a broader context by also considering layer thickness and type of layers. In the commonly employed current-perpendicular-to-plane geometry (CPP), all four types of scattering (type A–D) are likely to occur within the magnetic layers, with some type of scattering particularly prevailing at interfaces between magnetic and nonmagnetic layers. Indeed, thinner free layers have been observed to result in no measurable bulk scattering (or type C scattering). Nevertheless, they sometimes play a role in interface scattering, especially when the latter is combined with bulk (type D). The situation is a bit different for spacer layers where it has been predicted theoretically and observed experimentally that the GMR response decreases exponentially with spacer thickness regardless of what type of scattering dominates, or what kind of spin valve it is (artificial antiferromagnetic, bottom, or dual spin valve). The GMR ratio is also found to approach extremely low values when the pinned layer becomes very thin, although it mostly happens when bulk scattering takes place. Concurrently, short mean free paths in antiferromagnetic layers lead to sharp declines in GMR, but this fact may as well be attributed to a simultaneous reduction of exchange bias which likely occurs in very thin layers of the antiferromagnetic type.

A study [24] on a simple spin valve with a top pinned structure of Ta (6 nm)/Ni_{81} Fe_{19}/$Co_{90}Fe_{10}$ (1 nm)/Cu (1.8 nm)/$Co_{90}Fe_{10}$(3.5 nm)/$Ir_{20}Mn_{80}$(8 nm)/Ta(6 nm) has validated some of these results. In particular, the GMR ratio declined considerably when the thickness of the soft magnetic layer in the form of $Ni_{81}Fe_{19}$ was increased from 6 to 7 nm. On the other hand, below 6 nm and above 7 nm, no significant changes were registered in the GMR response. The performance of the spin valve was at its best when the thickness of other layers was optimized as well, such as the layers made of $Ir_{20}Mn_{80}$ (11 nm) and Ta (3 nm). In this case, the optimum thickness for the $Ni_{81}Fe_{19}$ layer was found to be 4.5 nm leading to GMR ratios of 9.15%, low coercivity (0.85 Oe), and a high sensitivity. It should be noted that in this study of this specific spin valve configuration, $Ni_{81}Fe_{19}$ was used to improve the softness of the free $Co_{90}Fe_{10}$ layer. At the same time, the second $Co_{90}Fe_{10}$ layer served as the pinned layer which coupled with the antiferromagnetic $Ir_{20}Mn_{80}$ causing the much needed exchange bias effect. The spacer layer provided by Cu which is nonmagnetic eliminated the magnetic coupling between the free and pinned layers so that the former was able to rotate freely. The two Ta films acted as buffer and capping layers to prevent oxidation of the underlying films. Lastly, a magnetic field of 50 Oe was used during layer fabrication to align the magnetic moments and induce an easy axis of magnetization. Similar studies performed in other research laboratories pointed out that the antiferromagnetic layer defines more of the properties of a spin valve than may be apparent at first. The coercivity of the ferromagnetic/antiferromagnetic system was found to depend on the growth order of these layers. As such, it matters whether the antiferromagnetic $Ir_{20}Mn_{80}$ layer is placed towards the top or bottom of the spin valve. If it is closer to the top, the coercivity is much lower than if the antiferromagnetic layer is at the bottom of the spin valve. At the same time, the 9.6% GMR ratio given by the spin valve with the top layer is superior to the 3% GMR obtained from the spin valve with the bottom antiferromagnetic layer. Both $Co_{90}Fe_{10}$ layers were observed to exchange couple ferromagnetically, as demonstrated through magneto-optical imaging. The improved GMR response of the top spin valve is attributed to the fact that the pinned layer in the top configuration has a large unidirectional anisotropy, whereas in the bottom version of the spin valve the same layer has only a partial and weak anisotropy that changes with the magnitude and orientation of the applied magnetic field. Finally, it should be mentioned that magnetotransport properties of spin valves are also defined by the spin structure as well as defects in the antiferromagnetic layer, proving once more that the latter plays a defining role in spin valve design.

6.3.4 Additional Interface Influences on Spin Valve Performance

Interface phenomena remain a high priority for the optimization of spin valves, in particular given the nanoscale dimensions of these structures. As a matter of fact, the configuration and cleanliness of interfaces is so essential, that unexpected effects have been observed in structures not normally associated with magnetic phenomena. Take for instance, heterostructures of nonmagnetic oxides where ferromagnetism and modified superconductivity have been demonstrated to coexist due to the presence of magnetic moments at the oxidized surface. Similarly, consider the presence of surface spins in nanoparticles which influence the GMR or cause an unusual magnetoresistance hysteresis. Surface defects caused by ion implantation have been observed to induce magnetism in silicon or carbon, which together with dangling bonds in the amorphous structure may be responsible for these effects. Recently [25], an unusually large magnetoresistance was recorded in epitaxial silicide nanowires, a type of material that is essential in microelectronics. This is even more so unexpected since these metallic compounds are largely nonmagnetic in their bulk state, although silicides are formed from the reaction of silicon with ferromagnetic metals such as Ni or Co. The unusual magnetoresistive effect observed in the silicide nanowires is attributed to the combined interactions between interfacial spins associated with dangling bonds. Theoretical models [25] have predicted a tendency for these interfacial spins to align in random or ordered configurations rendering magnetic properties to otherwise nonmagnetic structures.

The role of the exchange bias and its effect at interfaces has been emphasized as being an important aspect in the fabrication of spin valves. Lately, thermomagnetic techniques have been employed to induce exchange coupling in $DyFe_2/YFe_2$ superlattices by quenching the magnetization in $DyFe_2$ and thereby biasing the unpinned YFe_2 reversal of magnetic moments [26]. Depending on the cooling field amplitude, interface domain walls can be induced that tend to modify the orientation of the pinning moments at interfaces. When the unpinned spins of a ferromagnet couple to pinned spins of an antiferromagnet, a unidirectional magnetic anisotropy is produced and exchange bias occurs. Then the magnetically hard material can play the role of the antiferromagnet in inhibiting the soft magnetization reversal. Both hard and soft systems exhibit pinned magnetization and interface domain walls that develop parallel to unpinned/pinned interfaces. However, pinned magnetic moments are only about a percent of the saturation magnetization. It is, therefore, challenging to detect pinning, and even quantify the effect. Nevertheless, pinning at interfaces may raise specific concerns in exchange-coupling the magnetically hard layer. A good knowledge of the spin structure in the antiferromagnet is, therefore, required, especially close to interfaces. Thus, interface phenomena have been and remain a priority in the study of spin valves with the goal of obtaining further improvements in these devices.

Example 6.5. Discuss the influence of a nanosized oxide layer on coupling between different spin valve layers, if the oxide layer is placed at different locations around or within the Cu spacer layer. Consider a current-in-plane (CIP) spin valve geometry.

Answer: When designing spin valves it is, of course, desired to obtain the best possible GMR response. As such, several factors need to be considered, many discussed throughout this chapter. Among design considerations is achieving just the right amount of interlayer coupling, for instance the much needed antiferromagnetic coupling. In a CIP spin valve, due to the direction of current flow, the coupling between layers is particularly challenging. The spacer layer, usually made of Cu, could be made very thin to enhance the giant magnetoresistance ratio; however, below a certain nanometer size, problems start to occur when the coupling between the free and pinned layers increases significantly so that bias cannot be controlled, thereby decreasing the sensitivity of the spin valve. An oxide layer of nanometer thickness, enough to make a difference could be inserted around the pinned or ferromagnetic layers. Such a study was undertaken by a group of researchers [27] who decided to place the nanooxide layer at different locations around or within the Cu spacer. They noticed that in a bottom spin valve of the type Ta/NiFe/IrMn/CoFe/Cu/CoFe/Cu/Ta, placing the oxide after the Cu spacer or after the Cu cap layers enhances the GMR, in the former case by as much as 10%. However, results were very much dependent on oxidation conditions. As such, for large oxygen pressures and exposure times, the GMR response decreased. The GMR ratio was also noticed to be reduced for other positions of the nanooxide layer, such as before the Cu spacer or in the middle of the Cu spacer. The study was purely experimental and no explanation in terms of interface phenomena and their effect on spin was given as to the reasons for this observed behavior.

To better understand the place that spin valves occupy nowadays among commercial nano- and microintegrated semiconductor systems, we need to see how these structures are used within a device, especially in nonconventional applications. As such, a few devices are discussed in the next sections where spin valves were incorporated in less familiar functions and design configurations than the magnetoresistive read head. Three examples are presented where different sensors relying on the GMR effect have been integrated with other components. The first example describes how a magnetic flux concentrator can be used to serve a typical spin valve to improve its performance by reducing noise. In contrast, integrated systems illustrated in the next two examples encompass spin valves whose high sensitivity to extremely small magnetic fields is put to good use.

6.4 Other Devices and Spin Valves

6.4.1 Magnetic Flux Concentrators

Magnetic sensor manufacturers have used magnetic flux concentrators to alleviate the effects of the $1/f$ noise. It is possible for these flux concentrators to exist in a microelectromechanical system (MEMS) design to make them fully integrated with the microelectronics device they are serving [28]. Figure 6.16 illustrates an example of a two part magnetic flux concentrator integrated onto a silicon-on-insulator (SOI) wafer platform. The magnetic sensor is contained within the flux confinement area of the two parts of the flux concentrator. SOI wafers were used in this example because they require fewer processing steps while also providing a flat surface for fabricating the magnetic sensor. An SOI wafer contains a thin device layer of Si separated from the handle layer, which is by contrast a thick layer of Si. The separation layer is made of a 1–5 µm layer of SiO_2. By shifting the operating frequency of the device, the $1/f$ noise is reduced by several orders of

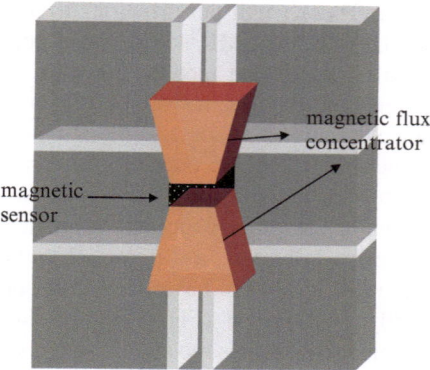

Fig. 6.16 Sketch of a two part magnetic flux concentrator integrated onto a MEMS device for reducing the $1/f$ noise of the magnetic sensor. The magnetic sensor is contained in the flux confinement area between the two parts of the concentrator. After [27]

magnitude. Some of the magnetic sensors that benefit from this technology are spin valves because magnetization fluctuations in magnetoresistive heads are a major limit on their signal-to-noise ratio.

The interesting aspect of this device idea is not the design of the spin valve itself, which may only have a classical configuration, but the fact that the field enhancement provided by the concentrator can improve the performance of the spin valve by reducing its noise. Usually, spin valves with the largest GMR ratio also tend to have considerable GMR noise. Hence, with proper design, the sensor noise can be reduced to a great extent. The reduction happens in this particular geometry when the two parts of the flux concentrator oscillate, modulating the enhanced field at the position of the sensor, and shifting the operating frequency of the device from the high $1/f$ noise region at low frequencies to higher frequencies where the sensor noise is orders of magnitude lower. For this concept to work, most of the $1/f$ noise has to come from the sensor element, and not another part of the system, and this provides a challenge considering the large size of the flux concentrator as compared to the small spin valve.

6.4.2 Vibration Detectors

Most vibration detectors are based on measuring velocity, displacement, or acceleration, and are not made of magnetic materials. However, low-power vibration detectors relying on measuring magnetic field variations can make use of GMR sensors. Consider, for instance, the Earth's magnetic field. Most machines contain a significant number of ferromagnetic parts, and when these vibrate, they disturb the Earth's magnetic field. These small perturbations in the field can be detected and measured with two half-bridge GMR sensors in a configuration where they are powered in chain, as demonstrated recently by a group of researchers [29]. In this case, the very sensitive magnetoresistance sensors convert the magnetic perturbations to resistance variations. By design, the sensitive directions of the two resistors are arranged such that they are opposing, so that when a time-varying magnetic field is present in the vicinity of the sensor, one resistor undergoes a decrease in resistance, while the other experiences an increase. Because of the voltage variation at the resistor dividers, a differential output signal is generated. Configurations with three GMR sensors have been tried, powered with the same current but independently conditioned (Fig. 6.17). In this particular design, a generalized impedance converter enables the supply of constant current while three different output voltages are measured. These three small voltage signals represent vibrations in the x-, y-, and z-directions, so they are amplified and filtered before being fed to an external acquisition system which allows visualization of the vibrations on the three different axes versus time and frequency.

The three GMR sensors used in this example can be integrated into microelectronics in configurations such as resistive dividers or a Wheatstone bridge. The sensors are placed in series, in an array with their sensing directions forming the orthogonal axes x, y, and z where vibrations are detected. They share a current of only 1 mA due to a generalized impedance converter mentioned earlier, thus reducing the power consumption which for conventional vibration sensing systems is about 5 mA per sensor. Nevertheless, as stated, a high gain (obtained through two amplifying stages) is still required as the vibrations are weak and the magnetic field strength decreases rapidly as the distance increases. This type of vibration detector is another example of an application where spin valves have been incorporated reliably wherever sensitivity to magnetic fields can be quantified and converted to electrical signals. Our next example will discuss another application where good

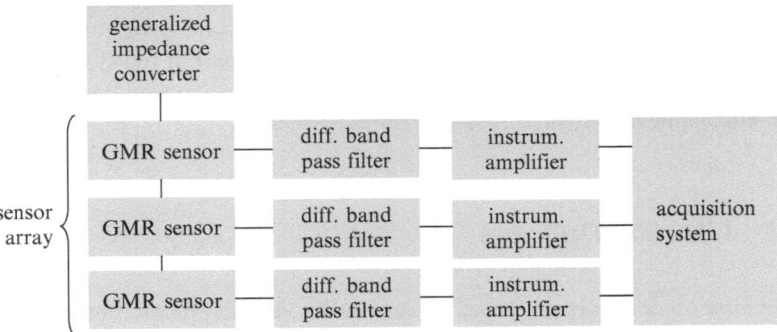

Fig. 6.17 Sketch of a vibration detector with a constant current source (generalized impedance converter) which powers three GMR sensors in chain. Each sensor detects vibrations in either x-, y-, or z-directions. After [28]

quality detection of small magnetic fields happens within the confinement of a microchip where a spin valve has been successfully integrated.

6.4.3 Microbridge Deflection Monitor

Extremely sensitive spin valves can detect deflections with nm precision in micro-electromechanical (MEMS) system bridges. These types of bridges could be part of a membrane/microbridge/cantilever system such as those microcantilevers that are increasingly being used in biochip applications as detectors or actuators. In a recently reported experiment, a spin valve sensor measuring $10 \times 2 \, \mu m^2$ and displaying a 6.5% GMR [30] was able to measure deflections of only a few nm in a Si:H/Al microbridge with a $1 \, \mu m$ air gap. The sensor was placed $3 \, \mu m$ away and $2.6 \, \mu m$ below the central point of the bridge where the deflection was about to occur, as shown schematically in Fig. 6.18. The movement of the bridge was controlled by an applied voltage on the gate. An rf-magnetron sputtered $Co_{78}Pt_{22}$ micromagnet was previously deposited on top of the bridge with the intention of creating a small magnetic field. The micromagnet was saturated in an external field. Because the magnet saturation occurred transverse to the bridge length, it created a transverse fringe field in the sensor, shifting the transfer curve. This fringe field of only 30–40 Oe was detected by the highly sensitive spin valve which responded with a GMR effect when changes in the fringe field occurred due to microbridge deflection. The sensor response was linearized with the help of an external bias field. The spin valve sensor had a typical top-pinned spin valve structure encompassing layers of Ta/NiFe/CoFe/Cu/CoFe/MnIr/Ta. The layers were deposited using ion beam deposition on a glass substrate, while pre-patterning was done by photolithography and ion-beam milling. Further improvements in the design and experimental conditions led to reported bridge deflections up to [31] $0.23 \, \mu m$. The sensing capability of spin valves has increased significantly in recent years so that these sensors can detect much smaller magnetic fields than earlier generation technologies.

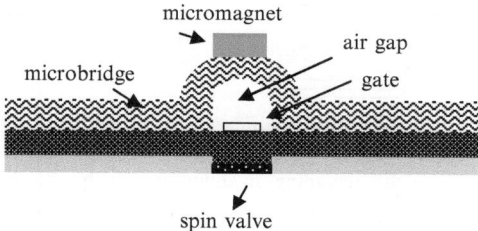

Fig. 6.18 Sketch of a semiconductor microbridge with an integrated spin valve and micromagnet. The spin valve senses deflections of the microbridge by detecting changes in the fringe field of the micromagnet. After [29]

6.5   Magnetoresistive Heads

6.5.1   The Role of the Magnetoresistive Head

A magnetoresistive read head detects the magnetic flux escaping from the transitions between two magnetized regions on a magnetic storage medium. On this medium, information has been previously stored by sequentially magnetizing regions in either of the two magnetization directions along the track (Fig. 6.19). The strength of the escaped magnetic flux is only a few tenths of an Oe; hence a spin valve sensor has to be extremely sensitive to be able to detect such small fields.

The magnetically hard layer of the spin valve has a fixed magnetization that will not rotate while it is in the proximity of the magnetic flux of the recording medium. As discussed previously, the hard layer is exchange coupled to a thin antiferromagnetic layer whose influence results in a shifted hysteresis loop to a higher magnetic coercivity. Hence, the layer becomes "hardened" displaying a larger coercivity that allows it to maintain its own magnetization. In contrast, the soft magnetic layer can still rotate its magnetization in the presence of a magnetic field. This field is provided by the escaped magnetic flux from the transitions between the two magnetized regions. Consequently, only the free layer of the spin valve "switches" in the low fields created by the magnetic regions on the medium. These magnetic regions each represent a *bit* that has been previously written (Fig. 6.19). The stored bits of information are recovered when the transitions between these magnetic bits are detected (Fig. 6.20). The spin valve displays a GMR effect when the magnetizations of the pinned and free layer change direction from parallel to antiparallel. The spin valve is the magnetoresistive head which reads the information stored on the magnetic medium as it moves along the track; however, heads incorporate both a reading and a writing function with different parts of the head performing different tasks (Fig. 6.21).

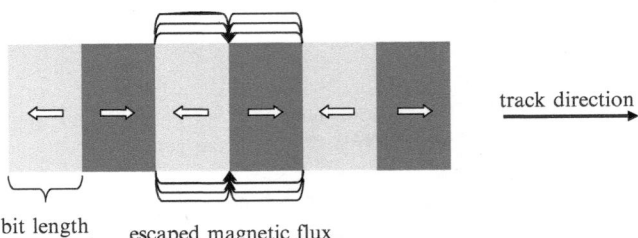

bit length   escaped magnetic flux

Fig. 6.19 Sketch of a magnetic recording medium on which information has been stored magnetically by magnetizing each region in either of the two magnetization directions along the track. Magnetic anisotropy within each region keeps the magnetization aligned with the direction of the track

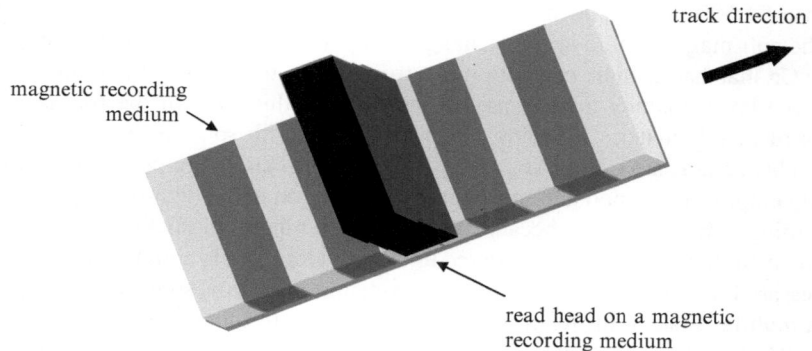

track direction

magnetic recording medium

read head on a magnetic recording medium

Fig. 6.20 Magnetoresistive head advancing along a track and reading information stored on the magnetic recording medium

Fig. 6.21 Sketch of a magnetoresistive sensor which is usually a read/write head serving for both reading and writing of information. The sensor contains a stack of several layers with different functions, as described throughout this chapter. Different parts of the sensor are responsible for each task

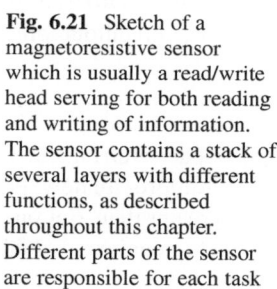

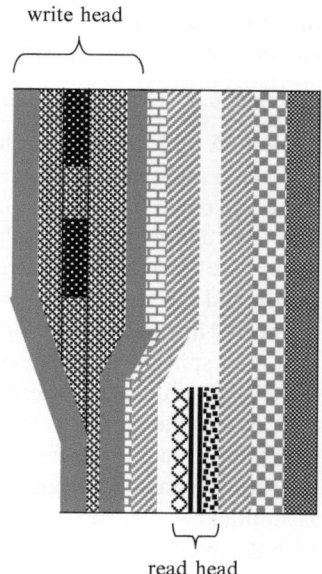

write head

read head

6.5.2 Material and Design Considerations for Magnetoresistive Heads

A variety of materials and material combinations have been investigated over the years for their suitability as layers for spin valve sensors, and some were already discussed in Sect. 6.2. Oxide pinning layers are sometimes preferred due to their increased corrosion resistance. As such, $CoFe_2O_4$ is often considered as a material for the pinning layer due to its high anisotropy constant; however, its magnetic properties are highly dependent on microstructure. GMR of 12.8% and pinning fields of 1,500 Oe have been observed in this material [32], keeping the properties of the soft magnetic free layer unchanged. In contrast, the hard magnetic material CoPtCr that has a high magnetic anisotropy leads to fringing fields that affect the free layer, making it necessary to increase the thickness of the free layer. In spite of it, this material remains a popular choice with spin valve manufacturers who choose it as a hard pinning layer. Antiferromagnetic IrMn or PtMn cannot be easily employed in small gap sensors that are required in the high-density magnetic recording industry. This is because PtMn needs a minimum thickness of 150 Å to become antiferromagnetically ordered upon annealing, while a minimum of 80 Å is necessary for IrMn to obtain an optimum exchange bias. Dual spin valves composed of a multilayer stack of, for instance, underlayer/IrMn [pinned layer/Ru/reference layer]Cu/free layer/Cu/[reference layer/Ru/pinned layer]IrMn/cap have become popular recently and are reported more often in the literature. However, a dual spin valve may pose a challenge from a magnetics perspective due to the dependence of the stability of the sensor on pinning of the top and bottom layers. Differences in pinning strength are unavoidable because of growth conditions, crystallographic and magnetic texture development, as well as roughness effects.

Example 6.6. Describe a spin valve with negative GMR.

Answer: A typical spin valve has the configuration ferromagnet/spacer/ferromagnet. If the spin asymmetry factor $\alpha = \rho_{\downarrow}/\rho_{\uparrow} \neq 1$ at both interfaces is approximately the same and equal to ~ 1, the GMR is positive. However, if α is greater than 1 at one interface, and less than 1 at the other interface, the GMR is likely negative. A dual spin valve structure can have a negative GMR because of both spin asymmetry factor, and the fact that the two spin valves act as two electrically distinct entities connected in parallel. A group of researchers [33] investigated such a negative GMR in a dual spin valve of the type CoFe/Cu/(CoFe/Ru/CoFe)/Cu/CoFe/IrMn, where the trilayer structure (CoFe/Ru/CoFe) formed the "artificial" antiferromagnet. They noticed that the thickness of the Ru layer leads to either negative or positive GMR, concluding that interlayer exchange coupling determined by the Ru layer thickness is one of the factors responsible for the differences observed. At the same time, Ru layer thickness can also reduce the bulk spin-dependent scattering to a value smaller than the interface spin-dependent scattering, and hence contribute to whether a positive or negative GMR is observed. The value for α may be determined by modifications to the density of states at the CoFe/Ru/CoFe interfaces. This is because Ru has a partially occupied d band at the Fermi energy, nearly equal to the Co minority d band so that the ferromagnetic CoFe can induce a magnetic moment on the Ru atoms near the CoFe/Ru/CoFe interfaces. In this experiment, measurements were done in the current-in-plane (CIP) geometry, so that most of the current flowed through the Cu spacer layers, so that the GMR originated mainly from the interfaces with the Cu layers. The two spin valves were essentially in parallel and the contribution from the different α cannot be distinguished from those of adjacent spin valve channels.

As already stated in previous sections, the current-perpendicular-to-plane (CPP) geometry is the one mainly used these days in the magnetic recording industry. For the recording head application, the spin valve sensor is placed between magnetic shields to improve linear resolution. In this shielded configuration, all demagnetizing fields are reduced. A current travels between two magnetic shields and perpendicular to the plane of the layers of the spin valve, but parallel to the air-bearing surface of the sensor [34]. It is the spacing between the two magnetic shields that determines the track resolution of the recorded bits; nevertheless, other physical parameters of the sensor determine its performance and magnetic stability. The electrical resistance of the total sensor is usually kept low to allow a high sensor bandwidth, as well as a high signal-to-noise ratio. As such, the sensor resistance is determined by the product *sensor electrical resistance × track width × sensor height*.

Example 6.7. Are organic semiconductors the future GMR read heads?

Answer: Organic semiconductors are very promising materials for spin valves, and therefore for GMR read heads. This is because of the large spin relaxation time in π-conjugated molecules, but also due to rather simple and low energy consuming device fabrication techniques. However, a big challenge facing organic spin valves at the moment is the necessity for low-temperature operation in most studied cases, which does not make them competitive in the magnetic recording industry. Nevertheless, we are witnessing ever increasing developments in their material properties and there is hope that more of these organic spin valves will be capable of room temperature functionality in the near future. Most studies of spin injection into organic semiconductors focus on the organic material known as Alq_3, or tris(8-hydroxyquinoline) aluminum. It would be interesting to have a look at other candidates, such as the perylene derivative PTCTE, or tetraethyl perylene 3,4,9,10-tetracarboxylate. Charge mobilities in organic materials are orders of magnitude smaller than in semiconductors; however, electron mobility in both Alq_3 and PTCTE is about the same order of magnitude. The newly acquired popularity of PTCTE as spin valve contenders is due to the fact that perylene molecules were used as electron transport materials in optoelectronic devices like organic electroluminescent diodes (OLEDs) and solar cells. When placed between ferromagnetic materials such as NiFe and Co, PTCTE spacer layers of 300 nm lead to overall GMR responses of up to [35] 3%. Assuming an exponential decay of spin polarization, this GMR values corresponds to $\sim$100–150 nm spin diffusion length which is a high reported value. Work out Problem P6.16 for a comparison to organic Alq_3 spin valves in which GMR responses as large as 40% were observed at low temperatures.

Digital data are stored these days mainly on magnetic hard disk drives of individual computers, but also on external magnetic memory drives. However, magnetic storage is not the only form of safekeeping of information. A certain class of solid-state devices uses the electronic state of transistors and capacitors to store data bits. Nevertheless, their memory is volatile, as they lose their data as soon as the computer is powered down. Hence, in spite of the perceived underperformance of magnetic storage which may seem slow in this era due to rotating disks and moving heads, the world still relies on hard disks. Not only has the density of information bits on the magnetic hard disk increased over the last decade, but reading the information is now faster than in previous years thanks to improved magnetoresistive head designs. But spin valves are not limited to magnetoresistive heads, and this is a point we tried to make in this chapter. Furthermore, many of the concepts touched upon here will come in handy in the next chapter which also builds a discussion around electron spin. We look forward in the near future to many more interesting applications, some of them already trying to break into markets. Among these novel applications, we count the bipolar spin switch [36], the micromagnetometer [37], and the micro-SQUID [38], while others have yet to emerge.

Tests, Exercises, and Further Study

P6.1 The influence of spin scattering on GMR is undergoing further study in many laboratories around the world. Which scattering is commonly viewed as being responsible for GMR: interface or bulk? See for instance [5] and also P6.7.

P6.2 Comment on the circumstances under which spontaneous phase separation occurs between ferromagnetic and ferrimagnetic phases. What role does exchange bias play in this case? Note: see [7].

P6.3 The BMPs are likely formed in $Zn_{1-x}Mn_xO$ compounds and contribute to a certain resistivity behavior, depending on the temperature range. Comment on the contribution of BMPs to the different transport mechanisms observed in $Zn_{1-x}Mn_xO$ in various temperature ranges. How many transport mechanisms can be identified? See [20] for further details.

P6.4 How is the spectral function $A(E)$ for $\uparrow$ spin and $\downarrow$ spin affected by the presence of a magnetic field acting perpendicular to the flow of current in the device channel? Compare results to those shown in Example 6.4 for a parallel field, and comment on any differences you may observe.

P6.5 Calculate the transmission as a function of energy for up-spin and down-spin electrons in a channel without impurities, while a magnetic field is acting parallel to the channel. Compare results to Fig. E6.2.

P6.6 Calculate the transmission as a function of energy for up-spin and down-spin electrons for a channel containing two scatterers located at a distance 1/3 and 2/3 with respect to the left contact. The second scatterer (at 2/3) should have an energetic value twice as high as the first (at 1/3).

P6.7 Describe four types of spin scattering.

P6.8 In a conventional spin-valve, a nonmagnetic metal is sandwiched between two magnetic layers. In this case, the magnetization of one magnetic layer (i.e., the pinned layer) is fixed by an antiferromagnetic FeMn layer (i.e., the pinning layer). However, the FeMn can be easily oxidized, creating instability. What can be done to overcome this challenge? Note: see for instance [39].

P6.9 Half-metallicity has been predicted theoretically in some Heusler alloys such as Co_2MnSi. This means a 100% spin-polarization of conduction electrons. These alloys have attracted much interest as ferromagnetic layers for current-perpendicular-to-plane GMR sensors. Extremely large GMR responses have been reported in Heusler alloys, sometimes exceeding 100%. The spacer layer plays a significant role in obtaining a certain GMR response. Describe the differences observed experimentally when Ag is interchanged with Cr in a spin valve structure of the type $Co_2MnSi/Ag/Co_2MnSi$, and $Co_2MnSi/Cr/Co_2MnSi$. Note: see for instance [40].

P6.10 The generation of pure spin current has been demonstrated at room temperature in graphene. Hence, graphene spin valves have emerged. However, these spin valves have been noticed to degrade rather quickly. What could be the possible reasons for this degradation, as compared to other spin valves made of different materials? Note: see for instance [41].

P6.11 Magnetic tunnel junctions are likely candidates for implementation into hard disk reading heads. With improvements in their fabrication process, the resistive insulating barrier has been lowered. This made the transfer of spin possible from a spin polarized current to a ferromagnet and subsequent magnetization manipulation without the help of magnetic fields. By inducing a steady magnetization precession, this magnetic dynamical regime can be converted into a microwave electrical signal through the GMR effect. It is therefore possible, at least in principle, to use GMR devices as a powerless nanoscale microwave frequency mixer. Describe how Georges et al. in [42] are proposing to do that.

P6.12 It has been reported that the giant magnetoresistive effect can also be induced by stress in certain types of spin valve structures. Describe the circumstances under which this can be observed. Note: see for instance [43].

P6.13 Magnetoresistive random access memory (MRAM) is based on the integration of the GMR effect and Si integrated circuits (IC). Termed a pseudo-spin valve MRAM, it is a nonvolatile memory technology with unlimited read/write endurance. Describe an example of such technology. Note: see for instance [44].

P6.14 Comment on the exchange bias observed in NiCoMnSb Heusler alloys as compared to BiCaMnTiO. Note: see for instance [45] and [46].

P6.15 What is the influence of Bloch wall scattering on magnetoresistance. Note: see for instance [47].

P6.16 Comment on GMR in organic Alq_3 spin valves. Compare to Example 6.5. Note: see for instance [48] and many other publications since.

P6.17 Corrosion resistant spin valves are important for the longevity of their technological applications. Describe such studies and their findings in terms of what can be done to improve the corrosion resistance of spin valves. Note: see for instance references [49] and [50].

P6.18 What are some of the fabrication advantages of magnetic memories as compared to semiconductor-based memories? Note: see for instance [51].

P6.19 Why are magnetic semiconductors preferred over metallic ferromagnets in the design of spin valves? Note: see for instance [52] and [53].

P6.20 A "transpinnor" is a magnetoelectronic solid-state device that incorporates spin valves. When the spin valves are arranged in a modified Wheatstone bridge configuration, their resistance can be changed by magnetic fields from currents flowing in one or more overlaying striplines that are not resistively connected. In this case, the device can be used as a transformer, an amplifier, or as logic gates. Describe an example of a "transpinnor" and its applications. Note: see for instance [54].

P6.21 The expected spin polarizations of NiFe and Co are, respectively, 25% and 35%. What is the maximum value of magnetoresistance according to Jullière's model?

P6.22 Comment on the role played by spin-dependent as well as spin-independent interface scattering in a symmetric spin valve. Also, discuss the antiferro-magnetic/ferromagnetic coupling in such a spin valve. See for instance [55].

P6.23 Discuss a core–shell spin valve structure fabricated from a free standing Ni nanowire serving as a platform on which different magnetoresistive layers are sequentially deposited. See for instance [56].

P6.24 Comment on the reliability of PtMn-based standard and artificial antiferromagnet spin valves for different thicknesses and compositions of the antiferromagnet. See for instance [57].

P6.25 Describe a spin valve with an artificial ferrimagnetic pinned layer whose impurity concentration is varied to control spin transport. See for instance [58].

Hints and Partial Answers to Problems in Chap. 6

The answers to these problems are in many instances just a hint as to where further information is to be found. They contain the main idea, but the reader needs to find the references that are sometimes recommended for each exercise and complete the answer through further bibliographic reading and study. Reference numbering follows the one used in that particular chapter.

P6.1 Scattering at interfaces is regarded as being mainly responsible for giant magnetoresistance effects. However, some scattering due to bulk effects can occur. Consult [5] and further literature for details.

P6.2 See referenced publication for details. A complex phase separation scenario is observed under certain fabrication conditions in $Nd_{1-x}Sr_xCoO_3$ that has been doped with holes in various amounts. With increased hole doping, ferromagnetic clusters form with long-range magnetic ordering. They coexist with ferrimagnetic clusters for $0.20 \leq x \leq 0.60$. An induced antiparallel ordering of the Nd spins in close proximity to the Co sublattice occurs, giving rise to a spontaneous phase separation between ferromagnetic and ferrimagnetic clusters. Further results indicate that the magnetization of the ferromagnetic phase can be reversed while that one of the ferrimagnetic phase cannot. This is indicative of the exchange bias phenomenon where one phase is reversible and the other one is rigid.

P6.3 According to [20], three different transport mechanisms can be identified in $Zn_{1-x}Mn_xO$. At temperatures below 50 K, the temperature dependence of the resistivity can be fitted by a modified Mott's variable-range hopping model of BMPs. Between 90 and 170 K, the nearest neighbor hopping of BMPs among impurity sites dominates the electrical conduction. In the temperature range of 170–300 K, electrical conduction is due to the combined effect of nearest-neighbor hopping of the BMPs and an activation above the mobility range described by the T^{-1} law. The nearest-neighbor hopping of BMPs is not very pronounced at low temperatures because of the reduced thermal energy.

P6.4 The mathematical treatment is similar to the one used in Example 6.4. Instead of taking a magnetic field acting parallel to the channel, a perpendicular field of the type used in Example 6.3 is considered. Figure P6.4 is obtained. The direction in which the field is applied matters, as more pronounced spectral density peaks are obtained when the field acts parallel to the channel, as opposed to the perpendicular along the z direction.

P6.5 Because no phase-breaking events take place during transport in the channel, the transmission for both ↑ spin and ↓ spin electrons occurs without impediments (see Fig. P6.5). The calculations were performed for a parallel magnetic field. Results are very similar to Fig. E6.2.

P6.6 Follow the approach described in Sect. 6.3.1, by introducing two self-energies $\Sigma_{\text{scatterer}}$ in the expression of the Green's function (6.11).

P6.7 See Sect. 6.3.3. (a) Pure interface scattering: transmission coefficients are spin-dependent, mean free paths are the same in all layers for both spin types. (b) Impure scattering: mean free paths are spin-independent but layer-dependent. (c) Bulk scattering: mean free paths are spin-dependent while the

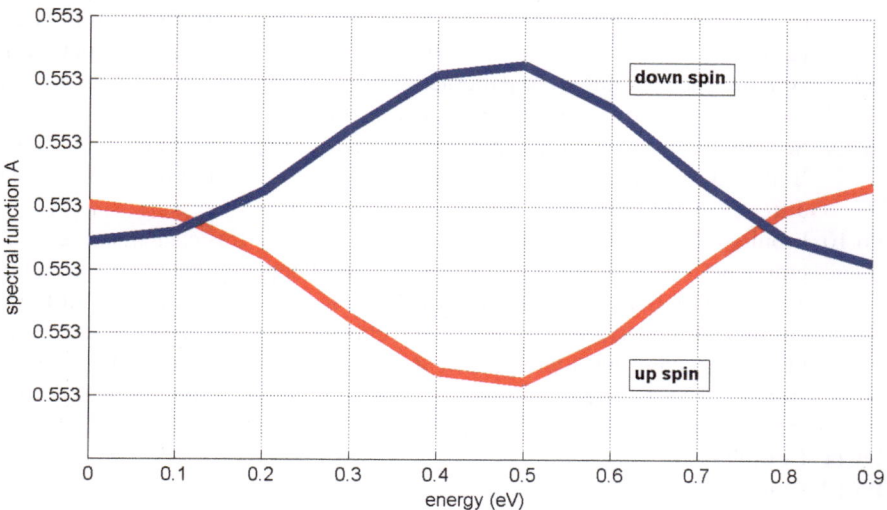

Fig. P6.4 The spectral density $A(E)$ for ↑ spin and ↓ spin electrons is affected by an applied perpendicular magnetic field. The direction of the applied field influences the height of the peaks. Compare this result to Example 6.4 while paying attention to the scale of the graph

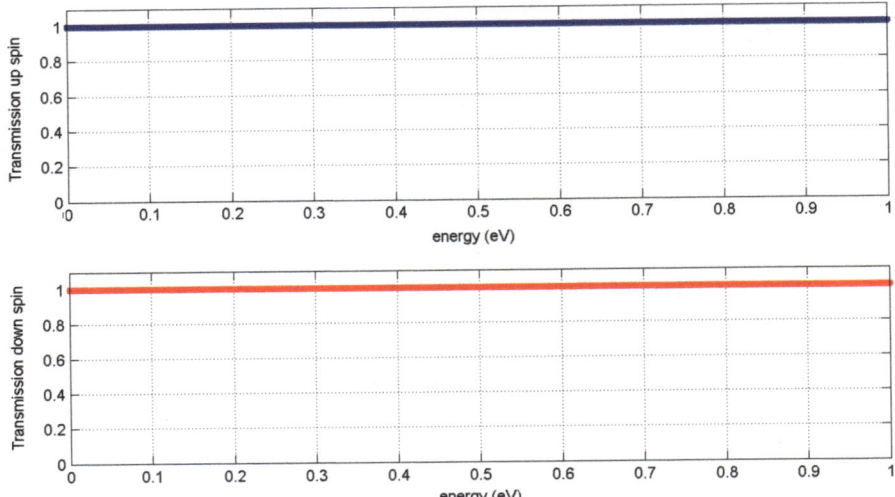

Fig. P6.5 Up-spin and down-spin electron transmission vs. energy through a channel without incoherent scattering, for a magnetic field acting parallel to the channel. Results are similar to Fig. E6.3 that was obtained for a perpendicular magnetic field

transmission coefficients do not depend on spin type. (d) Interface and bulk scattering: transmission coefficients and mean free paths are spin-dependent.

P6.8 From [39]: the spontaneous formation of a single-domain fixes the magnetization of one magnetic layer. This occurs as a result of nanopatterning, thereby pinning layers (e.g., FeMn) are no longer necessary. Magnetic multilayer samples with a NiFe(10 nm)/Co(1 nm)/Cu(13 nm)/Co(10 nm)/NiFe (2 nm) structure were deposited on a SiO_2 substrate using a dc sputtering system. The samples were patterned using nanoimprint lithography and an ion milling process. Consult [39] for further details.

P6.9 See for instance [40] for a detailed and interesting experimental analysis. Bulk and interface spin-asymmetry effects are responsible for the observed differences in GMR.

P6.10 From [41]: the degradation is not due to any intrinsic characteristic of graphene. The deviation from the linear dependence of spin signals occurs at high drain bias voltages which are necessary for spin injection. At these currents, the interface between the ferromagnet and graphene degrades due to two mechanisms: Joule heating and the existence of a tunneling barrier which is broken at a certain voltage, causing nonlinearity in signal. These mechanisms do not occur simultaneously.

P6.11 Consult [42].

P6.12 Consult [43].

P6.13 Consult [44].

P6.14 A large exchange bias is observed in $Ni_{50-x}Co_xMn_{38}Sb_{12}$ with $x = 0, 2, 3, 4,$ and 5. It increases with Co concentration. This exchange bias is attributed to the coexistence of two magnetic phases, ferromagnetic and antiferromagnetic in the metallurgical martensitic phase. A similar explanation is given for the exchange bias observed in $Bi_{0.4}Ca_{0.6}Mn_{1-x}Ti_xO_3$. A superposition of ferromagnetic and antiferromagnetic matrices leads to exchange bias which can be controlled through Ti and Mn doping. See [45] and [46] for further details.

P6.15 An electron with a certain spin polarization will traverse a Bloch wall and experience an exchange field which can cause the spin to flip. Thus Bloch walls produce a decoherence of electrons such that wall nucleation increases conductivity. The latter will fluctuate as walls themselves move. For further details consult [47] and bibliography therein.

P6.16 Consult [48].

P6.17 Spin valves were made of Ta/NiFe/CoNiFe/CuAu/CoNiFe/PtMn/Ta layers and built into conventional shield-type giant magnetoresistive heads. They were tested using a helical-scan tape drive. The spin valves showed a preservation of magnetic properties over 20 days. This occurred despite being exposed to corrosive gas, an estimated equivalent of environmental pollution a personal computer experiences over 10 years. See [49] and [50] for further details.

P6.18 Magnetic memories are fabricated with fewer processing steps than semiconductor-based memories. See [51].

P6.19 Spins leaving a magnetic semiconductor depend on the magnetization direction in the magnetic semiconductor. Thus, higher percentages in spin injection can be achieved. For further details consult [52] and [53].

P6.20 Some applications of "transpinnors" are nonvolatile data storage, and logic circuits for all-metal GMR magnetic random access memory (MRAM) chips (SpinRAM). A SpinRAM memory chip could be designed using only GMR "transpinnors" for word selection, digit selection, and differential sense amplifiers. See [54] and bibliography therein.

P6.21 According to Jullière's model, the maximum magnetoresistance is $2P_1P_2/(1 - P_1P_2) \approx 19\%$, if $P_1 = 25\%$ and $P_2 = 35\%$. In practice, smaller values are obtained due to reasons discussed in this chapter.

P6.22 A symmetric spin valve was investigated that had a typical configuration: unpinned ferromagnet/exchange-pinned ferromagnets on either side separated by two nonmagnetic spacers. It was observed that a 50% increase in GMR occurred, as compared to single spin valves. This response is attributed to the fact that the number of spin-dependent scattering interfaces was doubled, while the spin-independent scattering at the outer interfaces was reduced. See [55] for further details.

P6.23 The chemical vapor deposited Ni nanowire serves as the core around which sequential layers of CoO/Co/Cu/Co are sputter deposited as shells, forming a functional spin valve with a GMR of 9%. See [56] for further details.

P6.24 Artificial antiferromagnetic spin valves maintain their GMR ratio with time and at elevated operating temperatures as compared to standard spin valves. Thinner PtMn layers and of higher Pt content also display superior performance. See [57] for further details.

P6.25 By adjusting the impurity concentration, a synthetic ferrimagnetic pinned layer controls spin-dependent transport through a CPP spin valve. Calculations are performed based on the two current model, and validated through subsequent experiments. See [58] for further details.

References

1. M.N. Baibich, J.M. Broto, A. Fert, F. Nguyen van Dau, F. Petroff, Phys. Rev. Lett. **61**(21), 2472 (1988)
2. G. Binasch, P. Grunberg, F. Saurenbach, W. Zinn, Phys. Rev. B **39**, 4828 (1989)
3. P. Grünberg, R. Schreiber, Y. Pang, M.B. Brodsky, H. Sowers, Phys. Rev. Lett **57**(19), 2442 (1986)
4. S.S.P. Parkin, Giant magnetoresistance and interlayer exchange coupling in polycrystalline transition metal multilayers, in *Ultrathin Magnetic Structures*, ed. by B. Heinrich, J.A.C. Bland, vol. II (Springer, Berlin, 1994), p. 148
5. D.K. Ferry, S.M. Goodnick, *Transport in Nanostructures* (Cambridge University Press, Cambridge, 1997)
6. S. Datta, *Quantum Transport, Atom to Transistor* (Cambridge University Press, 2005)
7. M. Johnson, R.H. Silsbee, Phys. Rev. B **37**, 3312 (1988)
8. D.M. Edwards, R.B. Muniz, J. Mathon, IEEE Trans. Magn. **MAG-27** 3548 (1991)
9. V.S. Gornakov, V.I. Nikitenko, W.F. Egelhoff, R.D. McMichael, A.J. Shapiro, R.D. Shull, J. Appl. Phys. **91**(10), 8272 (2002)
10. W.L. Brown, M.F. Jarrold, R.L. McEachern, M. Sosnowski, G. Takaoka, H. Usui, I. Yamada, Nucl. Instrum. Method **B59/60**, 182 (1991)
11. C.Y. Duan, B.F.L. Ma, Z.Z. Wei, Q.Y. Zhang Jin, Chin. Phys. B **18**(6), 2565 (2009)
12. S. Maat, M.J. Carey, E.E. Fullerton, T.X. Le, P.M. Rice, B.A. Gurney, Appl. Phys. Lett. **81**(3), 520 (2002)
13. M. Patra, M. Thakur, S. Majumdar, S. Giri, J. Phys. Condens. Matter **21**, 236004 (2009)
14. L. Kong, Q. Pan, M. Li, B. Cui, S.Y. Chou, 56th Ann. Device Res. Conf. Digest 50 (1998)
15. A. Maeda, T. Tanuma, M. Kume, Mat. Sci. Eng. A **217**, 203 (1996)
16. F. Matsukura, E. Abe, H. Ohno, J. Appl. Phys. **87**(9), 6442 (2000)
17. M.L. Reed, N.A. El-Masry, H.H. Stadelmaier, M.K. Ritums, M.J. Reed, C.A. Parker, J.C. Roberts, S.M. Bedair, Appl. Phys. Lett. **79**(21), 3473 (2001)
18. T. Dietl, A. Haury, Y. Merle d'Aubigné, Phys. Rev. B **55**, R3347 (1997)
19. S. von Molnár, H. Munekata, H. Ohno, L.L. Chang, J. Magn. Magn. Mater. **93**, 356 (1991)
20. Y. Satoh, D. Okazawa, A. Nagashima, J. Yoshino, Physica E **10**, 196 (2001)
21. T. Dietl, H. Ohno, F. Matsukura, J. Cibert, D. Ferrand, Science **287**, 1019 (2003)
22. X.D. Peng, T.F.W. Zhu, W.G. Wang, Z.H. Huang Cheng, Chin. Phys. B **18**(6), 2576 (2009)
23. O.G. Heinonen, E.W. Singleton, B.W. Karr, Z. Gao, H.S. Cho, Y. Chen, IEEE Trans. Magn. **44**(11), 2465 (2008)
24. B.J. Qu, T.L. Ren, H.R. Liu, L.T. Liu, Z.J. Li, IEEE Sens. J. **5**(5), 905 (2005)
25. T. Kim, R.V. Chamberlin, P.A. Bennett, J.P. Bird, Nanotechnology **20**, 135401 (2009)
26. K. Dumesnil, C. Dufour, S. Fernandez, M. Oudich, A. Avisou, A. Rogalev, F. Wilhelm, J. Phys.: Condens. Matter **21**, 236002 (2009)
27. J. Qiu, P. Luo, K. Li, Y. Zheng, L.H. An, Y. Wu, IEEE Trans. Magn. **40**(4), 2260 (2004)
28. A.S. Edelstein, G.A. Fischer, M. Pedersen, E.R. Nowak, S.F. Cheng, C.A. Nordman, J. Appl. Phys. **99**, 08B317 (2006)

29. J.P. Sebastia, J.A. Lluch, J.R.L. Vizcaino, J.S. Bellon, IEEE Trans. Instrum. Meas. **58**(3), 707 (2009)
30. H. Li, J. Gaspar, P.P. Freitas, V. Chu, J.P. Conde, IEEE Trans. Magn. **38**(5), 3371 (2002)
31. H. Li, M. Boucinha, P.P. Freitas, J. Gaspar, V. Chu, J.P. Conde, J. Appl. Phys. **91**(10), 7774 (2002)
32. M.J. Carey, S. Maat, P. Rice, R.F.C. Farrow, R.F. Marks, A. Kellock, P. Nguyen, B.A. Gurney, Appl. Phys. Lett. **81**(6), 1044 (2002)
33. C. Fowley, B.S. Chun, J.M.D. Coey, IEEE Trans. Magn. **45**(6), 2403 (2009)
34. J.R. Childress, M.J. Carey, M.C. Cyrille, K.N. Carey, J.A. Smith, T.D. Katine, A.A.G. Boone, S. Driskill-Smith, K. Maat, C.H. Mackay Tsang, IEEE Trans. Magn. **42**(10), 2444 (2006)
35. J.F. Bobo, B. Warot-Fonrose, C. Villeneuve, E. Bedel, I. Séguy, IEEE Trans. Magn. **46**(6), 2090 (2010)
36. M. Johnson, Mater. Sci. Eng. B **31**, 199 (1995)
37. V. Cros, S.F. Lee, G. Faini, A. Cornette, A. Hamzic, A. Fert, J. Magn. Magn. Mater. **165**, 512 (1997)
38. C. Chapelier, M. El Khatib, P. Perrier, A. Benoit, D. Mailly, in *Superconducting Devices and Their Applications*, ed. by H. Koch, H. Lubbig (Springer, Berlin, 1991), p. 286
39. L. Kong, Q. Pan, M. Li, B. Cui, S.Y. Chou, Dev. Res. Conf. Digest **56**, 50 (1998)
40. T. Iwase, Y. Sakuraba, S. Bosu, K. Saito, S. Mitani, K. Takanashi, Appl. Phys. Exp. **2**, 063003 (2009)
41. K. Muramoto, M. Shiraishi, N. Mitoma, T. Nozaki, T. Shinjo, Y. Suzuki, Appl. Phys. Exp. **2**, 123004 (2009)
42. B. Georges, J. Grollier, A. Fukushima, V. Cros, B. Marcilhac, D.G. Crete, H.K. Kubota, J.C. Yakushiji, A. Mage, S. Fert, K. Yuasa Ando, Appl. Phys. Exp. **2**, 123003 (2009)
43. Q. Li-Jie, X. Xiao-Yong, H. Jing-Guo, Chin. Phys. B **18**(6), 2589 (2009)
44. S. Tehrani, M. Durlam, M. DeHerrera, E. Chen, J. Calder, G. Kerszykowski, Int. Nonvolatile Memory Techn. Conf. 43 (1998)
45. A.K. Nayak, K.G. Suresh, A.K. Nigam, J. Phys. D: Appl. Phys. **42**, 115004 (2009)
46. X. Zhu, Y. Sun, D. Shi, H. Lei, S. Zhang, W. Song, Z. Yang, J. Dai, S. Dou, J. Phys. D: Appl. Phys. **42**, 185001 (2009)
47. J.P.h. Ansermet, J. Phys.: Condens. Matter **10**, 6027 (1998)
48. Z.H. Xiong, D. Wu, Z.V. Vardeny, J. Shi, Nature **427**, 821 (2004)
49. M. Sekine, T. Watanabe, Y. Tamakawa, T. Shibata, Y. Soda, IEEE Trans. Magn. **42**(10), 2321 (2006)
50. D.W.D. Williams, Trans. Compon. IEEE, Hybrids Manuf. Technol. **11**(10), 36 (1988)
51. M. Plummer et al. (ed.), *The Physics of High Density Recording* (Springer, Berlin, 2001)
52. R. Fiederling, M. Keim, G. Reuscher, W. Ossau, G. Schmidt, A. Waag, L.W. Molenkamp, Nature **402**, 787 (1999)
53. R.P. Borges, C. Dennis, J.F. Gregg, E. Jouguelet, K. Ounadjela, I. Petej, S.M. Thompson, M.J. Thornton, J. Phys. D **35**, 186 (2002)
54. E.J. Torok, S. Zurn, L.E. Sheppard, R. Spitzer, S. Bae, J.H. Judy, W.F. Egelhoff Jr., P.J. Chen, IEEE Int. Magn. Conf. INTERMAG Europe, Digest Tech. Papers p AV8 (2002), ISBN 0–7803–7365–0
55. T.C. Anthony, J.A. Brug, S. Zhang, IEEE Trans. Magn. **30**(6), 3819 (1994).
56. K.T. Chan, C. Doran, E.G. Shipton, E.E. Fullerton, IEEE Trans. Magn. **46**(6), 2209 (2010)
57. S. Prakash, K. Pentek, Y. Zhang, IEEE Trans. Magn. **37**(3), 1123 (2001)
58. H. Oshima, K. Nagasaka, A. Jogo, T. Ibusuki, Y. Shimizu, A. Tanaka, IEEE Trans. Magn. **41**(10), 2929 (2005)

Chapter 7
Introduction to Spintronics

Abstract The past decade has witnessed a significant growth of new types of memory for nonvolatile data storage, as well as experimental devices for more efficient information processing [1]. Quite a few of these are based on an intangible property of electrons peculiar to quantum mechanics [2], and, for this reason, such devices are identified as *quantum devices*[3]. Many of these devices confine electrons to quantum-scale dimensions, creating what are known as quantum dots. The quantum mechanical property known as *electron spin* not only is key to the functionality of quantum devices but also underlies permanent magnetism [4]. Aligned spins suggest magnetism, whereas non-aligned ones indicate that the material is not magnetic. An applied current generating a magnetic field can align or randomize spins, allowing spin-based control.

Spin can be traced back to the magnetic moment of atoms, since atomic magnetic moments are determined by *electronic angular momenta*. Accordingly, each individual electron within an atom has an *angular momentum associated with its orbital motion*, and an intrinsic, or *spin angular momentum*. The latter plays a major role in today's emerging information processing technologies [5]. If, so far, electronic transistors have relied on the presence or absence of current to encode the ones and zeros of digital logic, a new type of transistors, dependent on spin, may do this more efficiently. Magnetic semiconductor quantum dots exploiting spin properties may lead to a new type of circuit based on magnetism instead of current flow.

The existence of spin or, more importantly, the statistics associated with it has permitted inventive designs of spin-based devices, particularly due to the *Pauli exclusion principle* that does not allow two electrons in an atom to occupy the same quantum state simultaneously [6]. This connection between spin degrees of freedom and electron charge imposed by the Pauli exclusion principle has added a different level of practicality to quantum devices, creating a new branch of electronics termed *spintronics* (i.e.,*spin electronics*). As an emerging field, spintronics has already had various degrees of experimental success in proving that information can indeed be encoded, transported, and stored using both electron charge and spin. It is hoped that a successful spintronic transistor will be capable of retaining its logic state in the absence of current while requiring less power to switch a bit. As such, it would

C.-G. Stefanita, *Magnetism*, DOI 10.1007/978-3-642-22977-0_7,
© Springer-Verlag Berlin Heidelberg 2012

reduce the electrical power required by a computer chip by what many believe to be as much as 99%.

Most practical challenges faced by spintronics, in particular those transistor-type applications, fall broadly into three categories: *injection of spin* in an "encoder", *control of spin transport* through a unity usually termed a "spacer", and *selective detection of spin* in the final element or "decoder". Another issue is temperature dependence and reliability. A practical device is expected to function at room temperature or above. However, many materials are ferromagnetic below certain temperatures. That means they retain their magnetic state in the absence of an applied field but lose that characteristic when they warm up. Therefore, many of these spintronic technologies are still in a research stage, and have yet to be commercialized. These topics will be addressed in this chapter, while particularly emphasizing developments where a significant amount of research seems to be concentrated nowadays.

7.1 Spin Injection

7.1.1 Ferromagnets as Data Encoders

Ferromagnetism originates in the magnetic moment of atoms with partly filled electron shells, where below a certain temperature termed the *Curie temperature*, uncompensated spins (i.e., in different quantum states) subject to a strong exchange coupling align parallel to each other. For instance, four uncompensated spins of the 3d electrons in an iron atom are mutually parallel, giving rise to an atomic magnetic moment. Figure 7.1 shows the schematic of an iron atom where the 26 electrons are divided into 4 principal electron shells, the innermost shell containing 2 electrons, the next shell 8, the next 14, and the outer shell 2. When completed, the first four shells of an atom contain 2, 8, 18, and 32 electrons, half of opposite spins. The spins of electrons in the three filled shells in an iron atom compensate each other such that these shells are magnetically neutral. However, in the third partly filled shell of the iron atom, there are four uncompensated electron spins that give rise to a permanent atomic magnetic moment. The atomic magnetic moments in a ferromagnet tend to align themselves parallel to each other and interact strongly with one another. This strong interaction is similar to an applied field, resulting in almost perfect alignment of the spins in spite of the thermal agitation, and, therefore, a spontaneous magnetization. The alignment of permanent magnetic moments in a ferromagnet is termed *ferromagnetic order* (Fig. 7.2a), and their effect is the presence of a strong inner magnetic field. When an external field acts on the ferromagnet, it changes the orientation of the spontaneous magnetization [7].

Paramagnetism is also attributed to unpaired electron spins; nevertheless, a different electron configuration allows these spins to be rather free to change their direction (actually, the spins can only take discrete orientations). For instance, electrons in the conduction band of many metals are considered to be responsible for

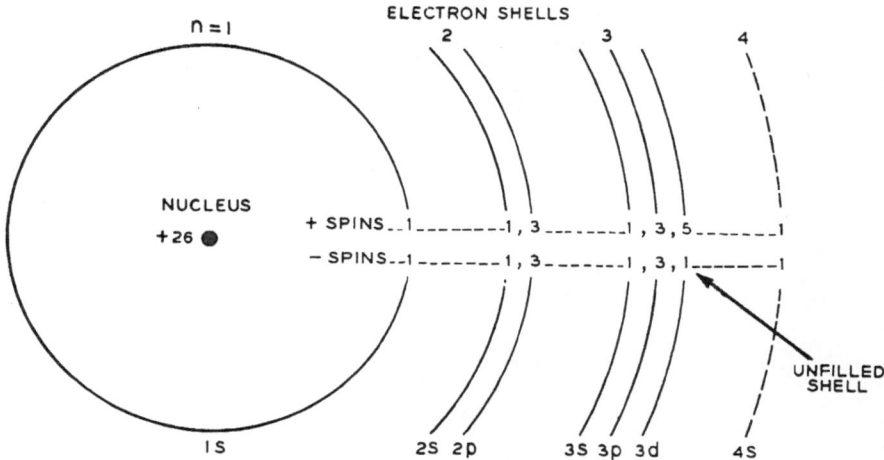

Fig. 7.1 Unfilled 3d electron shells in an iron atom giving rise to a magnetic moment. The 4 s electrons of isolated atoms become "free" in a metal. The 26 electrons are divided into 4 principal shells, the innermost shell containing 2 electrons, the next shell 8, the next 14, and the outer shell 2. Reprinted with permission from [7]

the paramagnetism displayed by these metals. Also, some complex salts containing Cr^{3+}, Fe^{3+}, Gd^{3+} ions, such as potassium chromium alum, ferric ammonium alum, and gadolinium sulfate octahydrate, are paramagnetic. The magnetic moments point in any direction and a strong thermal energy keeps them randomly oriented. Only an external magnetic field rotates the magnetic moments into the direction of the field, but the resulting magnetization is very small because it is opposed by strong thermal agitation. Only extremely high fields and very low temperatures come close to obtaining perfect alignment of moments and, therefore, saturation. The fields required to magnetize ferromagnetic substances are considerably weaker than those required for paramagnetic ones.

Example 7.1. The atomic magnetic moment is due to two sources, viewed classically as the orbital motion of the electron around the nucleus, and the spin motion of the electron about its own axis. It can be shown that the magnetic moment due to orbital motion and that produced by the spin of the electron differ in the ratio 1:2, where the factor $g = 1$ or $g = 2$ is called the *gyromagnetic factor* or the *g*-factor. Specifically, the magnetic moment M is given by

$$M = -g\frac{\mu_0 e}{2m}\Omega, \tag{E7.1.1}$$

where μ is the magnetic permeability of vacuum, e is the electron charge, m is the mass of the electron, and Ω is the angular momentum of the electron. By determining the ratio M/Ω, we can determine the value of g if it is close to 1 (orbital motion) or 2 (spin motion). Although classical physics cannot give a complete

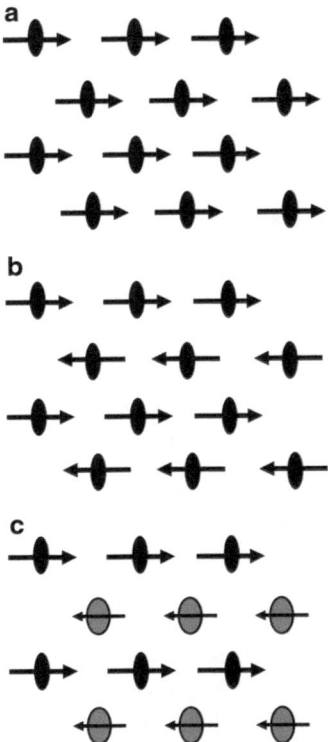

Fig. 7.2 Differences in spin ordering between: (**a**) ferro-, (**b**) antiferro-, and (**c**) ferrimagnetic materials below their Curie or Néel temperature[1]

description of the motion of the electron and the magnetic moment it produces, one can devise a simplified experiment based on classical principles and analogies (e.g., spinning top in a gravitational field) that shows which mechanism is responsible for ferromagnetism: electron orbital motion or electron spin. Show how you determine the *gyromagnetic factor* (*g*-factor) from this experiment. Historical experiments can be used as a reference.

Answer: For example, Chap. 3 in [8] can be used for guidance, and this approach is described below. The behavior of an atomic magnetic moment with angular momentum (whether orbital or spin) placed in an external magnetic field is similar to the behavior of a spinning top located in a gravitational field where it starts to precess (Fig. E7.1). This is the *gyromagnetic effect*. The moment of the couple which acts on a spinning top in a gravitational field is given by

---

[1] *These simplified sketches illustrate the basic idea, and should not be taken as accurate or complete representations of complex concepts.*

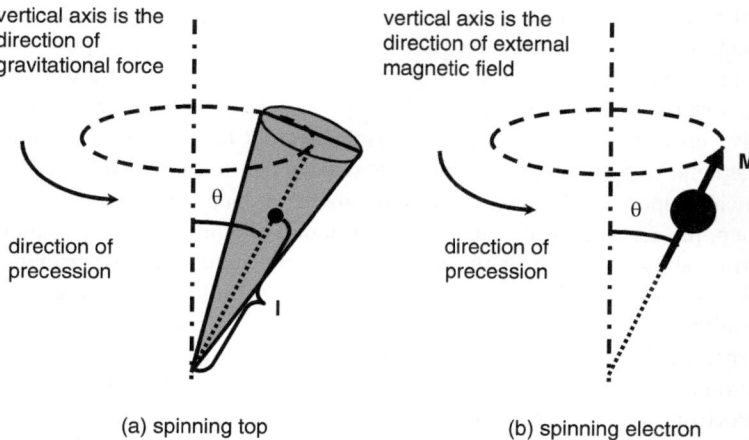

(a) spinning top   (b) spinning electron

Fig. E7.1 Precession of (a) spinning top in a gravitational field, and (b) spinning electron in a magnetic field

$$\Sigma = mgl \sin \theta, \qquad (E7.1.2)$$

where m is the mass of the top, l is the distance between the center of gravity and the bottom end of the axis, θ is the angle of inclination of the axis from the vertical direction, and g is the acceleration of gravity. Similarly, for a magnetic moment M in a magnetic field H, the moment of the couple is

$$\Sigma = -MH \sin \theta. \qquad (E7.1.3)$$

The two equations have the same dependence on θ.

The moment of the couple of forces acting on the top or electron magnetic moment is equal to the rate of change of angular momentum Ω

$$\frac{d\vec{\Omega}}{dt} = \vec{\Sigma}. \qquad (E7.1.4)$$

Whether the top is placed in a gravitational field or the electron is located in a magnetic field, the direction of Σ is always perpendicular to the axis of the top or the electron magnetic moment, and is also perpendicular to the direction of gravity or the magnetic field. The tip of the vector Σ draws a circle of radius $\Omega \sin \theta$ around the external agent, and the angular velocity ω of the precession is

$$\omega = \frac{\Sigma}{\Omega \sin \theta}. \qquad (E7.1.5)$$

When the top or the electron stop spinning, then they stop precessing and move in the direction of the external force. On the other hand, when a damping force acts during the precession motion, the top or the electron become partially influenced by that external force. Eventually, the top or electron will stop precessing as they give up to the damping agent a certain amount of angular momentum. In a gyromagnetic experiment, the g-factor is determined by exploiting the described dynamical properties. If the carrier of a magnetic moment, such as a ferromagnetic specimen, has an angular momentum, the rotation should induce an inclination of the carrier about the axis of the applied magnetic field. Any experiment where variables can be controlled such as to arrive at a measurable value for the ratio M/Ω based on the above relationships will allow determination of g. See, for example, the Barnett method, or the Einstein–de Haas experiment.

Contrary to ferromagnetism, antiferromagnetic materials exhibit an antiparallel electron spin arrangement, resulting in opposing total magnetic moments on different sublattices, thus giving rise to a total zero magnetization in the absence of an external field (Fig. 7.2b). This arrangement exists below a certain temperature termed *Néel temperature*, above which antiferromagnetic materials become paramagnetic. However, if a magnetic field is applied, the antiferromagnetic material displays a total nonzero magnetization that is similar to ferrimagnetism. This happens because the total net magnetization on one sublattice differs from that of the other sublattice. In contrast to antiferromagnets, a spontaneous magnetization exists in ferrimagnetic materials in the absence of an external field. This spontaneous magnetization is due to opposing moments on two different sublattices being unequal, for instance when the sublattices contain different materials or ions (Fig. 7.2c). Nevertheless, in a ferrimagnetic material, there is a certain point below the Néel temperature termed *magnetization compensation point* when the net magnetic moment resulting from both sublattices adds up to zero.

Example 7.2. A commonly employed magnetic material is permalloy $Ni_{78}Fe_{22}$ with $g \cong 2.07$ as determined from ferromagnetic resonance experiments. A magnetic field $H = 100\,A/m$ is required to magnetize a long, thin permalloy rod. Determine the angular velocity that is necessary to be achieved to magnetize the rod to saturation, if the rod is rotated around its axis (Fig. E7.2).

Answer: See previous example and relationships: $M = -g\frac{\mu_0 e}{2m}\Omega$; $\Sigma = -MH\sin\theta$; $\omega = \frac{\Sigma}{\Omega\sin\theta}$. From these relationships we arrive at

$$\omega = \frac{g\mu_0 eH}{2m} \tag{E7.2.1}$$

from which we determine $\omega = 7.2864 \times 10^6\,rad/s$ which corresponds to a frequency $f = 1.16\,MHz$.

Note: $e/m = 1.76 \times 10^{11}\,C/kg$; $\mu_0 = 4\pi \times 10^{-7}\,H/m$.

Magnetite is ferrimagnetic, while the protein ferritin and the related compound ferrihydrate found in sediments in nature display antiferromagnetic behavior

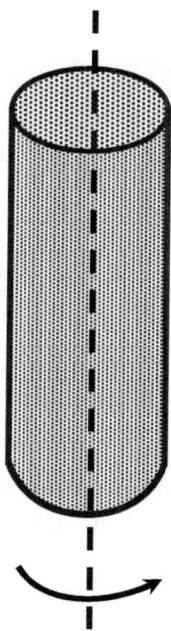

Fig. E7.2 A permalloy rod rotated about its axis in an external magnetic field that is applied along the vertical axis

[9]. The spin differences between ferromagnets, ferrimagnets, paramagnets, and antiferromagnets determine the role these materials play in spin-based quantum devices, and the choice for their employment is based on the amount of information available nowadays about these materials. For instance, magnetization can be thermally induced in some antiferromagnetic materials, a magnetization that increases further with temperature up to about the Néel temperature, and is dependent on particle size [10]. This is attributed to the fact that at low temperatures, the thermal energy is insufficient to induce a significant magnetization; therefore, the magnetic anisotropy energy dominates, which is not the case at higher temperatures. Additionally, thermoinduced magnetizations of antiferromagnetic nanoparticles are significantly larger than those of their bulk counterparts [11]. Some thermally induced magnetic anomalies reported in antiferromagnetic nanostructures are believed to be a result of changes with temperature in the uniform precession of the sublattice magnetization vectors about the net magnetization, as well as to transitions between the precession states causing magnetic moment fluctuations. In ferromagnetic nanoparticles, precessions occur with magnetic moments from electrons with parallel spins from one lattice, while in ferrimagnetic nanoparticles, precessions take place with magnetic moments from antiparallel electron spins from the two sublattices. In contrast to these, in antiferromagnetic nanostructures, the precessions of the two sublattice magnetizations M_{s1} and M_{s2} are at slightly

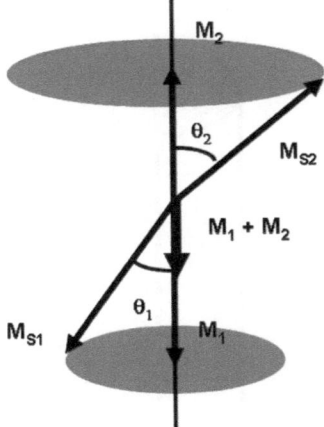

Fig. 7.3 Precession of sublattice magnetization vectors M_{s1} and M_{s2} about the net magnetization $M_1 + M_2$ with different precession angles θ_1 and θ_2 when thermally excited. After [9]

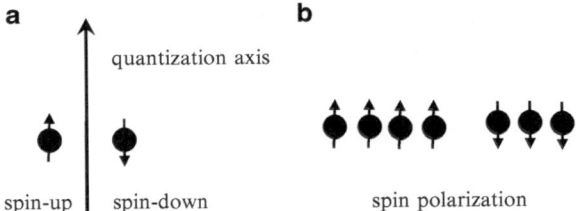

Fig. 7.4 Overly simplified sketches of: (**a**) spin-up and spin-down projection onto a given quantization axis, and (**b**) spin polarization, which shows that a certain number of spins of a particular orientation become available*

different angles θ_1 and θ_2 with respect to the net magnetization $M_1 + M_2$, when the nanostructures are thermally excited (Fig. 7.3). As a result, their precession frequencies are much higher than the precession frequencies of ferro- and ferrimagnetic resonances [12]. Furthermore, because the two angles are different when the uniform mode is thermally excited, antiferromagnetic nanostructures have a net magnetic moment. In bulk antiferromagnets, the probabilities for $\theta_1 < \theta_2$ and $\theta_1 > \theta_2$ are equal, and a large number of positive and negative contributions are added together so that the resulting net magnetization is vanishingly small. These facts have implications for the magnetic properties of antiferromagnetic nanostructures and their subsequent employment in today's devices. Even a very small amount of a strongly magnetic impurity phase can have a decisive influence on the magnetization of antiferromagnetic nanostructures because of their small magnetic susceptibility.

Spin permits an immediate differentiation between electrons, categorizing them into "spin-up" and "spin-down", depending on their $\pm 1/2$ spin projection onto a given quantization axis (Fig. 7.4a). This property is the basis for *spin polarization* (Fig. 7.4b), which is loosely a process through which a certain number of spins of a particular orientation become available, before they are injected through the quantum device. It is not only important to distinguish between spins, but also to ensure that a sufficient number of the same projection are present, as spin polarization is critical to encoding information. Yet, another process known as *spin flipping* can take place and cause several spins to change their orientation towards the opposite direction. Spin flipping is due to a variety of reasons, such as a fluctuating magnetic field caused by a varying electron density in the leads, or changes in the hyperfine coupling with nuclei, coupling influenced by lattice vibrations. Some of the reasons for spin flipping are discussed in subsequent sections. In any case, in many systems, the main source of spin flipping is the spin–orbit interaction which depends strongly on the crystal symmetry [13].

Example 7.3. The demagnetizing field H_{demag} due to a magnetization field density B normal to the axis of a ferromagnetic rod (see Fig. E7.3.1) is

$$H_{demag} = N_{demag}\frac{B}{\mu_0}. \tag{E7.3.1}$$

The demagnetizing factor is $N_{demag} = 1/2$ for a very long cylinder. In a ferromagnetic resonance experiment, a microwave with a 1 cm wavelength is applied perpendicular to the axis of precession of the atomic magnetic moments to maintain its precession. How strong should a magnetic field be, applied along the vertical axis of the ferromagnetic rod in order to obtain ferromagnetic resonance while taking into account demagnetization effects? Assume the rod is permalloy 78Ni 22Fe with $g \cong 2.07$ and $B_s = 1.1\,\text{Wb/m}^2$.

Answer: If the frequency of the microwave coincides with the precession frequency, ferromagnetic resonance is achieved. However, if demagnetizing effects take place, they need to be taken into account, and a magnetic field larger than the demagnetizing field needs to be applied. The demagnetizing field $(B \sin\theta)/2\mu_0$ produced

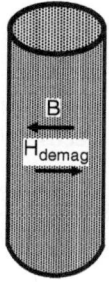

Fig. E7.3.1 Demagnetizing field H_{demag} in a long, ferromagnetic cylinder that occurs in response to an applied magnetic field density B acting perpendicular to the axis of the rod

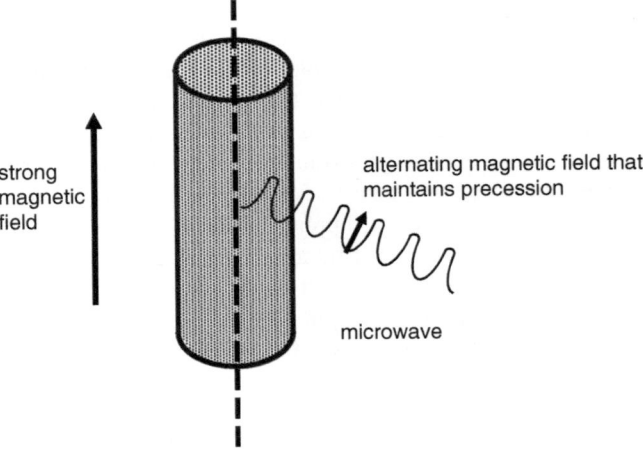

Fig. E7.3.2 In a ferromagnetic resonance experiment, in order to maintain the precession motion of the atomic magnetic moments about the vertically applied magnetic field, it is necessary to supply energy to the system. This is because the crystal lattice dampens the precession motion. An alternating magnetic field is applied normal to the axis of precession to ensure that the atomic magnetic moments keep precessing about the vertical magnetic field

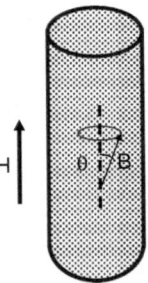

Fig. E7.3.3 A demagnetizing field B is produced during the ferromagnetic resonance experiment in the permalloy rod inclined at an angle θ with respect to the vertical axis. For this reason, a stronger magnetic field H needs to be applied to compensate for the demagnetization effects

in the permalloy rod increases the couple of force acting on the atomic magnetic moments. Hence, relationship 5P 6.3 needs to be amended as follows (Fig. E7.3.2):

$$\Sigma = -MH \sin \theta - \frac{MB_s}{2\mu_0} \cos \theta \sin \theta, \tag{E7.3.2}$$

where B_s has been used as it is the saturation, and, therefore, the maximum magnetic field density that can be attained in the ferromagnetic rod. Since θ is very small

$(\theta \ll 1 \text{ rad})$, we have $\cos \theta \cong 1$ (Fig. E7.3.3); therefore,

$$\Sigma = -M \left(H + \frac{B_s}{2\mu_0} \right) \sin \theta. \qquad (E7.3.3)$$

The resonance frequency becomes

$$\omega = g \frac{e\mu_0}{2m} \left(H + \frac{B_s}{2\mu_0} \right), \qquad (E7.3.4)$$

where the relationships from previous examples have been used. This yields $H = 21.49 \times 10^5 \text{ A/m}$, which is a strong magnetic field. Thus, demagnetizing effects are really significant and require a strong field to be overcome.

Ferromagnetic metals have the useful property of possessing available "spin-up" and "spin-down" electrons; nevertheless, they exist in two different channels (or sub-bands). This property is described theoretically in Mott's model [14] that dates back to the 1930s. It was since established experimentally that spin flipping occurs between the two spin channels within the ferromagnet, resulting in spins losing their initial orientation. Even so, spin flipping between channels is often neglected, given the short time scales of all the other processes in the system [15]. Nonetheless, the density of spin states in the two spin channels of the ferromagnet differs, so that the corresponding electrons do not equally contribute to the electrical current. Research shows that the electrical current that flows through the device originating in the ferromagnetic material is primarily due to those electrons with a lower density of spin states at the Fermi level. The low density of spin states leads to these electrons being known as *minority spin carriers* [16], as opposed to *majority spin carriers* corresponding to those with a higher density of states. Minority spin carriers constitute a fair supply of spins when the current passes from a ferromagnet to a nonmagnetic material (i.e., the spacer).

The natural imbalance at the Fermi level in the number of "spin-up" and "spin-down" electrons encountered in ferromagnets allows encoding of information in these materials. Therefore, the first unit in a quantum device, the encoder, is

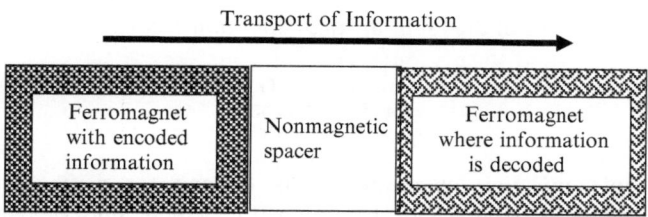

Fig. 7.5 Simplified sketch of transport of information through a trilayer structure consisting of two ferromagnets and a nonmagnetic spacer (usually a semiconductor). Notice the presence of the two interfaces which are actually two heterojunctions, if the first and third unit are metallic, and if the middle unit is a semiconductor

made of a ferromagnet (Fig. 7.5). The coding can be changed by remagnetizing the ferromagnet containing the information. Encoding is done through entities termed *qubits* (i.e., *quantum bits*) which are coherent superpositions of "spin-up" and "spin-down" states. A qubit involves two Zeeman sublevels of the same orbital level. During spin transport, qubits travel from the ferromagnet that serves as the encoder through a second unit, which is a nonmagnetic spacer that carries this information further. The qubits are detected and decoded in a third unit. Because of the superposition of quantum states existent in qubits, the data in spin-based quantum devices can be handled a few orders of magnitude faster than in existing microprocessors centered solely around charge [17], assuming that a large number of qubits are accessible. Therefore, hope exists that quantum computing will outperform classical computers once the experimental challenges associated with implementing quantum dots into electrical circuits are overcome.

7.1.2 Spin Injection Rate

To transport spin through a quantum device, it is necessary to first inject it. Spintronic devices rely on the injection of spin from a ferromagnet into a nonmagnetic material (usually a semiconductor), and this injection occurs at a certain rate at the interface. The *spin injection rate* at a single (i.e., one interface) ferromagnetic/semiconductor heterojunction is given by the ratio of the *net magnetization current* j_M to *electric current* j_e passing through the interface under an applied voltage, where μ_B is the Bohr magneton:

$$\frac{j_M}{j_e} = \eta_M \frac{\mu_B}{e}. \tag{7.1}$$

Relationship (7.1) is valid under the assumption that no spin-flip scattering or no spin precession takes place at the interface. The degree of spin polarization of the net electron flux through the interface is described by the interfacial transport parameter known as the *spin injection efficiency* η_M. The spin injection efficiency can be written in terms of the *"spin-up"and "spin-down"conductances* $G_\uparrow$ and $G_\downarrow$:

$$\eta_M = \frac{G_\uparrow - G_\downarrow}{G_\uparrow + G_\downarrow}. \tag{7.2}$$

A similar *spin injection efficiency* η'_M is defined for a *double* ferromagnetic/semiconductor/ferromagnetic heterojunction, taking into account the *total spin conductances* $G_\uparrow^{tot}$ and $G_\downarrow^{tot}$:

$$\eta'_M = \frac{G^{tot}_\uparrow - G^{tot}_\downarrow}{G^{tot}_\uparrow + G^{tot}_\downarrow}. \tag{7.3}$$

The elastic multiple scattering at the interfaces leads to *"spin-up"and "spin-down"transmission* probabilities $T_\uparrow$ and $T_\downarrow$ that result [13] in spin injection efficiency η'_M:

$$\eta'_M = \frac{T_\uparrow - T_\downarrow}{T_\uparrow + T_\downarrow - T_\uparrow T_\downarrow}. \tag{7.4}$$

A net spin current flows through the two interfaces when $T_\uparrow \neq T_\downarrow$. It is possible to experimentally observe spin injection by using a *spin valve* configuration, as discussed further below.

It should be mentioned that there is more than one possibility of achieving injection of spin. This is because electronic carriers of a material can transfer their angular momentum to a ferromagnet through a spin-polarized current originating in the first material, even without an external field. Because it results in a reorientation of magnetization direction in the ferromagnet, this effect is equivalent to applying a torque on its spins; therefore, it is termed [18] a *spin transfer torque*. The spin torque is created by the spin-polarized current of the first material, and is a reaction to a process known as [19] *spin filtering*, because spin-polarized electrons interact selectively with the ferromagnet. Spin filtering implies that incoming electrons with spin components perpendicular to the magnetic moment in the ferromagnet are being filtered out. As a result, this perpendicular spin is transferred as a torque. The latter is exerted on the moment due to conservation laws that have to take place. Thus, the mechanism for spin transfer involves a spin filtering process without which the transfer would not occur. In any case, the ferromagnet has to be thin, as the requirements for this process to take place are not fulfilled in bulk materials [20]. If an occurrence known as *spin transmission resonance* takes effect, the magnetic moment in the ferromagnet does not change, leading to no spin transfer.

Efficient spin injection is sometimes hindered because of a loss of spin polarization at the two material interfaces where spin flipping is likely to occur [21]. Although the two materials may be closely lattice-matched, the interface is, in most cases, still far from ideal. Unfortunately, even if only a few ferromagnetic atoms diffuse into the nonmagnetic side of the interface, they can already pose problems, as the diffused atoms scatter electrons between the two spin channels. Scattering is likely to occur because the diffused ferromagnetic atoms will carry a local magnetic moment randomly oriented with respect to the magnetization direction. That moment will interact with the moment of the electrons crossing the interface.

One strategy for avoiding spin flipping is to employ a type of material known as a *half-metallic ferromagnet*, as this material does not provide final states for spin flipping. Alternatively, an intermediate layer of some material may be introduced between the ferromagnet and the spacer. This intermediate layer could maintain the spin polarization, and pass it on to the actual spacer layer. The additional layer may just be a tunneling barrier or a Schottky barrier whose depletion zone constitutes the intermediate layer. If the Schottky barrier is sufficiently extended, it can reconcile the dissimilarities in electrochemical potentials between the metal and semiconductor, thereby improving spin injection efficiency. Conversely, if a nonmagnetic tunneling barrier is used, the voltage drop across the barrier controls the injected current and its polarization. Spin flipping may still occur; nevertheless, the injected polarization is, in this case, independent of the diffusion constants [22].

Rashba [23] suggested using a process termed *spin tunnel injection* to avoid the interface problems related to the continuity of the electrochemical potential at the metal/semiconductor interface. In this case, a tunnel contact between a ferromagnet and a semiconductor would act as a spin selector, based on the spin polarization of the density of states. The interface barrier spin polarizes the tunnel current sustaining the difference in chemical potential between the "spin-up" and "spin-down" bands at the interface. Nevertheless, the possibility exists that a very thin or discontinuous tunnel barrier could worsen the efficiency of the spin injection.

7.1.3 The Concept of Spin Diffusion Length

Obtaining spin polarization for spin injection implies generating first a nonequilibrium spin population [24]. This is quite often achieved through optical techniques [25] where polarized photons transfer their angular momentum to electrons. On the other hand, proponents of spintronics prefer an electrical spin injection realized by connecting a magnetic electrode to a specimen. In this case, a current from the electrode sends spin-polarized electrons to the device, bringing about what is known as a nonequilibrium *spin accumulation*. However, spins have the tendency to revert back to an equilibrium state through a process termed *spin relaxation*. Spin relaxation results in an exponential spin accumulation *decay* away from the interface where spins had first accumulated.

The process of spin relaxation away from the interface takes place over a period of time of picoseconds to microseconds [26, 27]. The distance into the device over which spin relaxation occurs is termed *spin diffusion length*. This spin diffusion length l_{sd} is estimated to be given by the following relationship [28]:

$$l_{sd} = \sqrt{\frac{v_F \tau_{\uparrow\downarrow} \lambda}{3}}, \tag{7.5}$$

where v_F is the Fermi velocity, $\tau_{\uparrow\downarrow}$ is the spin-flip time, and λ is the mean free path.

For spin accumulation to happen at all, it is necessary to counterbalance the spin relaxation by a particular spin accumulation rate at the interface [29]. This rate has been calculated and was found to be expressed in terms of the spin density n at distance x from the interface:

$$n = n_0 e^{-x/l_{sd}}, \tag{7.6}$$

where n_0 is the spin density at the interface. A volume integration of (7.6) gives the total number of spins in the accumulation region with a spin accumulation rate of $n_0 l_{sd}/\tau_{\uparrow\downarrow}$. It should be noted that a total accumulated spin density of $10^{22}\,\mathrm{m}^{-3}$

is estimated in a quantum device, compared to a total available electron density of about 10^{28} m^{-3}.

The above results show that the available spin density stands for only one part in a million of the total electron density; thereby, measuring spin density is quite a difficult process. Furthermore, the spin diffusion length is reduced by impurities that shorten the mean free path and the spin-flip time. These facts explain part of the problems associated with a successful spin injection process, and why so much research is dedicated to this particular area of spintronics. In the end, the transport of data through a device is possible only if polarized spins are injected successfully into the device, and the length of the device can accommodate the short spin diffusion lengths. To complicate matters further, other conditions also need to be met, such as maintaining *spin phase coherence* during the voyage of the spins.

Since spins are "attached" to mobile electrons, they travel together along conduction paths. However, based on earlier observations regarding spin diffusion lengths, and the necessity of maintaining spin phase coherence for the electronic wave function to be preserved, a key requirement for quantum devices is to be built on short-length scales. Spin transport is influenced by various spin decoherence mechanisms because, unlike charge, spin is a nonconserved quantity. The spin polarization of electrons or holes is maintained as long as the electrons or holes do not come across a magnetic impurity [30] or interact with the lattice via spin–orbit coupling. At the same time, collisions with magnons may occur, which cause momentum transfer between spin channels, a process often referred to as *spin mixing*. Nevertheless, the interaction between spin and the transport environment needs to be controlled to maintain spin polarization. Solely, *mesomagnetic* length scales of somewhere between tens and thousands of angströms allow control of spin-polarized electrical conduction and therefore satisfy these requirements. The term *meso* indicates lengths that are large compared to the atomic scale; however, they are smaller than the scale on which the Boltzmann transport theory is applied. Nanotechnology holds promise of constructing functional devices on these scales; nevertheless, more research is necessary [1, 31].

7.2 Choice of Materials for Spintronic Devices

7.2.1 Integrating New Techniques into Traditional Manufacturing

A careful selection of materials for the production of spintronic devices is required, as not only their type, but also the material processing routes will determine their functionality. Additionally, adapting manufacturing to existing fabrication methods ensures a more likely transition from the traditional semiconductor industry to novel technologies. For this reason, magnetic semiconductors belonging to group IV have caught researchers' attention, as this unique class of semiconductors is potentially compatible with Si-based devices. Yet, complex fabrication requirements involving high doping levels for the magnetic semiconductors or elevated growth temperatures could create undesired processing problems not normally encountered for more traditional materials. For instance, phase separations may happen in magnetic semiconductors, causing them to be inhomogeneous. On the other hand, doping is unavoidable, as it is crucial to acquiring certain magnetic characteristics. Some new properties may be gained as well. In general, doping magnetic semiconductors with several elements is preferred over doping with a single one, because any introduced element alters the local kinetic and strain conditions. As a matter of fact, X-ray diffraction measurements show that strains developing in the grown magnetic semiconductor film are linearly dependent on the concentration of Co and Mn atoms [32]. Other studies suggest that Co in Ge contracts the lattice, while doping with Mn expands it [33]. Nevertheless, strain has been noticed to stabilize growth in Co-rich samples because of the small lattice mismatch. Furthermore, electric and magnetic properties of Co-rich $(Co_a Mn_b)_x Ge_{1-x}$ epitaxial films are seen to be controlled by the doping concentration, while both electrons and holes are likely to be spin polarized. If double spin polarization is further confirmed, it would open new avenues for heterojunction device applications.

Traditional fabrication methods notwithstanding, some lesser known and usually small-laboratory-confined growth techniques have started to gain steadily more ground, and enter the spotlight. They offer advantages hard to ignore. For instance, *oxygen-plasma-assisted-molecular-beam-epitaxy*-grown $Co_x Ti_{1-x} O_{2-x}$ usually results in a high concentration of oxygen vacancies that allow thermally driven redox reactions to go on [34]. Consequently, Ti becomes fully oxidized at the expense of Co, and Co nanoparticles form. Additional spectroscopic investigations show O vacancies to be effectively bound to substitutional Co, while maintaining its $+2$ oxidation state. The advantage of the material fabricated this particular way is its great thermal stability, demonstrated by the fact that it does not change its magnetic properties when annealed in vacuum at 825 K. It stands in contrast to *pulsed-laser-deposition*-grown $Co : TiO_2$ that displays reduced thermal stability.

Another technique termed *combinatorial-laser-molecular-beam-epitaxy* has been specifically developed for the rapid synthesis of samples with compositional variations. Differences in composition can easily be obtained in this high-throughput synthesis method, facilitating an efficient search for advanced functional materials, while identifying structure–composition–property relationships not otherwise easily obtainable. The technique was applied to the study of new magnetic oxide semiconductors, such as ZnO and TiO_2 doped with 3d transition metals. Results showed that none of the 3d transition metals doped into ZnO films gave rise to ferromagnetic behavior. Conversely, some studies showed that Co-doped ZnO grown on sapphire substrates is slightly ferromagnetic up to [35] $\sim$350 K, while Co mole fractions of up to 0.35 were obtained. Too much Co doping would result in losing those ferromagnetic properties. As an interesting fact, oxygen vacancies were observed to be responsible for n-type conductivity. Co(II) was found to substitute for Zn(II) in the ZnO lattice. The related Co-doped TiO_2 compound was observed to display the highest saturation and remanent magnetization, as well as the highest Curie temperature ($\sim$700 K) among all half-metallic ferromagnetic oxides. Furthermore, measurements showed transparency in the visible and near-infrared regions for a $Co_{0.06}Ti_{0.94}O_2$ film.

7.2.2   Half-Metallic Ferromagnets

There is a group of materials known as *half-metallic ferromagnets* [36] that has the interesting property of possessing only one spin channel with unfilled states at the Fermi level, even in the absence of a magnetic field. The other spin channel has an energy that is not at the Fermi level, and the spin channel is either entirely full or completely empty [37]. The classification of half-metallic ferromagnets contains subgroups, such as *Heusler alloys* [38], *ferromagnetic oxides* [39], and some *diluted magnetic semiconductors* (DMS), each with its own characteristics.

In the first mentioned subgroup, the Heusler alloys have a cubic crystallographic structure with four interpenetrating *fcc* sublattices, an arrangement that gives them certain magnetic properties. This class of materials is considered to include compounds such as Co_2MnSi, $NiMnSb$, Co_2MnGe, and Co_2FeSi. On the other hand, the discovery of unexpected properties in the second mentioned subgroup, the half-metallic ferromagnetic oxides, has created a general interest in these materials all over the world. Half-metallic ferromagnetic oxides have itinerant electrons as is the case for CrO_2 and Sr_2FeMoO_6, or localized electrons as in Fe_3O_4, rendering them unique characteristics. Of course, the type of fabrication technique is key in further determining what properties a certain compound will have.

The third subgroup, the DMS containing compounds such as the ferromagnetic phase of (Ga,Mn)As [40] or (In,Mn)As [41], is characterized by the spontaneous splitting of the valence band into "spin-up" and "spin-down" sub-bands [42]. The DMS are semiconductors, initially nonmagnetic, subsequently doped with a few percent of magnetic impurities. Depending on the percentage of the dopant and the complex processing involved, these materials can exhibit ferromagnetism at and above room temperature. However, what sometimes appears to be ferromagnetism in a DMS is, in fact, just an effect given by a secondary magnetic impurity phase.

7.2.3   Magnetic Semiconductors

Semiconductors present advantages in that they can be processed to high purity, the equilibrium carrier densities can be varied through doping, and electronic properties are tunable by gate potentials. Additionally, depletion zones, voltage blocking, diffusion currents, the tunnel effect, as well as conduction in two-dimensional electron gases [43] (2DEG) can sometimes be exploited in the design of spintronic devices. For all these reasons, magnetic semiconductors are considered promising candidate materials for spintronic applications. Recent advances in the fabrication of high-quality magnetic semiconductors, as well as improvements in interface matching, have led to semiconductor compounds being preferred over ferromagnetic metals. Injection of spin-polarized carriers from a magnetic semiconductor into a "normal" semiconductor results in a polarization of the injected carriers that depends on the magnetization direction in the magnetic semiconductor. A better injection is obtained with carriers from a semiconductor than from a ferromagnetic metal. However, magnetic elements have a low solubility in III–V semiconductors, only of the order of [44] $\sim 10^{18}\,cm^{-3}$. Their concentrations need to be well controlled to ensure that spin polarization is obtained [45]. An undesired effect is the formation of additional phases that impede achieving adequate spin polarization. Techniques such as molecular beam epitaxy or laser ablation have been used to precisely introduce transition metals into the semiconductor matrix.

(Ga,Mn)As [46] and manganite perovskites [47] have demonstrated in some studies a high spin polarization assumed to be due to Mn ions. The latter are electrically neutral in II–VI compounds, but act as an acceptor in III–V semiconductors because of the large density of states in the valence band [48]. Consequently, holes contribute to spin injection in spite of their lower mobilities and shorter lifetimes as compared to electrons. In a study by Tanaka et al., tunneling magnetoresistance values of $\sim 70\%$ were demonstrated in GaMnAs/AlAs/GaMnAs structures [49]. However, these high TMR values were only attainable at very low temperatures (8 K).

The large spin polarization of holes due to the strong spin–orbit coupling in the valence band of (Ga,Mn)As is assumed responsible for the *giant planar Hall effect* observed in an epitaxial thin film when an in plane magnetic field is applied (Fig. 7.6). Two abrupt changes are observed in the planar Hall resistance at distinct magnetic field values, while the longitudinal resistance also experiences small sudden changes [50]. These changes seem independent of sample size or geometry. In any case, the smaller specimens show Barkhausen jumps (Fig. 7.6f) because the propagation of domain walls is limited by specimen geometry. Also, a spontaneous transverse voltage has been noticed when a longitudinal current flows through a (Ga,Mn)As specimen, even if no field is applied.

Another way of inducing spin polarization in magnetic semiconductors is through applied electric fields, by controlling ferromagnetic interactions due to spin–orbit coupling. An electron spin polarization has been induced electrically in n-type ZnSe layers, polarization that depended on doping concentration [51]. In this case, an in plane spin polarization and a transverse spin accumulation were produced by using solely electric fields, an effect known as the *spin Hall effect*. Spin polarization was determined by measuring the Kerr rotation angle with an in plane magnetic field applied perpendicular to the laser propagation direction. An electron g-factor of 1.1 and a spin coherence time of up to 50 ns were measured at low temperatures (~ 5 K) and zero field. The low value for the g-factor indicates a strong orbital contribution to the magnetic moment. Also, it was determined that the spin mobility was over an order of magnitude smaller than the electron mobility for that ZnSe sample.

Most methods proposed for controlling electron spin rely on time-dependent magnetic fields, which is not always convenient. Therefore, the idea of using electric fields for spin manipulation is particularly appealing, since it is more practical. In a semiconductor quantum well, electron spins couple to an electric field because of

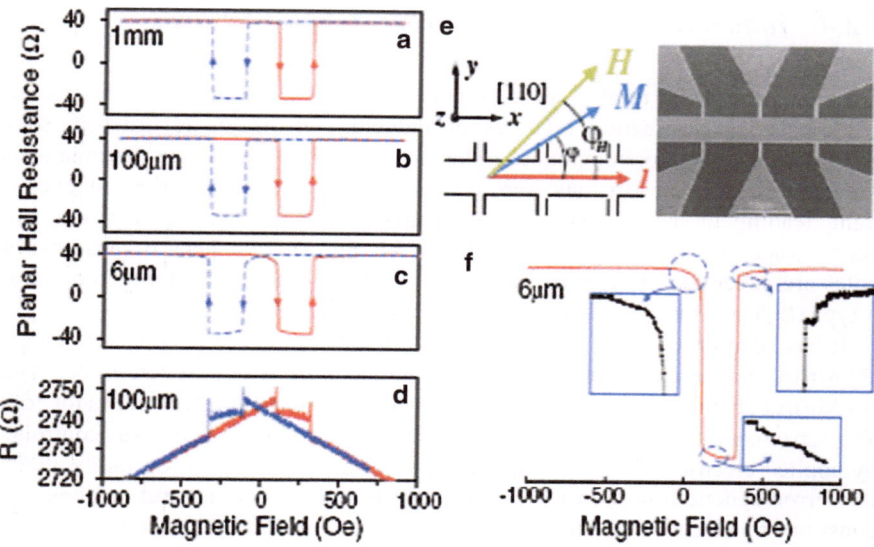

Fig. 7.6 (**a**) to (**c**) Planar Hall resistance in three specimens of different size obtained at 4.2 K as a function of an in-plane magnetic field at 20° fixed orientation; (**d**) field-dependent sheet resistance of the 100 μm wide Hall specimen; (**e**) Relative orientations of sensing current I, external field H, and magnetization M. An SEM image of the 6 μm specimen is also shown; (**f**) Barkhausen jumps observable in the 6 μm specimen near the resistance transitions. Reprinted from [50] with permission from the American Physical Society. URL: http://link.aps.org/abstract/PRL/ v90/e107201;DOI:10.1103/PhysRevLett.90.107201

the spin–orbit interaction, as Rashba et al. [52, 53] predicted theoretically. In their calculations, they showed that an external in-plane electric field as small as 0.6 V/cm can be used to efficiently manipulate electron spin. In addition, this coupling should be significantly stronger than the coupling to an ac magnetic field; however, these predictions have yet to be confirmed practically.

A challenge faced by spintronics is how to maintain spin polarization at room temperature [54]; therefore, any results regarding room temperature ferromagnetism are important. Encouraging findings have been reported by a few research groups that found high-temperature ferromagnetism in (Ga,Mn)N [55, 56]. Experimen-tal findings have been supported by theoretical predictions of a Curie point above room temperature for Mn doped p-type semiconductors such as (Ga,Mn)N [57]. To raise the Curie temperature, several research groups have undertaken to grow magnetic III–V/II–VI semiconductor material superlattices [58] believed to exhibit superior magnetic properties. Despite the limited functionality given by low operating temperatures and other challenges posed by materials problems, magnetic semiconductors hold promise as spin injectors in the near future. The fact that ferromagnetism can be induced using Mn doping in zinc-blende III–V compounds, or II–VI semiconductors has opened up new branches of materials studies.

7.2.4 Inducing Ferromagnetic Order in Some Systems

Ferromagnetic order implies an alignment of uncompensated spins. Various laboratories are investigating methods of inducing ferromagnetic order in some popular semiconductor compounds or heterostructures. For instance, ferromagnetic interactions between Mn ions were observed to intensify under the influence of light, leading to ferromagnetic order in III–V diluted magnetic semiconductor heterostructures such as (In,Mn)As/GaSb [59, 60]. This happens because excess holes are generated by light in the (In,Mn)As layer, making it p-type. In contrast, n-type (In,Mn)As is paramagnetic [61].

It has been observed that electric fields are able to vary hole concentrations in (In,Mn)As and reversibly change the Curie temperature. This was determined in an (In,Mn)As channel that was part of an insulating-gate field-effect transistor [62]. The ferromagnetic exchange interaction among localized Mn spins was controlled by applied electric fields that varied the hole concentration and reversibly altered the ferromagnetic transition temperature. Such findings are hopeful, as they may constitute one step forward toward the integration of ferromagnetic semiconductors with nonmagnetic III–V semiconductor devices such as lasers.

Mn-doped β-FeSi$_2$ single crystals display similar magnetic characteristics as ferromagnetic semiconductors, judging by the data obtained on the magnetic field dependence of the magnetization and of the Hall resisitivity [63]. These characteristics are more visible at lower temperatures (<40 K), and some vanish at room temperature, as is the case with the remanent magnetization. A small anomalous Hall contribution was still observable at about room temperature. In contrast, low temperatures showed large anomalous Hall coefficients. As far as the observation of an anomalous Hall effect in the Mn-doped β-FeSi$_2$ single crystals is concerned, it should be noted that the latter can occur in any material containing large localized magnetic moments, and is not a characteristic of ferromagnetic materials only. This is because the anomalous Hall effect is attributed to scattering of charge carriers upon irregularities in the arrangement of localized magnetic moments. On this basis, the anomalous Hall coefficient provides some insight into the processes responsible for the asymmetric scattering of conducting carriers. Consequently, the differences observed in the behavior of the anomalous Hall coefficient for β-FeSi$_2$ single crystals with various concentrations of Mn atoms led to the conclusion that long-range ferromagnetic ordering takes place in the specimens with 0.42 at%, as opposed to specimens with lower Mn concentration (e.g. 0.11 at% Mn). The β-FeSi$_2$ single crystals were p-type, showing once more that p-type materials can be magnetic.

Some concepts that were discussed in preceding sections are applicable to *spin transport* as well, because they are related to the idea of manipulating spin dynamics. A variety of properties of spintronics materials are proof that transport of spin can occur, and furthermore, it can be controlled. The sections that follow constitute a closer look at the manipulation aspect of spin transport without which the development of spintronics devices would not be possible.

7.3   Control of Spin Transport

7.3.1   Polarize/Analyze Experiments

Various geometrical arrangements of layers are being used in spintronics to study the efficiency of spin injection, filtering, transport and detection. Among these arrangements are the "polarize/analyze" geometries consisting in principle of two magnetized ferromagnets separated by a nonmagnetic spacer layer. A change in voltage occurs when the magnetizations in the two ferromagnets reorient from parallel to antiparallel. This is caused by the transport of spin through the layers. The first ferromagnet serves as a "polarizer" of spin, while the second is a spin "analyzer"; however, the way the layers are arranged leads to significant differences in what kind of results are obtained. Julliére's study [64], reported in 1975, was the first demonstration of "polarize/analyze" experiments without using superconductors. His arrangement employed two metal ferromagnets and a semiconductor spacer, Fe/Ge/Co, while measurements took place at temperatures of 4.2 K.

The more frequently utilized "polarize/analyze" geometry is *current-perpendicular-to-plane* (CPP) (Fig. 7.7 bottom). This geometry is the most referred to in mathematical treatments [19], while it also constitutes the discussion basis for many of the relationships mentioned earlier. Nevertheless, the *current-in-plane* (CIP) (Fig. 7.7 top) is the geometry that was historically employed first due to practical reasons [65]. Spin accumulation cannot be achieved in a CIP geometry; therefore, the *critical length scale* is not the spin diffusion length as seen in the

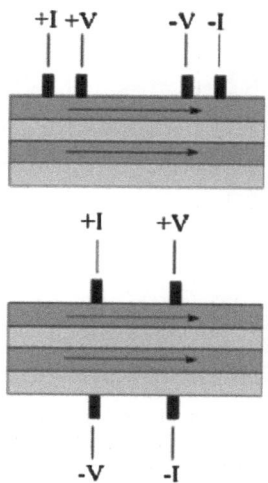

Fig. 7.7 *Top*: current-in-plane (CIP); bottom: current-perpendicular-to-plane (CPP) geometry. Reprinted from [28] with permission from the Institute of Physics

previous section, but the electron *mean free path*. This is because the drop in resistance measurable in a CIP sample when the magnetizations are aligned in the two ferromagnetic layers is attributed to different mobilities for the "spin-up" and "spin-down" electrons, and not to spin accumulation. The drop in resistance in a CIP geometry can be further explained as follows. A very thin nonmagnetic spacer layer results in electrons passing through, and reaching a ferromagnet where their momentum undergoes successive scattering events. However, there are differences between how "spin-up" and "spin-down" electrons scatter, and this fact leads to resistance changes. When the magnetizations of the ferromagnetic layers are aligned, one spin type is significantly scattered, while the other type remains relatively unscattered [19]. In the case of antiparallel magnetization of the layers, neither spin type has a high-enough mobility due to heavy scattering in the ferromagnets. It is highly desirable to observe the effect termed *giant magnetoresistance* [66] (GMR) in "polarize/analyze" geometries, also discussed in the previous chapter. However, for GMR to take place in CIP samples, the thickness of the spacer layer would have to be less than the mean free path of about 20–30 Å. These restrictions make CIP geometries rare in current research studies.

The *parallel resistor model* is a theoretical concept, shown schematically in Fig. 7.8, in which the five series resistors represent the resistances of each magnetic trilayer configuration. This model is used to describe the "spin-up" and "spin-down" transport differences obtained in the two spin channels by using resistors instead of

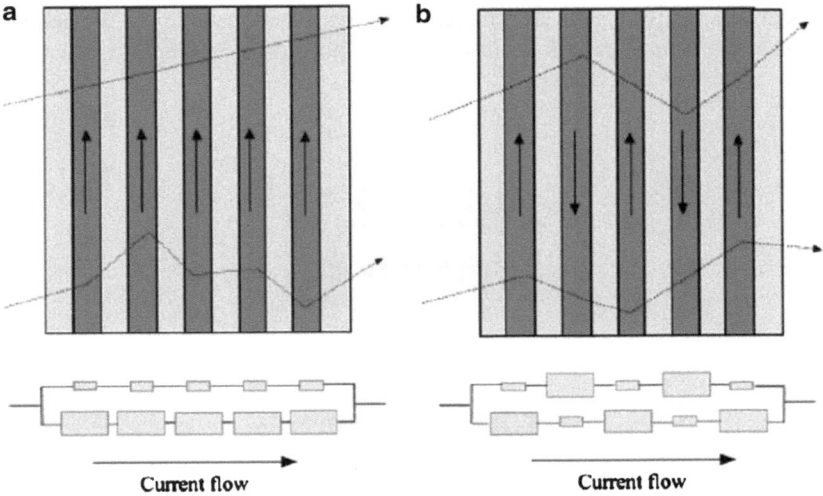

Fig. 7.8 Two-resistor model provides a simple explanation for the low (**a**) and high (**b**) resistance configurations of magnetic multilayers. In (**a**) the layers have parallel magnetizations allowing spins in one channel to pass through with no scattering, whereas spins in the other channel experience scattering while traversing the layers. In (**b**), the magnetizations of adjacent layers are antiparallel, and spins in both channels are scattered. The resistors in each path represent the resistances the spins experience in each layer. Reprinted from [28] with permission from the Institute of Physics

spin in each channel. A high resistance corresponds to one spin type, for instance, 10Ω, and a low resistance of, for example, 1Ω to the other. Thus, a parallel magnetization in the layers corresponds to a minimum resistance state with a total resistance of 4.5Ω, while an antiparallel magnetization in the layers gives rise to a higher resistance of 13Ω. Nevertheless, the model fails to explain what occurs at adjacent interfaces, and does not take into account spin accumulation, diffusion or relaxation.

7.3.2   *Transport in GMR and TMR Structures*

Giant magnetoresistance (GMR) [67, 68] and tunneling magnetoresistance (TMR) [69, 70] are two important transport mechanisms for many quantum device applications. TMR and GMR display some similarities not only in terms of current flow, but also when it comes to challenges encountered with their configurations. Aside from spacer layer composition and measurement geometries, effects at adjacent interfaces in GMR and TMR devices lead to certain results for the effective resistance of the whole multilayer or trilayer structure.

In a GMR multilayer stack where some of the layers are ferromagnets, an unequal distribution of "spin-up" and "spin-down" electron states occurs and is due to a spin splitting of the Fermi level in the ferromagnetic metals. If the GMR stack is an "ohmic" and non-tunneling CPP configuration while also Mott's model applies, the spins in one channel pass through with no scattering. At the same time, spins in the other channel experience scattering at interfaces when the magnetizations in all layers are parallel. For the antiparallel alignment of magnetizations of neighboring magnetic layers, spins experience maximum scattering due to spins in both channels being scattered while traversing the layers. Bloch walls are also known for scattering spin because of *adiabatic passage*, an effect encountered in magnetic resonance spectroscopy. In this case, the exchange field acting on the electron spin changes the orientation of the latter while it passes through the Bloch wall. Spins experience a potential barrier as the spin precession does not quite follow the exchange field. A *fast passage* occurs when the spin precession about the applied field is rapid in comparison to the angular velocity of the rotation of the magnetic field [71]. Also, a *sudden passage* can occur when the electron passes through magnetic multilayers or a tunnel barrier, as the electron is suddenly subjected to another magnetic field [72]. Scattering at Bloch walls is assumed to be responsible for colossal magnetoresistance (CMR) in manganate perovskites [73].

TMR devices are based on an electrical current passing through an insulating barrier by quantum mechanical tunneling between two ferromagnetic layers. Depending on the relative orientations of the magnetizations in the ferromagnetic layers, two-terminal spin tunneling junctions can show large changes in tunnel resistance. The rate of tunneling is proportional to the product of the electron densities of states at that particular energy on each side of the barrier (Julliére's model). A low resistance for the device is obtained when electrons with majority spins tunnel if the two half-metallic ferromagnets have aligned magnetizations. On the other hand, the product of the initial and final densities of states is zero when the magnetizations are antiparallel, therefore no current flows in this case. The type of insulator material, as well as barrier height and width, plays an important role in TMR. Additionally, TMR varies with barrier impurities, temperature, and bias voltage. One of the reasons is that spin-flip scattering increases with the content of magnetic impurities in the barrier, and it probably also increases with temperature. Some research [74, 75] indicates that increases in temperature bring about a reduction in the overall magnetization in the ferromagnet because of excitations of magnons. Because magnons are *spin* 1 quasi-particles, the excitations of magnons increase electron–magnon scattering and, therefore, electron spin flipping [76], reducing the TMR response.

The type and combination of ferromagnet/insulator materials is a determining factor in obtaining a certain TMR value. Spin polarization is not controlled just through the ferromagnet alone, but also through the type of interface between the ferromagnet and the insulator. For instance, the Fermi surfaces of Co and Cu match well for the majority spin electrons in Co, with a poorer match for minority spins [77]. The unequal alignment of sub-band structures in the Co/Cu system leads to a spin asymmetry of the tunneling current [78]. The Co/Cu configuration is very sensitive to the interface conditions between Co and Cu, necessitating smooth and sharp interfaces between the magnetic and nonmagnetic layers. It was noticed that a reduced quality of the interface results in "hot spots" or "pinholes". In the former case, electrons tunnel preferentially through localized regions of low barrier thickness, whereas the latter situation implies direct contact between the ferromagnets [79]. Both GMR and TMR are examples of spintronic effects where the flow of spin-polarized electrons is manipulated with the help of magnetic fields.

7.3.3 Nonmagnetic Spacer Layers

To more efficiently manipulate spin, transmission needs to be extended over longer distances, particularly during spin voyage in the spacer layer between the encoder and decoder of a spintronics configuration. The remnant magnetization in the encoder, which is usually a ferromagnet, renders a polarization to the electron spins. This polarization is carried (after some interface spin scattering) by the electron current into the spacer layer; however, the polarization decays spatially over the spin diffusion length. The spin polarization of electrons usually persists over longer lengths than that of holes. When the separation between the encoder and the decoder is smaller than the spin diffusion length, a voltage is induced. This spin-induced voltage is believed to be due to the increase in "spin-down" and decrease in "spin-up" population in the spacer layer. A difference between electrochemical potentials is likely to exist between the spacer and the decoder; therefore, the electrochemical potential of the decoder must rise to match that of the spacer layer. At the same time, a non-equilibrium magnetization exists in the spacer layer that depends on several factors: the partial polarization at the Fermi level, the differences in Fermi velocities, and a partial or inefficient spin transfer across the interface.

It was mentioned several times that nonmagnetic spacer layers are employed to interface with magnetic layers; therefore, some discussion needs to be dedicated to them. A popular choice these days are organic semiconductors because of the wide range of properties they display, as well as their fabrication versatility. Due to the *proximity effect* [80], a spin polarization exists at the interface between an organic semiconductor such as Alq_3 [tris-(8-hydroxy-quinolinolato) aluminum] and the ferromagnet from which the spin injection takes place. This spin polarization extends over a few lattice constants while a small Schottky barrier is also formed at the organic/ferromagnet interface. The spin-polarized carriers tunnel through the first interface into the organic material. As the carriers reach the second interface through which they also tunnel, their spin polarization has decayed exponentially in the spacer layer (Fig. 7.9). A current is detected in the second ferromagnet. Based on the Julliére model [81], the spin diffusion length in Alq_3 is estimated to be $\sim$4 nm, while it is fairly independent of the Alq_3 layer thickness [82]. This value was obtained with spin polarizations in nickel and cobalt at their Fermi energies of 33%, and respectively 42%, while the Alq_3 layer thickness was 33 nm. The obtained spin diffusion length is almost an order of magnitude smaller than the spin diffusion length reported in other studies for thin film layers. Since the above quoted value is for nanostructures, it implies that the spin relaxation occurs at a higher rate in these structures than in thin films.

Organic semiconductors have been used in the fabrication of spin valve structures, whether in thin film or nanowire form [83]. It was determined in the latter that using Alq_3 spacers leads to long spin relaxation times ranging from a few milliseconds to over 1 second at low temperatures ($\sim$1.9 K). Both values are higher than in inorganic semiconductors. These long times are believed to be due to the Elliot–Yafet spin relaxation mechanism [84] that seems to be dominant in nanostructures. Previous and current studies have shown that trilayer configurations containing an organic spacer can be responsive to magnetic fields. Figure 7.10 shows an example of a hysteresis curve for $Ni/Alq_3/Co$ trilayer nanowires obtained at

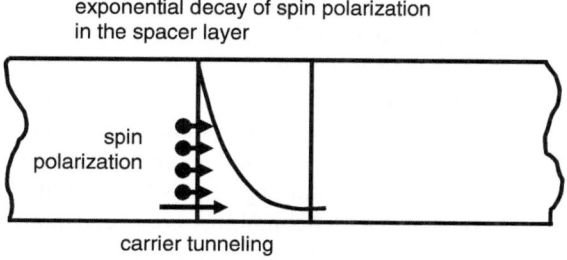

Fig. 7.9 Simplified schematic of a trilayer magnetic structure encompassing an organic spacer layer. A spin polarization is present at the first interface and it decays exponentially from the interface. The dimensions of the spacer layer are exaggerated

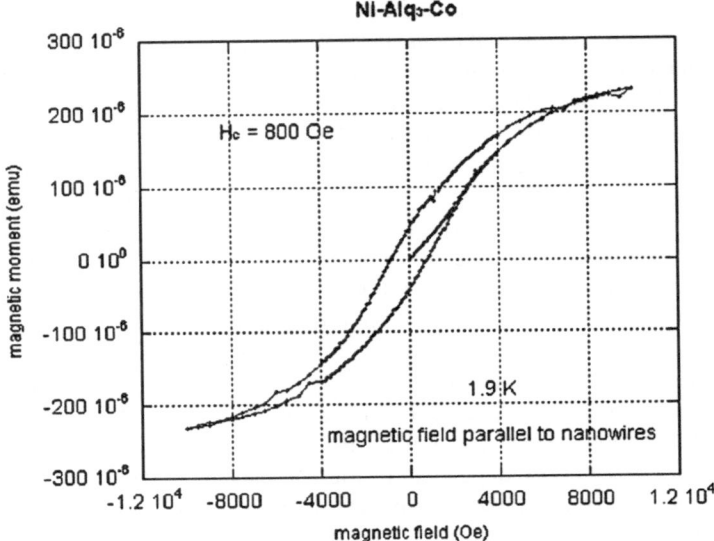

Fig. 7.10 Hysteresis curve for Ni/Alq$_3$/Co trilayer nanowires obtained at 1.9 K using a physical parameter measurement system (quantum design). The trilayer nanowires were fabricated in porous alumina obtained by anodization [C.-G Stefanita, S. Pramanik, S. Bandyopadhyay]

1.9 K. These results are important because the state of the electron spins needs to be controlled by an external magnetic field for the spin valve to be functional. GMR values as high as 40% have been obtained for thin film spin valves at low temperatures. Half-metallic ferromagnets with near-100% spin polarization give better GMR results when they are used as spin injectors.

It is not surprising that organic semiconductors are being intensely studied as potential replacements for metal and inorganic semiconductor spacers. One of the reasons is the unwanted diffusive electron transport that takes place in metals and inorganic semiconductors caused by shorter momentum scattering lengths as compared to spin scattering lengths. The distinctive properties that organic semiconductors display show promise for spintronics applications. These properties may be partly attributed to the strong intramolecular rather than inter-molecular interactions. Conjugated polymers are quasi-one-dimensional systems; therefore, weakly screened electron–electron interactions result. Another distinct effect encountered in organic semiconductors is that electrons and the lattice are strongly coupled. Therefore, unlike in inorganic semiconductors, charge carriers in these organic materials are positive and negative polarons, rather than holes and electrons [85]. It was observed that carrier mobility is reduced in Alq$_3$ because of the additional Coulomb scattering caused by the charged surface states [86]. However, an advantage is that charge injection into organic semiconductors happens mainly through tunneling [87], thus circumventing a likely mismatch in conductivity often experienced at ferromagnetic/inorganic semiconductor interfaces where spins can switch from "spin-up" to "spin-down" and conversely. This spin switching results in an electrochemical potential that produces variations in conductivity.

Carbon nanotubes have also been considered as spacers for spin transmission due to the fact that they display very long electron and phase scattering lengths [88]. The construction of carbon nanotubes is partly responsible for the observed properties. A single-walled nanotube (SWNT) is formed by wrapping around one graphene sheet to build a 1–3 nm diameter cylinder, whereas multiwalled nanotubes contain several concentric cylinders of diameters up to 80 nm. The chirality of the wrapping in a SWNT determines whether it behaves as a semiconductor or a metal. The reduced spin-flip scattering is to some extent attributed to the lightness of the carbon atom.

7.3.4 The Advantage of Using Quantum Dots

Quantum dots constitute a particular "structural geometry" suited for quantum devices, as carriers are confined in all three space dimensions of the ferromagnet, limiting the dot size to that of the Fermi wavelength. For this reason, quantum dots have properties not normally encountered in structures of larger size where electrons are not confined. Theoretical considerations for spins of confined electrons point out a fact that leads to unusually low spin-flip rates in GaAs quantum dots. The localized character of the electron wave function in a dot suppresses the intrinsic spin-flip mechanism attributed to the absence of inversion symmetry in GaAs-like crystals [89, 90]. Quantum dots are known as well for their ability to allow manipulation of electrons. For instance, a magnetic field applied perpendicularly to the structure can alter the electronic states in the quantum dots. Also, by changing the gate voltage the shape and size of quantum dots can be varied. Through this latter process, the number of electrons in the quantum dots is tuned.

Example 7.4. Give an example of a method of obtaining magnetizations of different orientation in two quantum dots.

Answer: Two quantum dots with different Curie temperatures T_C can be heated above their Curie temperatures, and then cooled down below their Curie temperatures while using two different external fields. By applying the first field and cooling initially below the higher Curie temperature T_{C1} for the first dot, the magnetization of that dot will become parallel to the external field and will be fixed. When cooling is continued, upon reaching the Curie temperature T_{C2} of the second dot, a second field is applied in a different direction that orients the magnetization of the second dot parallel to the second field without changing the magnetization of the first dot *(After* [1]*Chap. 8).*

One of the main practical challenges facing quantum device fabrication is maintaining an adequate isolation compromise between quantum dots. After all, it is desired to realize individual quantum entities. On the other hand, a weak coupling of dots to each other is necessary to ensure a reduced spin decoherence (i.e., dephasing), usually determined through the *transverse decoherence time T_2*.

It is very important to keep coherence on sufficiently long time scales. One way the T_2 can be measured experimentally is from the line width of the resonance peak obtained through electron spin resonance [91]. This decoherence time should not be confounded with the *spin relaxation time T_1*. It should also be noted that *charge decoherence* is not the same as *spin decoherence*. This is because superpositions of "spin-up" and "spin-down" states become decoherent over different times than charge (i.e., orbital) electron states. Actually, both theory and experiment indicate that spin coherence times significantly exceed charge coherence times [92]. Nevertheless, more experiments are necessary to completely validate these results. Once proven as undoubtedly true, these longer coherence times would make spin-based quantum devices more resilient than charge-based devices. Thus, using spin over charge for data transport through quantum devices may have more advantages.

GaAs is a semiconductor material that has displayed remarkable properties, and, therefore, it has been actively studied in many laboratories. These days, this material continues to be used as part of experimental device structures in which the number of electrons in the conduction band of a quantum dot has been restricted. Interesting effects, such as tunneling between neighboring dots and Coulomb blockade have been observed in coupled quantum dot arrays controlled by electrical gating [93,94]. Figure 7.11 shows such an array where quantum dots are formed by manipulating the gate voltage and confining one electron to each dot. The electron in each quantum dot is subject to a distinct Zeeman splitting under the influence of a local magnetic field. A single spin can couple to an effective Zeeman field if the dot carrying the spin is exchange coupled to a ferromagnetic layer as in Fig. 7.11, or to a ferromagnetic dot. The ferromagnetic layers or dots are made of either magnetic metals or magnetic semiconductors. Individual detection of spins occurs by applying an additional ac magnetic field of corresponding Larmor frequency, and recording electron spin resonance pulses.

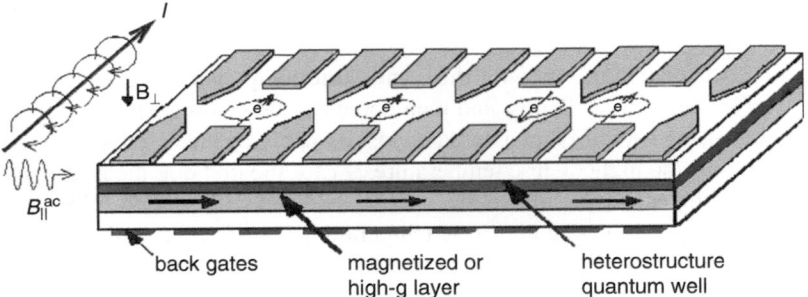

Fig. 7.11 Electrical gating (through the dark gray electrodes) leads to the formation of an array of quantum dots (shown by *circles*). Under local magnetic fields created by a current-carrying wire, distinct Zeeman splitting occurs for the electron in each quantum dot. Consequently, spins can be detected individually by applying an additional ac magnetic field of corresponding Larmor frequency, and recording electron spin resonance pulses. The rightmost two dots are depicted as tunnel-coupled. Reprinted with permission from [90] Copyright 2002 Springer

7.4   Spin Selective Detection

7.4.1   *"Reading" Spin*

Assume for a short time that spintronic devices are only meant for quantum computation, although in reality, there are many more spin-based devices being developed nowadays, as will be discussed below. In any case, when reading information after an operation known as quantum computation involving coherently entangled spins or qubits, it is necessary to identify the state to be read that is needed for the interpretation of results. In particular, it is important to detect whether the state after the computation is "spin-up" or "spin-down". Fortunately for us, no details need to be known about the coherent superposition of spins contained in the qubit. This convenient situation arises because the superposition of spins is only important during the execution of the algorithm, while the result needs to be only a classical bit, either an "up" or a "down" spin. Although this simplifies the task significantly, the challenge of detecting the magnetic moment of a single spin still remains. With today's electron spin resonance microscopy capability, the smallest volume element that can be imaged must contain 10^7 electron spins [95], which poses a problem for single spin detection. Fortunately, there is a technique with high enough sensitivity that has the ability to detect an individual electron spin. Magnetic resonance force microscopy (MRFM) is based on detecting a magnetic force between the spins in the sample and a ferromagnetic tip (Fig. 7.12). For instance, a mass-loaded silicon cantilever has been used to measure unpaired electron spins in silicon dioxide specimens [96]. Attached to the cantilever was a 150-nm-wide SmCo magnetic tip

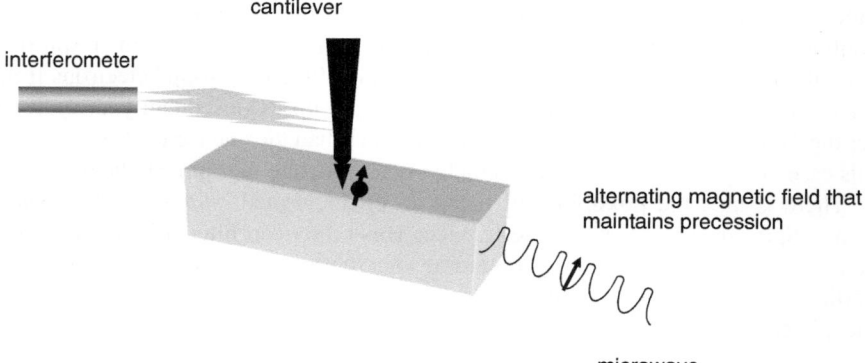

Fig. 7.12 Sketch of single-spin MRFM measurement using an ultrasensitive cantilever beam and a perpendicular magnetic field oscillating at microwave frequency. The perpendicular field is necessary to maintain precession of the electron spin. The specimen contains unpaired electron spins. A spin interacts with the field from the magnetic tip of the cantilever beam, producing a deflection sensed by the interferometer

that sensed the force from the electron spin. Ultrasensitive cantilever-based force sensors are the latest development in MRFM, made specifically for the purpose of facilitating single-spin detection [97]. The MRFM technique allows imaging of spins as deep as 100 nm below the surface. However, in spite of its promises, MRFM is still inconvenient due to the fact that the force from an electron spin is only a few attonewtons, thereby $\sim 10^6$ times smaller than the forces detected by atomic force microscopy (AFM), a widespread and more affordable technique.

There are ways to bypass the need for single-spin detection, by using charge to deduct information about spin. This is basically a problem of reducing the spin measurement to a charge or current measurement. A technique employed by many research groups involves spin filtering, where either "spin-up" or "spin-down" electrons can tunnel through a barrier; hence, the coined name "filtering". Electrons tunneling through with spin of a particular orientation are then detected, with no detection for the other spin state which does not make it through. Another technique uses a reference electrode with 100% spin polarization [98]. For this situation, we have a tunneling current flowing from the first electrode, which is also the material being investigated. This current reaches the reference electrode whose spin polarization depends only on the choice we make for the reference material. Unfortunately, it was noticed that the interface electrode/insulator influences the result; therefore, it is important to employ a combination that minimizes defects at interfaces, preventing spin-flipping [99].

A Coulomb blockade may prove useful in filtering spin states. As such, consider two materials with different effective g-factors, one for a quantum dot and the second one for the current leads attached to the dot. In this case, the two different g-factors cause the spin-degeneracy to be lifted because of the different Zeeman splittings in the leads and dot [100]. If the Zeeman splitting on the dot is much larger than in the attached wires, the quantum dot acts as a spin filter. As a result, Coulomb blockade peaks are obtained, which belong to a particular spin state of the electron confined to the quantum dot. In the converse case, when the larger Zeeman splitting is on the fully spin-polarized wires, they can only absorb "spin-up" electrons from the dot if the polarization of the wires is also "spin-up". Consequently, the current for the "spin-down" states on the dot will be larger than the one for the "spin-up". In this case, a "spin to charge" conversion has occurred, and the spin can be read out.

There is also the possibility of using an effect termed *quantum interference*, in principle similar to optical interference. This effect can filter out the unwanted component of spin current by obtaining destructive interference only for that particular component. However, this technique poses experimental challenges and, therefore, is not usually employed.

7.4.2 The Datta and Das SFET

It was stated before that quantum computation is not the only area of spintronics which benefits from the manipulation and detection of spin. Studies about the interaction of spin with its environment have kept researchers focused for nearly thirty years on what are nowadays a variety of branches of spin electronics. Therefore, the next section of spin-based devices introduces the reader to a range of applications that rely on spin, and, for that reason, are likely to deal with the necessity of detecting spin (or spin states) as part of their functionality. However, at the moment we will briefly discuss what is now considered a historical turning point in the general field of spintronic devices.

In 1990, Datta and Das [101] published the scheme of a possible spintronic device, a *spin field effect transistor* (SFET) shown schematically in Fig. 7.13. Two ferromagnets, the first acting as a source or *spin injector*, and the second as a drain or *spin detector*, have parallel magnetic moments. The electron current through the transistor is made up of polarized spins traveling with the electrons through a quasi-one-dimensional channel. The spins are unavoidably exposed to the Rashba effect, which occurs because the electron velocities in devices are of the order of $\sim 10^6$ m/s, and the magnetic component of the local fields in the depletion layer is likely to influence the transport of these relativistic electrons. For this reason, the stationary electric field already present in the material is experienced by the relativistic electrons as partially magnetic. At the same time, the gate voltage at the top of the channel generates an effective magnetic field that causes the spins to precess. Depending on the value of the voltage, spins can be parallel or antiparallel with respect to the source and drain. The gate voltage tunes the magnetic field so that the "on" state of the transistor occurs when the spins are parallel to the magnetizations in the source and drain. This is when a large current is obtained and conductance is high. On the other hand, when the spins are antiparallel to the magnetizations in the ferromagnets, the spins are scattered and the transistor is in the "off" state with very low conductance. Although the Datta and Das spin field effect transistor was not put into practice, it has significantly influenced the development of spin-based devices, as it illustrates some of the principles on which these devices are expected to work.

One of the reasons it is believed that the Datta and Das SFET cannot be functional is the influence of the high magnetic field created in the channel by the magnetizations of the two ferromagnets, source and drain. This magnetization is usually neglected, although studies [102] indicate that it should be taken into

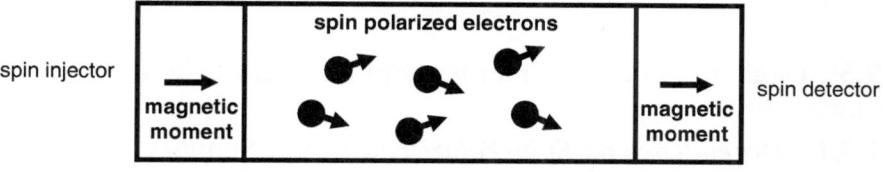

quasi-one-dimensional channel

Fig. 7.13 Simplified sketch of the Datta and Das spin field effect transistor illustrating a source and drain for spin-polarized electrons. A gate voltage at the top of the quasi-one-dimensional channel tunes the magnetic field about which the spins precess

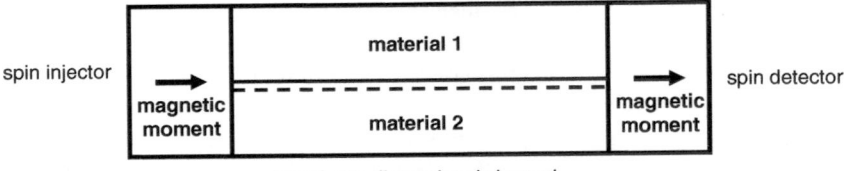

quasi-one-dimensional channel

Fig. 7.14 Simplified sketch of an electron spin interferometer with a two-material channel. The horizontal dashed line highlights the formation of a quasi-one-dimensional electron gas formed at the semiconductor interface where the two materials come into contact. The magnetization in the source and drain is assumed along the horizontal direction

account. If this is the case, then the conductance modulation in the channel of these transistors is dominated by what are known as Ramsauer or Fabry–Perot-type transmission resonances, rather than the expected Rashba effect. Because of its similarity to an interferometer, this type of transistor has been termed a *gate-controlled electron spin interferometer.* Figure 7.14 illustrates a sketch of this device with a two-material channel. But what is the Ramsauer effect? When interface potential barriers are large where the semiconductor and the other material make contact, and the effective mass mismatch between them is considerable, transmission resonances occur due to reflections at the interferometer's contacts [100]. When the gate voltage is varied, the Fermi level moves up or down with respect to the conduction band edge. However, when it finally lines up with the resonant energy levels above the barrier between the contacts, the transmission through the channel reaches a maximum. These conductance oscillations that occur every time the Fermi level sweeps through the resonant levels are generally known as the *Ramsauer effect.* The effect is believed to be responsible for controlling the spin precession in the channel instead of the Rashba effect as was assumed previously, because oscillations are very pronounced for one-dimensional channels.

Unfortunately, the control of spin transport is not similar to that of charge, as additional problems need to be considered that can put a stop to building functional spin-based devices. In spite of numerous difficulties that still need to be resolved, spintronics seem to have a bright future for applications such as switching, amplification, or non-volatile memory storage, to name just a few. The next section should provide some details.

7.5 Challenges Facing Emergent Spin-Based Devices

7.5.1 Dependence on Spin-Polarized Current Supplies

Success for spintronic devices hinges significantly on obtaining steady supplies of spin-polarized currents, and much attention is still given to this research area. There is a magnetic imaging technique capable of sub-nanometer scale resolution called *spin-polarized scanning tunneling microscopy (SP-STM)*, or its closely related version *spin-polarized scanning tunneling spectroscopy (SP-STS)*, both very useful in analyzing the spin polarization state on either the surface, or in the bulk of a sample. Several spin-polarized surface states have been analyzed for Fe(001), Cr(001), Gd(0001), Mn(001), and Co(0001), given the large GMR responses reported at junctions between these materials and MgO barriers. Nevertheless, the efficiency with which spin is injected and detected depends not only on the spin polarization of the ferromagnetic electrode, but also on interfacial phenomena, as well as solving the conductivity mismatch problem due to the interface resistance at the ferromagnet/semiconductor junction. A tunneling barrier at the interface may offer some advantages with respect to the latter. The combination of ferromagnets such as CoFeB with the popular and constantly improving MgO tunnel barriers shows the most promising results (*see for instance, problem P7.13*). When connected to a doped GaAs channel, it seems that at an effective depletion width of $n^+ - GaAs$ can become thin enough for electron tunneling near the Fermi level, so that the conductivity between CoFeB and the doped GaAs increases sharply. This, in turn, leads to a resistance area product of the CoFeB/n^+-GaAs/MgO junction strongly dependent on the bias voltage, and, hence, the ability to manage the conductivity mismatch problem.

The spin valves discussed in the previous chapter also depend on the injection of a coherent spin current as, without it, there would not be a conductance modulation when spin traverses the different GMR layers. Hence, numerous research groups are dedicated to finding innovative ways to generate and manipulate spin-polarized currents that have the required stability to be reliably injected and controlled while propagating through spin-based devices. The following example illustrates a case attempting to resolve these problems.

Example 7.5. A recent report of a possibly feasible spin current transistor proposes a way to obtain an enhanced and controllable spin polarization within such a device. The authors suggest that a single-mode device can be designed by considering a two-dimensional electron gas (2DEG) channel. Electrons pass through a series of magnetic–electric barriers built into the device. The number and distribution of barriers can be varied to control transmission of the spin-polarized current [103]. Describe briefly this type of spin current transistor.

Answer: The proposed design of this spin current transistor relies on a 2DEG channel along which external electric and magnetic fields are applied at certain locations along the channel. These fields take on the role of barriers across the conduction path while they operate through ferromagnetic gates. The periodicity of the ferromagnetic gates defines the type of barriers controlling electron flow. Each of the ferromagnetic gates is magnetized in the perpendicular direction. Aside from the magnetic field of the gates, an electric potential is also applied at the gates. The electric potential induces a change in the electrostatic potential barriers in the 2DEG channel. By applying these external magnetic and electric fields, it is expected that the conductance of the spin transistor channel can be modulated enough to control

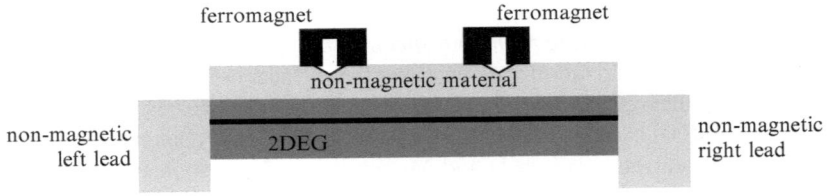

Fig. E7.5 Schematic of a proposed spin current transistor with two external ferromagnetic gates along a channel containing a two-dimensional electron gas (2DEG). After [103]

the passage of an electron and its spin. Figure E7.5 shows a schematic of such a spin current transistor.

The previous chapter discusses a technique in which these gates can be modeled as Büttiker probes. This is one method, but not the only one that can be followed for describing spin transport through the gated channel. However, in the example illustrated here, the authors adopt an approach similar to the Büttiker probes method. The modeling of these probes takes into account the potential barrier in the channel, and the effect of the magnetic field from the ferromagnetic gates. The effective potential barrier in the channel is taken to depend on both the gate bias, as well as the Fermi energies of the leads and 2DEG. The single-mode device regime is obtained by considering transport only over the lowest wavevector sub-band, thus making the summation over all wave vectors unnecessary and thereby simplifying calculations. The gates are treated as expanding waves due to the broadening of the magnetic fields at the barriers. This assumption is in contrast to previous calculations, where the magnetic fields were treated as Dirac delta functions. In this example, however, the broadening magnetic fields lead to a linearly increasing vector potential across the conduction channel. This results in a current with enhanced spin polarization which can be further manipulated by the gates. The authors believe that the reasons behind the enhanced spin polarization are, on one hand, this increasing vector potential, and, on the other hand, the low source-drain bias which allows conductance modulation. Therefore, they suggest that their device is suitable for use as a tunable source of spin current supply. In the end, the supply depends on the number and physical distribution of ferromagnetic gates along the channel. The reader is encouraged to explore the Büttiker probes method and to try to reproduce the results obtained in this example.

Another interesting idea on controlling spin-polarized current supplies is that of employing a novel type of spin polarizer. This particular spin polarizer is based on using a quantum dot array within which the spin degeneracy of the carriers in the quantum dots has been lifted with the help of a magnetic field [104]. A theoretical model has been developed to prove the feasibility of such a design. The quantum dot array is regarded as coupled to leads and their effect is taken into account in the generalized Hamiltonian expression. Only small quantum dots are studied where the separation between energy levels is large, so that only one level is considered per quantum dot. A Green's function approach is used for calculations of spin-dependent transmissions obtained by varying different parameters of the geometry of the system (*Note: some examples were given in Chap. 6*). The authors [104] base their study on previously published research, pointing at the similarity to the Fano and Dicke resonances in optics, also found to be present in quantum dot configurations. In principle, these resonances could be used to generate spin polarization of transmitted carriers.

7.5.2   *Superconductivity and Spintronic Devices*

As we have noticed for a while, spintronic devices come in different configurations, not only in "traditional" designs as spin field effect transistors. As a matter of fact, a ferromagnetic single electron transistor has been conceived [105] by building on knowledge acquired from single electron transistors, while capitalizing on new developments in magnetic tunnel junctions and superconductivity. Consider the case of two ferromagnetic materials separated by a superconductor so that a spin accumulation occurs at the interface within the superconductor when the magnetizations in the ferromagnets are antiparallel. Furthermore, when not only one, but two ferromagnet/superconductor interfaces are present, a double tunnel junction system is obtained, as shown schematically in Fig. 7.15. For such structures, the spin accumulation in the superconductor is due to the imbalance in the non-equilibrium spin density of "spin-up" and "spin-down" tunneling electrons. Because of spin accumulation, the superconductivity is then suppressed, while the conductance of the system increases when the bias voltage is increased. However, a different situation is obtained when the superconductor is placed in a magnetic field. This leads to the Meservey–Tedrow tunneling observations brought about by the excitation level of "spin-up" and "spin-down" electrons shifting by $\mu_B B$ due to the Zeeman splitting. Although conductance changes as well in this case, a distinction should be made as to the cause of this change, whether it is due to spin accumulation or the Meservey–Tedrow effect. Fortunately, in case of the Meservey–Tedrow conductance variation, an opposite sign change is observed under reverse bias, thereby making it possible under a proper bias configuration to separate the two situations, and identify which conductance changes are due to spin accumulation only.

One group of researchers [105] has undertaken to prove that there is a difference, and had a simple experiment designed, involving a basic spintronic single electron transistor with a double ferromagnet/superconductor/ferromagnet tunnel junction (Fig. 7.15). The spin polarization of the electrons injected at one junction into the superconductor decays over the spin diffusion length l_{sd} according to relationship (7.5). If the superconductor is longer than the spin diffusion length within it, any effects of spin accumulation or magnetoresistance on the current can be neglected. When a positive bias is applied, for instance, the electrode on the right (Fig. 7.15) is

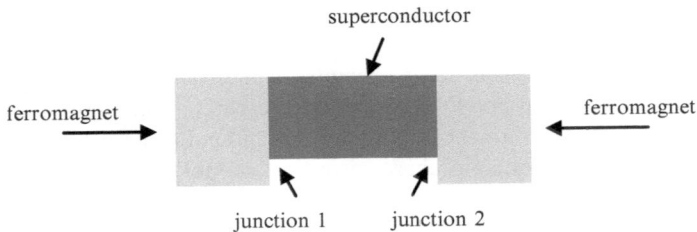

Fig. 7.15 Schematic of a double ferromagnet/superconductor/ferromagnet tunnel junction where the polarity of the bias can be changed. Spin accumulation can be obtained at interfaces within the superconductor. This system constitutes a basic spintronic single electron transistor. After [105]

at positive potential, and electrons tunnel from left to right. The converse is true for reverse bias.

If a magnetic field is also applied, either parallel or anti-parallel to the magnetization within the ferromagnets, the outcome is as expected (and discussed in Chap. 6) in terms of which magnetization follows the field and which switches at a certain field value. Thereby, four different situations can be obtained, depending on the relative orientation of the magnetic field with respect to the magnetizations in each ferromagnet. The researchers calculated the currents for all four situations and used their results for comparison to experimental data. Findings indicate that typical single electron tunneling behavior is observed, with a Coulomb blockade and a superconducting energy gap obtained at low temperatures. Above the critical temperature, only the Coulomb blockade is detected.

Measurements performed under the influence of the magnetic field show that for parallel alignment to the magnetizations in the ferromagnets, there is an indisputable effect of the field on the superconducting gap. However, under the positive applied bias, it was expected that the current would decrease as soon as anti-parallel alignment of the ferromagnets' magnetizations was achieved. Instead, a peak in current was measured for field values corresponding to magnetizations switching from parallel to anti-parallel alignment. In an anti-parallel configuration, the majority spin electrons in the source ferromagnet tunnel easier to the superconductor, and encounter a higher resistance when they tunnel through the second junction, from the superconductor to the other ferromagnet. As a result, a spin imbalance occurs in the superconductor, leading to a shift in the chemical potential of the electrons with different spins. If the superconductor were a normal metal, the chemical potential imbalance would translate into a voltage drop across the device and a spin polarization in the ferromagnets. However, since it is a superconductor, the consequence of the imbalance reduces the formation of Cooper pairs and, therefore, suppresses the superconductivity. Thus, spin accumulation in itself leads to a reduction of the superconducting energy gap, a fact distinguishable from consequences of the Meservey–Tedrow effect [105].

7.5.3   MTJs within Spin-Based Devices

An abundance of reported experimental achievements in spintronics is brought to light every month, in which researchers manage through persistance and ingenuity to finally break through what once appeared to be impenetrable barriers. With frequent and simultaneous improvements in materials properties and device design, long spin diffusion lengths in semiconductors have been, and continue to be, obtained, while the spin orbital strength is now controlled through gates with various degrees of success, although some researchers may not entirely agree with these results. Nevertheless, we expect to see in this decade more and more spin-based devices in the form of sensors and transistors, as well as logic, memory, and optoelectronic applications powered by spin currents. Already, proposals for spin diffusion transistors, spin valve transistors, magnetic bipolar spin transistors, spin metal-oxide-semiconductor FET, spin single electron transistors, and spin torque transistors have emerged.

For instance, a spin transistor driven by a current induced magnetic field was recently [106] demonstrated to have a power amplification capability of 5.6. The basic unit of such a device consists of a magnetic tunnel junction (MTJ) of low resistance-area product, and an electrically insulated metallic wire. This metallic wire produces a magnetic field that can control the magnetization of the free layer in the MTJ, thus resulting in a resistance change. When this resistance variation happens, another metallic wire connected to the MTJ changes its power consumption, leading to a magnetization switching in a second MTJ cascaded next to it. The basic unit of this device is driven when the difference in the output power ΔP_{out} between the parallel and anti-parallel magnetization configurations in the first MTJ is larger than the input power P_{in} to the first wire. The power gain of the basic unit is then $\Delta P_{out}/P_{in}$ which is further cascaded to more similar units. The MTJs can be MgO-based, of the form: buffers layers/PtMn(15)/CoFe(2.5)/Ru(0.85)/CoFeB(3)/MgO(1)/CoFeB(2)/cap layers on thermally oxidized Si substrates. The resistance area product of such a fabricated structure [106] is $3.7\Omega\,\mu m^2$, while the magnetoresistance change was reportedly as high as 129% for current-in-plane tunneling measurements, although these may vary in a more practical application setting. The output power was noticed to strongly depend on the bias voltage applied to the MTJ. The mentioned power gain of 5.6 was obtained for the high bias of 0.4 V.

In general, MgO-based MTJs enjoy a lot of attention because of their potential to be implemented into hard disk reading heads or as bit cells in magnetic random access memories (MRAMs). Improvements in their fabrication, for example obtaining a low resistance insulating barrier, have made it possible to obtain a transfer of spin angular momentum from a spin-polarized current to a ferromagnet. This phenomenon termed *spin transfer torque* has enabled the manipulation of magnetization without a magnetic field, thus achieving a much more promising magnetic dynamical regime. As such, a steady magnetization precession can be induced by injection of a dc current, which can then lead through the magnetoresistive effect to the conversion into a microwave electrical signal [107]. This particular type of realization has applications in telecommunications, especially when a low frequency ac current added to the dc current enables a standard feature like frequency modulation. Thus, the new generation of devices, better known as spin transfer nano-oscillators, may become the promising next generation high frequency synthetizers. An excitation frequency much closer to the natural one of such a device can result in frequency locking, which, in turn, can lead to phase locking in arrays of such coupled spin transfer nano-oscillators. Hence, injecting a spin-polarized current of microwave frequency into a spin based nanodevice can make new applications possible beyond what has been initially imagined when the Datta and Das spin field effect transistor was designed.

The key to a workable spintronic device remains a high efficiency injection of spin-polarized current. This should be followed by a spin coherent propagation, an induction of controlled spin precession, and finally a spin selective detection. The good news is that over the last two decades, tremendous progress in GMR research has advanced several areas within the field of spintronics. Many people anticipate that ultimately quantum computers with spintronic components will perform very high speed logic and memory operations at a fraction of the power of conventional charge-based electronic devices. We will, therefore, look at one more spintronic device before closing this chapter, especially since it is one implemented in memory applications, which are discussed in more detail in the next chapter. We should point out that in spite of the many challenges that remain, it is expected that further developments in spintronics will be determined by advances in their materials. For instance, it is expected that all-semiconductor structures will be preferred to hybrid configurations with metallic ferromagnets. Nevertheless in the end, progress will depend largely on our ability to fabricate high-quality interfaces with nonmagnetic materials.

Example 7.6. A device called ternary content addressable memory (TCAM) has a parallel data processing capability and, as such, is employed in network routers, parallel image processors, and CPU caches. These devices are usually fabricated using conventional complementary metal oxide semiconductor (CMOS) structures and principles. However, with shrinking area requirements that worsen leakage currents, especially in nanoscaled CMOS-based cell circuits, researchers are looking at replacing some of the CMOS structures in TCAM cells. By using the same principles, but different semiconductor nanostructures, specifically, spin injection writable magnetic tunnel junctions (MTJs), a more compact and yet non-volatile TCAM cell circuit chip can be fabricated that performs the same bit-level logic operations as well as data storage. Give an example of such an MTJ implementation in a TCAM cell.

Answer: A spin injection writable magnetic tunnel junction (MTJ) device was fabricated in the following configuration [108]: $Co_{20}Fe_{60}B_{20}(2)/Ru(0.8)/Co_{20}Fe_{60}B_{20}(1.8)$ as an "artificial" ferrimagnetic free layer, separated through a 1 nm thick magnesium oxide (MgO) tunnel barrier from a $Co_{20}Fe_{60}B_{20}$ fixed layer. The magnetic tunnel junction resist patterns were drawn using e-beam lithography, each pattern having dimensions of $50 \times 200\,\text{nm}^2$. Such a junction annealed at $325°\text{C}$ has reportedly resulted in a tunneling magnetoresistance of 167%. In this configuration, switching is observed at a critical current density, hence the ability of this MTJ to function as a logic element and store information. The developers of this device are proposing to use a stack of magnetic tunnel junctions as a ternary content addressable memory (TCAM) cell.

The disadvantage of a conventional CMOS implementation of a TCAM cell is that storage elements itself are large and *placed next* to logic elements on a CMOS layer. Hence, the area of a conventional TCAM is dominated mainly by two-dimensional cell arrays. A comparison study of MTJ implementation in a TCAM cell with this conventional CMOS-based memories shows that the MTJs are significantly smaller, and when *stacked over* a CMOS layer, they result in a reduced cell size. This gives them a space advantage without losing anything in performance, because the stacked MTJ devices are used not only as storage elements, but also as logic elements.

In an MTJ implemented TCAM cell, two series-connected n-type MOS transistors and two series-connected MTJ devices work together in parallel to perform the 1 bit match operation between the external search input bit and the corresponding stored bit. Programming data in the TCAM cell is achieved through the two MTJ devices that store two types of logic values, while the TCAM current indicates a match result. Actually, a match operation in a word circuit is basically a comparison between the amount of current flow in the TCAM cells and a reference cell. There are also peripheral circuits such as search-line/word-line/bit-line and output drivers. Fully parallel equality searches are performed in the TCAM between a search word and each pre-stored word. As such, word-line/bit-line drivers are activated, while a write-enable signal is sent. When a stored word is equal to the search word, the output current becomes high, otherwise it is low. Data needs to be programmed into each TCAM cell prior to a match operation. Thus, this example shows how a TCAM cell can be compactly fabricated by using MTJ devices implemented with MOS transistors, resulting in about 1/6 cell size as compared to an entirely CMOS-based TCAM cell.

Tests, Exercises, and Further Study

P7.1 How is spin decoherence of a single electron affected if the electron is confined to an isolated quantum dot, and the electron spin is subjected to the hyperfine interaction of the electron spin with the nuclei? *Note: See for instance* [109].

P7.2 Describe three electronic, magnetic, or optical properties of quantum dots not normally encountered in structures of larger size made of the same material. Consult appropriate research literature and reference it accordingly.

P7.3 Some challenges in spintronics are related to interfacing different materials so that spin injection is obtained. Describe three techniques (and add appropriate references) that may help in overcoming these problems encountered in spin injection at the interface. The problems include loss of spin polarization, spin flipping, and mismatch in electrochemical potentials.

P7.4 Comment on the effects of doping and strain in magnetic semiconductor processing. Reference any experimental results such as appropriate concentrations, or type of materials where influences are more noticeable.

P7.5 Low dimensional electron systems "handle" spin polarization differently than 3D structures. For instance, it has been observed that a spin polarization exists in a 2D system even in a zero magnetic field (*Note: See, for instance [111]*). Comment on the reasons of what may cause this spontaneous spin polarization and what effects accompany it.

P7.6 Name three factors that modify the free-electron value of the g-factor.

P7.7 How does the spin react, from a mathematical point of view, to Lorentz transformations and magnetic fields? In summary, the answer would be through "Wigner rotations". Consult reviews such as for instance [112] and [113] and comment on spin's behavior under such circumstances.

P7.8 Quantum computation with superconducting qubits encounters several practical difficulties, among which we count the much discussed decoherence. Spin decoherence occurs through spin relaxation and spin dephasing. These may be induced when the qubit state is manipulated. In circuits employing superconducting qubits for quantum computation, the qubit's state is read in the presence of a microwave field. However, this driving field could contribute to the decoherence of the spins entangled (or superposed) in the qubit. Comment on research performed on the driving field-induced decoherence of spins in qubits and what can be done to minimize or avoid it. *See for instance* [114].

P7.9 A semiconductor quantum dot is significantly larger than a simple, organic molecule. If semiconductor quantum dots were replaced with organic molecules, the overall size of the information storing unit would shrink significantly. Furthermore, organic molecules offer the fabrication advantage of self-assembly. Also, unpaired electrons in organic molecules couple to each other by means of intra-molecular exchange interaction, while intra-molecular ferromagnetic coupling is functional at room temperature. All this makes ferromagnetic organic molecules good candidates as magnetic storage units. But how can they be read? Nanoscale paramagnets can be ordered on a surface as probe systems, and an RF resonance tunneling current between the tip of an STM microscope and the probe systems could flow through a ferromagnetic molecule and read its spin state. The technique is termed ESR-STM (electron spin resonance – scanning tunneling microscopy). Describe such an experiment and comment on the conditions the candidate molecule needs to satisfy for the experiment to be viable. *See for instance* [115].

P7.10 Describe other types of Büttiker probes and the experimental configurations to which they apply.

P7.11 Discuss the scaling of the device presented in [106], and compare measured power gains to theoretical predictions.

P7.12   Describe a frequency converter based on nanoscale MgO magnetic tunnel junctions.

P7.13   Discuss the spin polarization and spin injection advantages in spintronic structures made of CoFeB ferromagnetic electrodes, MgO tunnel barriers, and GaAs channels. *See for instance,* [116].

P7.14   Spin injection and band transport of injected spins have been reported to occur only at low temperatures of up to $\sim 120\,\mathrm{K}$ in Fe/MgO/Si spintronic structures. Describe in some detail the method used to measure spin injection into such a silicon structure by using a MgO tunnel barrier. *See for instance,* [117].

P7.15   It has been recently reported that the perpendicular magnetic anisotropy can be controlled by a voltage in Au(001)/bcc-Fe(001)/MgO(001) junctions. Building on these results and combining them with the possibilities offered by MTJs, researchers were able to control the magnetization in a spintronic device by employing only a voltage, as opposed to using a current induced magnetic field. This method would offer a technique for writing in magnetic memory devices that requires a low critical current density for magnetization switching. Describe these achievements and the advantages they would offer for magnetic writing. *See* [118] *and references therein for a start.*

P7.16   Highly spin-polarized ferromagnetic materials are in high demand as a source of spin current. They are of crucial importance for the success of spintronic devices. Comment on how ferromagnetic resonant tunneling diodes with split quantum levels in a ferromagnetic quantum well could be used for such purpose. *See* [119] *for instance, and references therein.*

P7.17   High-density spin torque magnetoresistive random access memory (Spin-RAM) requires a spin-polarized current for magnetization switching, although voltage-controlled rather than current-controlled methods are also being explored (See problem P7.15). Give an example of a current controlled method for this type of application. *See for instance,* [120].

P7.18   Magnetic tunnel junctions (MTJs) are basically two-terminal devices. Due to the large amount of research leading to TMR responses getting higher while at the same time decreasing resistance area products, MTJs have also evolved towards becoming multi-terminal devices. Give an example of a three-terminal device incorporating an MTJ used for a spintronic application. *See for instance,* [121].

Hints and Partial Answers to Problems in Chap. 7

The answers to these problems are in many instances just a hint as to where further information is to be found. They contain the main idea but the reader needs to find the references that are sometimes recommended for each exercise and complete the answer through further bibliographic reading and study.

P7.1 Spin decoherence time T_2 is shorter than the nuclear spin relaxation time arising from the dipole–dipole interaction between nuclei. This is partly because spin decoherence is caused by the spatial variation of the electron density that corresponds to a particular discrete quantum state in the quantum dot. This spatial variation gives rise to a broad distribution of the local values of the hyperfine interaction. Consequently, the electron spin perceives a nuclear magnetic field of about 100 gauss due to the hyperfine interaction, and the decay of the spin precession amplitude is of a *power law type*, and not exponential. *See* [109] *for further details.*

P7.2 From various research papers, a few examples are:

Magnetic. Co quantum dots display a different coercivity than bulk Co. *(Note: See for instance research papers by L. Menon and S. Bandyopadhyay).*

Electronic. Nanometer-scale Co structures exhibit a unique standing wave pattern that arises from scattering of surface state electrons off defects in the substrate on which the Co nanostructures were grown. *(Note: See for instance [110].)*

P7.3 See text for a brief description of the techniques and corresponding references.

P7.4 See text for a starting point in the discussion. Search appropriate literature in a science database.

P7.5 See the mentioned paper for additional details.

P7.6 a. A common cause for a modified free-electron g-factor is the spin–orbit coupling.
   b. The free electron g-factor can be drastically enhanced by doping a semiconductor with magnetic impurities.
   c. Different "structural geometries" such as quantum wells, wires, and dots confine electrons and, hence, modify the free electron g-factor. Furthermore, in these cases, the g-factor is affected by an external bias voltage.

P7.7 See [112] and [113], as well as references therein.

P7.8 When we think of spin decoherence, we usually attribute it to "free decay" brought about by the properties of the material itself, or the material combinations with layer interfaces, not the external conditions. However, quantum computation with qubits implies the existence of some external circuits. It is this coupling to external circuits that was noticed to result in short decoherence times for the entangled spins in the qubit, in particular, large driving fields had the most effect. A theoretical study that includes the influence of the driving field and the electrical circuit, found a relaxation time of, for instance, 142.24 ns, and a decoherence time of 140.01 ns for a normalized field strength of 5×10^{-4}. By comparison, the same qubit in free decay has a calculated relaxation time of 109.28 ns, and a decoherence time of 202.67 ns. This results in a dephasing time of 2.790 μs. Calculations were repeated for other field strengths. At small field strengths, relaxation and decoherence times are independent of field strength. On the other hand, stronger fields result in longer relaxation times, but shorter decoherence times. The SQUID qubit can be optimized by varying the external circuit parameters. *See* [114].

P7.9 Candidate molecules need to be attached on a conductive surface, their magnetic moment needs to be maintained and not suppressed during tunneling, and they must have an intrinsic conductivity for STM tunneling currents to be allowed and manipulated. *See* [115] *for details about such an experiment.*

P7.10 Other types of Büttiker probes include, for instance, electrostatic potentials and "edge" magnetic fields applied on top of a 2DEG channel of a high-electron-mobility transistor. In these situations, the spin current is polarized parallel to the "edge" magnetic fields. *See appropriate published experiments for further details, including references within* [103].

P7.11 See [106].

P7.12 See [107].

P7.13 See [116] for further details.

P7.14 *See* [117] *for further details.* Authors use a non-local measurement method consisting of a four-terminal probing system with an electromagnet. Their measurement range spans 4.2 K to room temperature, while the magnetic field is swept from -80 mT to 80 mT.

P7.15 Manipulating the magnetization in nanoscale ferromagnets by using a voltage rather than a current would lead to a reduction in power consumption. One key aspect for achieving voltage-controlled magnetization switching is to have perpendicularly magnetized ferromagnetic materials as opposed to in plane. Nevertheless, a small assisting magnetic field is also necessary. *See* [118] *for further details.*

P7.16 The ferromagnetic quantum level split depends on electron spin and, as such, allows a resonant tunneling diode to operate as a spin filter or energy filter. The resonant tunneling effect accompanied by a usually large tunneling magnetoresistance ratio has already been studied in semiconductor GaMnAs ferromagnetic double barrier heterostructures. Similarly, Fe-based ferromagnetic heterostructures are currently under investigation due to their relatively high Curie temperature. In particular, $Fe_3Si/CaF_2/Fe_3Si$ magnetic tunnel junctions on Si(111) substrates show promising results due to the fact that they can be matched to the Si lattice, but also on account of their large conduction band discontinuity, allowing electron tunneling and the observation of negative differential resistance. *See* [119] *for further details.*

P7.17 A more efficient current-controlled spin transfer magnetization switching that also works at room temperature has been reported. It uses FePt/Au/FePt nanopillars with perpendicular anisotropy, structures that allow a lower current density, a key property for obtaining higher density applications. It also offers superior thermal stability. *See* [120] *for further details.*

P7.18 A new three-terminal spintronics device has been proposed featuring an MTJ, and based on the current-induced magnetic vortex dynamics. This device shows promise as a highly sensitive sensor for the vortex core dynamics, but also as an RF amplifier with positive gain. The first two terminals are input and output terminals, while the third terminal is similar to the bias voltage terminal. *See* [121] *for further details.*

References

1. D.D. Awschalom, D. Loss, N. Samarth (ed.), *Semiconductor Spintronics and Quantum Computation* (Springer, Berlin, 2002)
2. P. Lambropoulos, D. Petrosyan, *Fundamentals of Quantum Optics and Quantum Information* (Springer, Heidelberg, 2007)
3. P.W. Shor, in *Proceedings of the 35th Symposium on the Foundations of Computer Science* (IEEE Computer Society Press, 1994), p. 124
4. N. Majlis, *The Quantum Theory of Magnetism* (World Scientific, Singapore, 2000)
5. M. Le Bellac, *A Short Introduction to Quantum Information and Quantum Computation* (Cambridge University Press, Cambridge, 2006)
6. D.J. Griffiths, *Introduction to Quantum Mechanics* (Prentice Hall, Upper Saddle River, 1995)
7. R.M. Bozorth, *Ferromagnetism* (IEEE, 1993), p. 435, Fig. 10–6
8. S. Chikazumi, *Physics of Magnetism* (Wiley, NY, 1966)
9. S. Mørup, C. Frandsen, Phys. Rev. Lett. **92**(21), 217201 (2004)
10. M.S. Seehra, V.S. Babu, A. Mannivannan, J.W. Lynn, Phys. Rev. B **61**, 3513 (2000)
11. A. Punnoose, H. Magnone, M.S. Seehra, J. Bonevich, Phys. Rev. B **64**, 174420 (2001)
12. C. Kittel, Phys. Rev. **82**, 565 (1951) *and subsequent work*
13. A.V. Khaetskii, Y.V. Nazarov, Phys. Rev. B **64**, 125316 (2001)
14. N.F. Mott, Proc. R. Soc. London, Ser. A **156**, 368 (1936)
15. F. Mireles, G. Kirczenow, Europhys. Lett. **59**(1), 107 (2002)
16. A. Barthelemy, A. Fert, Phys. Rev. B **43**, 13124 (1991)
17. E. Svoboda, IEEE Spectrum, January 2007, p. 15
18. J.C. Slonczewski, J. Magn. Magn. Mater. **159**, L1 (1996)
19. L. Berger, Phys. Rev. B **54**, 9353 (1996)
20. W. Kim, F. Marsiglio, Phys. Rev. B **69**, 212406 (2004)
21. J.P.h. Ansermet, J. Phys.: Condens. Matter **10**, 6027 (1998)
22. B.T. Jonker, S.C. Erwin, A. Petrou, A.G. Petukhov, Mater. Res. Soc. Bull. **28**, 740 (2003)
23. E.I. Rashba, Phys. Rev. B **62**, R16267 (2000)
24. D. Hagele, M. Oestreich, W.W. Ruhle, N. Nestle, K. Ebert, Appl. Phys. Lett. **73**, 1580 (1998)
25. D.D. Awschalom, J.M. Kikkawa, Phys. Today **52**(6), 33 (1999)
26. F.X. Bronold, I. Martin, A. Saxena, D.L. Smith, Phys. Rev. B **66**, 233206 (2002)
27. R.S. Britton, T. Grevatt, A. Malinowski, R.T. Harley, P. Perozzo, A.R. Cameron, A. Miller, Appl. Phys. Lett, **73**, 2140 (1998)
28. J.F. Gregg, I. Petej, E. Jouguelet, C. Dennis, J. Phys. D: Appl. Phys. **35**, R121 (2002)
29. R. de Sousa, S. Das Sarma, Phys. Rev. B **68**, 155330 (2003)
30. J. Korringa, Can. J. Phys. **34**, 1290 (1956)
31. C.G. Stefanita, F. Yun, M. Namkung, H. Morkoc, S. Bandyopadhyay, in *Handbook of Electrochemical Nanotechnology*, ed. by H.S. Nalwa. Electrochemical Self-Assembly as a Route to Nanodevice Processing (American Scientific, 2006)
32. F. Tsui, L. He, D. Lorang, A. Fuller, Y.S. Chu, A. Tkachuk, S. Vogt, Appl. Surf. Sci. **252**(7), 2512 (2006)
33. Y. Matsumoto, H. Koinuma, T. Hasegawa, I. Takeuchi, F. Tsui, Y.K. Yoo, Mater. Res. Bull. **28**, 734 (2003)
34. S.A. Chambers, R.F.C. Farrow, Mat. Res. Bull. **28**, 729 (2003)
35. K. Ueda, H. Tabata, T. Kawai, Appl. Phys. Lett. **79**, 988 (2001)

36. R.A. de Groot, F.M. Mueller, P.G. van Engen, K.H.J. Buschow, Phys. Rev. Lett. **50**, 2024 (1983)
37. Y.K. Yoo, F.W. Duewer, H. Yang, Y. Dong, X.D. Xiang, Nature **406**, 704 (2000)
38. C. Palmström, Mater. Res. Soc. Bull. **28**, 725 (2003)
39. T. Dietl, H. Ohno, F. Matsukura, J. Cibert, D. Ferrand, Science **287**, 1019 (2000)
40. S.J. Potashnik, K.C. Ku, S.H. Chun, J.J. Berry, N. Samarth, P. Schiffer, Appl. Phys. Lett. **79**, 1495 (2001)
41. T. Omiya, F. Matsukura, T. Dietl, Y. Ohno, T. Sakon, M. Motokawa, H. Ohno, Physica E **7**, 976 (2000)
42. T. Dietl, H. Ohno, Mater. Res. Soc. Bull. **28**, 714 (2003)
43. N. Biyikli, J. Xie, Y.T. Moon, F. Yun, C.G. Stefanita, S. Bandyopadhyay, H. Morkoç, I. Vurgaftman, J.R. Meyer, Appl. Phys. Lett. **88**, 142106 (2006)
44. H. Ohno, H. Munekata, T. Penny, S. von Molnár, L.L. Chang, Phys. Rev. Lett. **68**, 2664 (1992)
45. H. Ohno, Science **281**, 951 (1998)
46. J.G. Braden, J.S. Parker, P. Xiong, S.H. Chun, N. Samarth, Phys. Rev. Lett. **91**, 056602 (2003)
47. J.H. Park, E. Vescovo, H.J. Kim, C. Kwon, R. Ramesh, T. Venkatesan, Phys. Rev. Lett. **81**, 1953 (1998)
48. T. Dietl, Physica E **10**, 120 (2001)
49. M. Tanaka, Y. Higo, Phys. Rev. Lett. **87**, 026602 (2001)
50. H.X. Tang, R.K. Kawakami, D.D. Awschalom, M.L. Roukes, Phys. Rev. Lett. **90**, 107201 (2003)
51. N.P. Stern, S. Ghosh, G. Xiang, M. Zhu, N. Samarth, D.D. Awschalom, Phys. Rev. Lett. **97**, 126603 (2006)
52. A. Bychkov, E.I. Rashba, J. Phys. C **17**, 6039 (1984)
53. E.I. Rashba, A.l.L. Efros, Appl. Phys. Lett. **83**(25), 5295 (2003)
54. A. Shen, H. Ohno, F. Matsukura, Y. Sugawara, Y. Ohno, N. Akiba, T. Kuroiwa, Jpn. J. Appl. Phys. **36**, L73 (1997)
55. M.L. Reed, N.A. El-Masry, H.H. Stadelmaier, M.K. Ritums, M.J. Reed, C.A. Parker, J.C. Roberts, S.M. Bedair, Appl. Phys. Lett. **79**, 3473 (2001)
56. H. Hori, S. Sonoda, T. Sasaki, Y. Yamamoto, S. Shimizu, K. Suga, K. Kindo, Physica B: Condens. Matter **324**(1–4), 142 (2002)
57. T. Dietl, F. Matsukura, H. Ohno, Phys. Rev. B **66**, 033203 (2002)
58. F. Matsukura, H. Ohno, A. Shen, Y. Sugawara, Phys. Rev. B **57**, R2037 (1998)
59. H. Munekata, T. Abe, S. Koshihara, A. Oiwa, M. Hirasawa, S. Katsumoto, Y. Iye, C. Urano, H. Takagi, J. Appl. Phys. **81**(8), 4862 (1997)
60. S. Koshihara, A. Oiwa, M. Hirasawa, S. Katsumoto, Y. Iye, C. Urano, H. Takagi, Phys. Rev. Lett. **78**(24), 4617 (1997)
61. S. von Molnár, H. Munekata, H. Ohno, L.L. Chang, J. Magn. Magn. Mater. **93**, 356 (1991)
62. H. Ohno, D. Chiba, F. Matsukura, T. Omyia, E. Abe, T. Dietl, Y. Ohno, K. Ohtani, Nature **408**, 944 (2000)
63. E. Arushanov, L. Ivanenko, H. Vinzelberg, D. Eckert, G. Behr, O.K. Rößler, K.H. Müller, C.M. Schneider, J. Schumann, J. Appl. Phys. **92**(9), 5413 (2002)
64. M. Julliére, Phys. Lett. **54**, 225 (1975)
65. B. Dieny, V.S. Speriosu, S.S.P. Parkin, B.A. Gurney, D.R. Wilhoit, D. Mauri, Phys. Rev. B **43**, 1297 (1991)
66. M.N. Baibich, J.M. Broto, A. Fert, F. Nguyen Van Dau, F. Petroff, Phys. Rev. Lett. **61**(21), 2472 (1988)
67. M.A.M. Gijs, G.E.W. Bauer, Adv. Phys. **46**, 285 (1997)
68. E. Hirota, H. Sakakima, K. Inomata, *Giant Magneto-Resistance Devices* (Springer, Berlin, 2002)
69. T. Miyazaki, N. Tezuka, J. Magn. Magn. Mater. **139**, L231 (1995)
70. J. Mathon, A. Umerski, Phys. Rev. B **63**, 220403 (2001)
71. G. Tatara, H. Fukuyama, Phys. Rev. Lett. **78**, 3773 (1997)

72. J.F. Gregg, W. Allen, S.M. Thompson, M.L. Watson, G.A. Gehring, J. Appl. Phys. **79**, 5593 (1996)
73. S.F. Zhang, Z. Yang, J. Appl. Phys. **79**, 7398 (1996)
74. J.S. Moodera, J. Nowak, R.J.M. van de Veerdonk, Phys. Rev. Lett. **80**, 2941 (1998)
75. A.H. MacDonald, T. Jungwirth, M. Kasner, Phys. Rev. Lett. **81**, 705 (1998)
76. S. Zhang, P.M. Levy, A.C. Marley, S.S.P. Parkin, Phys. Rev. Lett. **79**, 3744 (1997)
77. J.E. Ortega, F.J. Himpsel, G.J. Mankey, R.F. Willis, Phys. Rev. B **47**, 1540 (1993)
78. J. Mathon, M. Villeret, R.B. Muniz, J. d'Albuquerque e Castro, D.M. Edwards, Phys. Rev. Lett. **74**, 3696 (1995)
79. T. Dimopulos, V. Da-Costa, C. Tiusan, K. Ounadjela, H.A.M. van den Berg, J. Appl. Phys. **89**, 7371 (2001)
80. Z.H. Xiong, D. Wu, Z.V. Vardeny, J. Shi, Nature **427**, 821 (2004)
81. G. Schmidt, D. Ferrand, L.W. Molenkamp, A.T. Filip, B.J. van Wees, Phys. Rev. B **62**, R4790 (2000)
82. E.Y. Tsymbal, O.N. Mryasov, P.R. LeClair, J. Phys. Condens. Matter **15**, R109 (2003)
83. S. Pramanik, C.G. Stefanita, S. Patibandla, S. Bandyopadhyay, K. Garre, N. Harth, M. Cahay, Nat. Nanotechnol. **2**, 216 (2007)
84. R.J. Elliott, Phys. Rev. **96**, 266 (1954)
85. M. Wohlgenannt, Z.V. Vardeny, J. Shi, T.L. Francis, X.M. Jiang, O. Mermer, G. Veeraraghavan, D. Wu, Z.H. Ziong, IEE Proc. Circuits Devices Syst. **152**(4), 385 (2005)
86. V. Pokalyakin, S. Tereshin, A. Varfolomeev, D. Zaretsky, A. Baranov, A. Banerjee, Y. Wang, S. Ramanathan, S. Bandyopadhyay, J. Appl. Phys. **97**, 124306 (2005)
87. H. Bassler, Polym. Adv. Technol. **9**, 402 (1998)
88. R. Saito, G. Dresselhaus, M.S. Dresselhaus, *Physical Properties of Carbon Nanotubes* (Imperial College, Singapore, 1998)
89. A.V. Khaetskii, Y.u.V. Nazarov, Phys. Rev. B **61**(12), 639 (2000)
90. G. Burkard, D. Loss, in *Semiconductor Spintronics and Quantum Computation*, ed. by D.D. Awschalom, D. Loss, N. Samarth. Electron Spins in Quantum Dots as Qubits for Quantum Information Processing (Springer, 2002), p. 234
91. H.A. Engel, D. Loss, Phys. Rev. Lett. **86**, 4648 (2001)
92. D. Loss, D.P. DiVincenzo, Phys. Rev. A **57**, 120 (1998)
93. I.J. Maasilta, V.J. Goldman, Phys. Rev. Lett. **84**, 1776 (2000)
94. S. Tarucha, D.G. Austing, T. Honda, R.J. van der Hage, L.P. Kouwenhoven, Phys. Rev. Lett. **77**, 3613 (1996)
95. A. Blank, C.R. Dunnam, P.P. Borbat, J.H. Freed, J. Magn. Reson. **165**, 116 (2003)
96. D. Rugar, R. Budakian, H.J. Mamin, B.W. Chui, Nature **430**, 329 (2004)
97. T.D. Stowe, K. Yasumura, T.W. Kenny, D. Botkin, K. Wago, D. Rugar, Appl. Phys. Lett. **71**, 288 (1997)
98. J.M. de Teresa, A. Barthelemy, A. Fert, J.P. Contour, F. Montaigne, P. Seneor, Science **286**, 507 (1999)
99. R.D.R. Bhat, J.E. Sipe, Phys. Rev. Lett. **85**, 5432 (2000)
100. P. Recher, E.V. Sukhorukov, D. Loss, Phys. Rev. Lett. **85**, 1962 (2000)
101. S. Datta, B. Das, Appl. Phys. Lett. **56**, 665 (1990)
102. M. Cahay, S. Bandyopadhyay, Phys. Rev. B **68**, 115316 (2003) *and references therein*
103. S. Bala Kumar, S.G. Tan, M.B.A. Jalil, G.C. Liang, Nanotechnology **20**, 365204 (2009)
104. J.H. Ojeda, M. Pacheco, P.A. Orellana, Nanotechnology **20**, 434013 (2009)
105. D. Wang, M. Zhang, G. Tateishi, G. Bergmann, J.G. Lu, in *Proceedings of the 20055th IEEE Conference of Nanotechnology* (2005)
106. K. Konishi, T. Nozaki, H. Kubota, A. Fukushima, S. Yuasa, M. Shiraishi, Y. Suzuki, Appl. Phys. Express **2**, 063004 (2009)
107. B. Georges, J. Grollier, A. Fukushima, V. Cros, B. Marcilhac, D.G. Crété, H.K. Kubota, J.C. Yakushiji, A. Mage, S. Fert, K. Yuasa Ando, Appl. Phys. Express **2**, 123003 (2009)
108. S. Matsunaga, K. Hiyama, A. Matsumoto, S. Ikeda, H. Hasegawa, K. Miura, J. Hayakawa, T. Endoh, H. Ohno, T. Hanyu, Appl. Phys. Express **2**, 023004 (2009)

109. A.V. Khaetskii, D. Loss, L.I. Glazman, cond-mat/0201303
110. O. Pietzsch, S. Okatov, A. Kubetzka, M. Bode, S. Heinze, A. Liechtenstein, R. Wiesendanger, Phys. Rev. Lett. **96**, 237203 (1996)
111. A. Ghosh, C.J.B. Ford, M. Pepper, H.E. Beere, D.A. Ritchie, Phys. Rev. Lett. **92**(11), 116601 (2004)
112. A. Chakrabarti, Fortschr. Phys. **36**, 863 (1988)
113. A. Chakrabarti, J. Phys. A: Math. Theor. **42**, 245205 (2009)
114. Z. Zhou, S.I. Chu, S. Han, IEEE Trans. Appl. Supercond. **17**(2), 90 (2007)
115. P. Messina, M. Mannini, L. Sorace, D. Rovati, A. Caneschi, D. Gatteschi, in 4th IEEE Conference on Nanotechnology (2004), p. 645
116. T. Inokuchi, T. Marukame, M. Ishikawa, H. Sugiyama, Y. Saito, Appl. Phys. Express **2**, 023006 (2009)
117. T. Sasaki, T. Oikawa, T. Suzuki, M. Shiraishi, Y. Suzuki, K. Tagami, Appl. Phys. Express **2**, 053003 (2009)
118. Y. Shiota, T. Maruyama, T. Nozaki, T. Shinjo, M. Shiraishi, Y. Suzuki, Appl. Phys. Express **2**, 063001 (2009)
119. K. Sadakuni, T. Harianto, H. Akinaga, T. Suemasu, Appl. Phys. Express **2**, 063006 (2009)
120. K. Yakushiji, S. Yuasa, T. Nagahama, A. Fukushima, H. Kubota, T. Katayama, K. Ando, Appl. Phys. Express **1**, 041302 (2008)
121. S. Kasai, K. Nakano, K. Kondou, N. Ohshima, K. Kobayashi, T. Ono, Appl. Phys. Express **1**, 091302 (2008)

Chapter 8
Some Magnetic Recording Developments

Abstract Magnetic recording has been for some time one of those areas of engineering that managed to capture many researchers' attention, and ultimately their imagination. One always seems to wonder how far we can still go in the pursuit of higher recording densities, and what kind of limitations are threatening to put an end to our magnetic storage dreams. Thus, we cannot leave this introduction to magnetism and magnetic materials without discussing some new developments in magnetic thin film systems, or improvements in writing/reading capabilities of magnetic devices that serve them. Additionally, we should also have a look at reasons behind some technological challenges encountered in the pursuit of higher recording densities. A few are mechanical, as they relate directly to the flying height of heads on the hard drive or contamination sensitivity issues. Others are problems inherent to the media, such as noise and data stability. Nevertheless, in spite of numerous obstacles, the magnetic recording industry somehow manages to push further its technological limits, ultimately overcoming numerous hurdles. Fortunately, the human mind has proven to be an endless source of innovation, and will likely find answers to many problems still awaiting to be unveiled.

The multilayer spin valve GMR sensors and the unique MTJ-based devices presented in previous chapters were once a curious novelty in spintronic technologies. Nevertheless, they are nowadays absolutely essential for further developments in magnetic recording. Furthermore, we have seen that through their design, spintronic devices rely on the flow of spin-polarized electrons, usually manipulated with the use of magnetic fields. However, magnetic fields are also needed to control the orientation of magnetic moments in magnetically engineered thin film systems. These are part of the magnetic hard disk drive, allowing magnetic storage and retrieval. Unfortunately, shape-dependent demagnetizing fields in micro and nanoscale magnetic structures raise the necessity of increasing the magnetic fields required to change the magnetic state of sensor devices. This means that sensors will be limited in what strengths of magnetic fields they can detect, as the low fields between transitions would become in this case undetectable. Consequently, demagnetizing fields and the control over them play a significant role in designing good sensors. Moreover, as areal densities increase and bit size correspondingly

C.-G. Stefanita, *Magnetism*, DOI 10.1007/978-3-642-22977-0_8,
© Springer-Verlag Berlin Heidelberg 2012

decreases, we cannot but ask ourselves how much more information can we store and extract from the ever decreasing volume of our magnetic media? We keep thus stumbling across added barriers that are putting at risk further developments in magnetic information storage.

Magnetic recording has evolved over a long and uneven road since the middle of the last century, with a few significant milestones along the way. From the economical magnetic tapes with their reduced per-bit cost and high reliability to the magnetically engineered nanostructures of the twenty-first century, this industry has known its ups and downs through numerous challenges. Many new developments would not be possible without advances in nanofabrication on which magnetic electronics have come to rely so heavily. Therefore, in this chapter, we will be reviewing some of these design and fabrication questions and solutions, in particular those related to the magnetic media themselves. We will finally end by having a look at new trends in magnetic memory types such as high-performance nonvolatile magnetic random access memories. As these open new doors for data storage, they arrive with an entire set of specifically new and sometimes unpredictable challenges waiting to be embarked upon.

8.1 Traditional Magnetic Tapes

8.1.1 Basic Structure of the Magnetic Film

The magnetic tapes of the kind many of us grew accustomed to over the decades are made of magnetically hard layers such as hexagonal closed packed (hcp) Co-based alloys. These alloys contain Pt, Cr, Ta, and B, and are fabricated by sputter deposition onto NiP-coated aluminum or glass substrates. The additive's role, in particular that of Pt, is to increase the magnetic anisotropy of Co. The deposited material forms a single sheet of fine, single-domain ferromagnetic grains similar to the sketch in Fig. 8.1. The magnetic easy axis of Co coincides with its hcp c-axis which for longitudinally oriented grains is in the plane of the sheet or tape. When data are stored on the tape by magnetizing the grains, some regions on the tape become magnetized in one direction, while for others the magnetization is in the opposite direction. The resultant magnetization of a group of grains thus forms a *bit-cell* of either "1" or "0," depending on the direction of magnetization. About 1,000 grains are contained within a bit-cell, whereas the size of the bit-cell is roughly 100 nm nowadays for perpendicularly magnetized (and *prepatterned* media).

Some problems usually arise with these magnetic tapes. For instance, magnetic grains within the tape are likely to have a random crystallographic orientation, while also being of irregular size. This inconvenient manufacturing aspect is attributed to the fabrication process itself. However, by making sure that the substrate is crystallographically textured, one step can be taken towards influencing the overall direction of the easy axis of the magnetic layer on top. Thus, a reorientation of magnetization becomes possible along some preferred direction. Nevertheless,

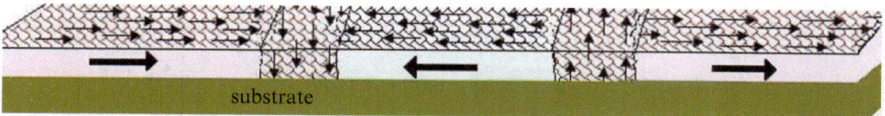

Fig. 8.1 Sketch of a traditional magnetic recording tape made of small granular areas magnetized in a longitudinal direction. One group of grains constitutes a bit-cell of either a "1" or a "0." A bit-cell will have the same magnetization direction; however, due to strong demagnetizing effects, some grains will spontaneously remagnetize. This happens particularly in a transition area of the tape where magnetization changes from one direction to the other, from a "1" to a "0," or converse. The grains magnetized up or down in the figure represent areas of remagnetization. Their overall magnetization is no longer aligned longitudinally; therefore, the varying magnetization orientation creates errors when reading the information on the tape

magnetic tape media with higher capacity have been persistently demanded over time. If tapes with a reduced number of grains have problems, imagine what happens when numerous and more minute grains are trying to occupy smaller and smaller spaces on the tape.

8.1.2 Thermal Effects

Manufacturers had to come up with new solutions before yielding further to market pressure for increased data storage densities. They had to resolve to fabricate tapes with smaller and more numerous grains, while avoiding problems of their predecessors. For instance, it was noticed that in order to maintain a sufficient signal-to-noise ratio, the number of grains per bit cannot be reduced indefinitely. Therefore, in order to pack more bits on the tape, the grain volume had to be reduced in order to reduce the size of the bits themselves. Unfortunately, increasing the grain density by reducing the grain volume in the magnetic layer leads to thermal effects that destabilize not only the magnetization of individual grains but also that of their neighbors. This is because each grain behaves as a single magnetic particle displaying a certain degree of exchange coupling to its neighbors. As the grains become smaller, their coupling increases. Thereby, when grain sizes of 10 nm or smaller start to exhibit thermal fluctuations, the spontaneous reversal of grain magnetization becomes hard to prevent. When one grain reverses its magnetization, other grains follow soon. In the end, if grains are too small, demagnetizing fields at bit transitions brought about by thermal effects will result in a loss of information. The spontaneous reversal of grain magnetization is termed the *superparamagnetic effect*, and is related to the fact that the magnetic energy per grain becomes too small, leading to thermally activated reversals (Fig. 8.2). It can be described by a *thermal stability factor* $\kappa_\alpha v / kT$, where κ_α is the magnetic anisotropy constant, v is the magnetic switching volume, and kT is the thermal activation energy. The thermal stability factor has to be larger than 60 to allow some control over the thermally activated reversal. However, the magnetic anisotropy constant κ_α cannot

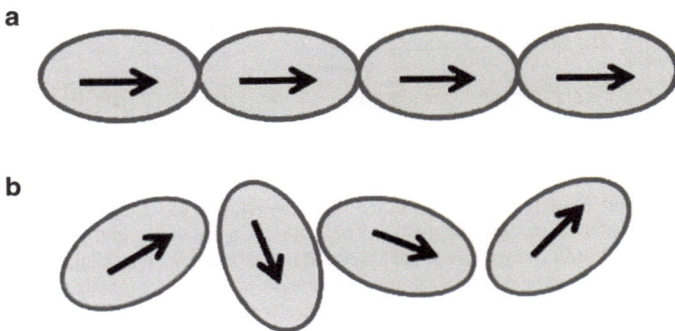

Fig. 8.2 The superparamagnetic reversal of magnetization is a spontaneous reversal of grain magnetization that occurs due to thermal fluctuations: (**a**) magnetic entities previously magnetized and not affected by superparamagnetic effects; (**b**) thermally activated reversals of magnetization due to superparamagnetic effects. The schematic shows only magnetization reversals, the grain does not rotate

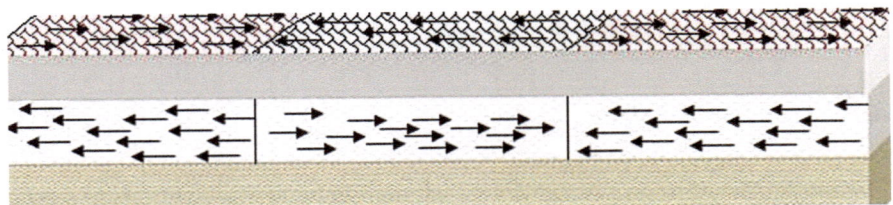

Fig. 8.3 The top magnetic layer on which information is recorded is separated from another ferromagnetic layer through a Ru coupling layer. The coupling layer provides formation of an artificial antiferromagnetic structure needed to overcome the superparamagnetic limit. The second ferromagnetic layer underneath the Ru layer stabilizes magnetic domains in the top ferromagnetic layer through its mirror-image magnetization. These mirror-image magnetic domains of opposite magnetization make it harder for the domains in the top layer to switch their moments when thermal fluctuations occur. The thickness of the Ru layer is exaggerated in this sketch. It is usually extremely thin, and so are the other layers

be increased too much, as this also increases the required magnetic field for writing a bit. Existing head designs handle only fields that do not exceed 15 kOe, and can write only on ∼8 nm diameter magnetic grains, but these values change in time, and the reader should consult appropriate literature for more current data.

The challenge brought about by superparamagnetic reversals of grain magnetization can be controlled through the use of a Ru layer. This layer is placed between the top magnetic film where the information is recorded in magnetized domains, and the bottom ferromagnetic layer which has oppositely magnetized mirror-domains. A sketch of this arrangement is shown in Fig. 8.3. The Ru layer provides a much needed *interlayer exchange coupling*, meaning that an exchange interaction exists between magnetic moments in the two oppositely magnetized layers. In this case, uncompensated interfacial spins are locked and cannot easily follow another magnetic field in alignment, ensuring that domains in the top layer will not switch their moments when thermal fluctuations occur. An added benefit of this configuration is that smaller net magnetic moments can be achieved per grain, allowing grains of reduced size.

Example 8.1. Discuss magnetization processes in transition metal (TM) layers that have been antiferromagnetically coupled through *interlayer exchange coupling* provided by a Ru layer.

Answer: Ru is very often employed as a spacer layer in the magnetic recording industry because it exhibits strong interlayer exchange coupling. The antiferro-magnetic exchange coupling provided by Ru in multilayer structures of the type Co/Ru/Co, Fe/Ru/Fe, and Fe/Ru/Co is particularly of interest for the fabrication of magnetic tapes. A multilayer configuration of the type TM/Ru/TM, where TM = Co or Fe was recently investigated [1] by a technique known as resonant X-ray magnetic reflectivity (RXMR). In Fe/Ru/Co multilayers, the thickness of the Fe layer is so adjusted that the total magnetic moment of the Fe layer is equal to that of the Co layer. That helps with magnetization curve measurements. The RXMR technique can detect both parallel and perpendicular components of magnetization vectors of the resonant element with a good depth resolution, information which cannot be obtained from usual magnetization measurements. As such, RXMR can reveal the magnetic behavior of the upper and lower layers in the multilayer configuration. The investigations performed by the researchers involved hysteresis measurements at Co and Fe K absorption edges, while the thickness of the Ru layer was tuned near the second maximum of the oscillatory interlayer exchange coupling. This type of coupling is expected for such reduced sizes of the Ru layer, and the right thickness of the layer must be chosen, depending on the coupling strength one wishes to obtain. Hysteresis measurements showed that upper and lower TM layers were indeed antiferromagnetically coupled. Magnetic reflectivity profiles were made by θ–2θ scans in the specular reflection condition while switching X-ray helicity every one second. An avalanche photodiode detector was used to count X-rays. Significant differences were obtained for Co/Ru/Co, Fe/Ru/Fe multilayers vs. mixed Fe/Ru/Co structures. Magnetizations rotated towards mutually opposite directions during a magnetization process. As such, the magnetization of the upper layer in Co/Ru/Co and Fe/Ru/Fe configurations rotates counterclockwise, while in mixed Fe/Ru/Co multilayer structures each magnetic layer contains clockwise and counterclockwise rotation domains with an equal probability. Such detailed information about magnetization rotation in layers helps determine the amount of antiferromagnetic coupling for different layer configurations.

8.1.3 Improving the Quality of the Tape

It was observed that increased deposition rates are one fabrication variable of longitudinal magnetic media that leads to improved quality of these tapes. Another fabrication aspect is the necessity to embed the magnetic grains in a SiO_2 matrix in order to reduce the magnetic grain coupling responsible for media noise. As such, the magnetic grains need to be isolated from each other, and the nonmagnetic phase seems to accomplish this goal. It turns out that there is an optimum deposition rate that produces well-defined SiO_2 networks, while minimizing coupling between magnetic grains. Thus, deposition rates of $\sim 10\,nm/s$ are usually beneficial for longitudinally magnetized films made of CoCrPt in a SiO_2 matrix.

Not long ago, areal recording densities for flexible media had reached 3 Gbits/in^2 for metal particulate media, and about 11.5 Gbits/in^2 for metal evaporated media. These densities are not sufficient for applications nowadays, therefore, to maintain further increases in data storage on traditional magnetic tapes, the granular film of the Co-based alloy would have to be reduced even more in thickness and grain size. However, to overcome the thermal challenges, it becomes necessary to replace the longitudinally magnetized granular tapes with perpendicularly magnetized granular tapes. If magnetic films with longitudinal magnetization have high thermal activation reversals predominantly at densities in excess of 100 Gb/in^2, media with perpendicularly magnetized grains have a higher magnetization stability with the onset of superparamagnetism occurring at much larger densities of about 0.5 to 1.0 Tbits/in^2. Only a handful of material systems exhibit perpendicular magnetic anisotropy, such as multilayers of Co combined with Pt, Pd, or Ni. A few Co amorphous films (e.g., TbFeCo and GdFeCo), and some FePt combinations may also have a perpendicular magnetic anisotropy. Unfortunately, many of these systems are difficult to fabricate at room temperature, or cannot withstand material processing. We will see in subsequent sections that a better solution to the magnetic stability problem is the *perpendicular bit media*. These display an enhanced thermal stability of single-domain islands, as well as a higher density potential due to larger head write fields.

Example 8.2. Discuss how an in plane magnetic anisotropy can be transformed into a perpendicular magnetic anisotropy.

Answer: Co layers exhibit an in plane magnetic anisotropy. However, a perpendicular magnetic anisotropy is highly desired in magnetic recording media. This is because perpendicularly magnetized grains have an increased thermal stability, thereby allowing larger grain densities. As such, a group of researchers [2] set out to study how a perpendicular magnetic anisotropy could be obtained in thin Co layers. They finally reported a strong and thermally very stable perpendicular magnetic anisotropy in a 20 Å thick Co layer grown on a Co/oxide/Pt multilayer structure. They were able to transform the in plane magnetic anisotropy of the Co layer to a perpendicular magnetic anisotropy by incorporating a few oxygen atoms in the Co layers of a Co/Pt multilayer structure. Co/Pt films are usually known to have an inherent limited thermal stability. Nevertheless, by exposing these films to a pure oxygen atmosphere, and allowing some native oxide to grow between the Co and Pt layers, they increased the thermal stability of the structure. Furthermore, they noticed that by annealing the Co (10 to 15 Å)/native oxide/Pt system, a Co layer previously grown on top of this system will align its magnetization perpendicular to the film plane after annealing. It is interesting to note that pure Co/Pt systems were found to display an out of plane magnetization only if the Co layer on top was below 7 Å, and that annealing leads to a deterioration of this perpendicular anisotropy due to species intermixing. On the other hand, Co/oxide/Pt systems with top layer Co thicknesses above 15 Å exhibit a perpendicular anisotropy and annealing significantly enhances it. In fact, the transition from an in plane to an out of plane magnetization after the sample has undergone annealing occurs around that value for the thickness of the top Co layer. The researchers believe that the oxidation and thermal annealing create an additional perpendicular anisotropy that would otherwise not be present in pure Co/Pt systems.

Granular structured CoCrPt-based magnetic alloys with an out of plane magnetization are commonly used as perpendicular recording media. The magnetic grains forming bit cells are separated from each other because of exchange coupling which allows us to achieve a high-density recording performance by reducing medium noise. This exchange coupling is controlled by adding oxides to the film or introducing oxygen gas during the sputtering process. The oxides then control the exchange coupling by forming boundaries between grains. Grains average in size between 1 and 10 nm. This fine grain structure influences the shape of the magnetic bit cells and their ability to handle recorded information. For this reason, it is highly desired to analyze the relationship between grain structure and magnetically recorded bits. As such, some techniques are available for this characterization. Dimensions and crystallographic orientations between grains can be evaluated with sub-nanometer resolution by transmission electron microscopy (TEM). Using statistical methods, an average size and dispersion can be estimated for these grains. On the other hand, the magnetic structure of the grain can be determined by magnetic force microscopy (MFM) with a resolution under 20 nm. However, the drawback of this approach is that magnetic information is erased on the disk during the characterization process. We shall discuss in the example below how to avoid erasure of recorded data while still being able to characterize grain structure and establish a correlation between this structure and the recorded bit cells.

Example 8.3. It was mentioned that the grain size as well as crystallographic orientations between grains can be evaluated by TEM with sub-nanometer resolution, while the magnetic structure of the grain can be determined by MFM. However, this approach has its drawbacks, because it destroys the magnetic information. Describe a safe method for analyzing the grains and the exchange coupling between them in order to assess the quality of this high-density recording medium.

Answer: An approach used by researchers [3] in Japan involves fabricating marks on the magnetic disk to be observed, prior to performing characterization measurements. These Ga irradiated marks are used for later guidance. However, obtaining guiding marks by Ga irradiation is tricky by itself, as irradiation of the center portion of the bits needs to be avoided, in order to preserve the recorded information. When marking is performed successfully, the so-called target marks identify the position of TEM and MFM observation, while an additional mark is used for confirming image rotation. An optical microscope is used to find the target marks. When

Fig. E 8.3 MFM image of
recorded bit patterns
including five guide marks.
Reprinted with permission
from [3]

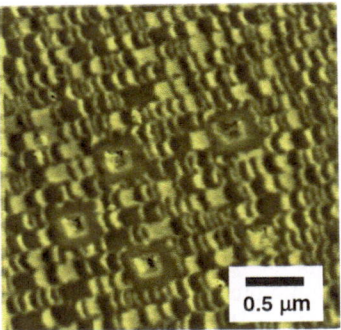

MFM imaging of the magnetic bits is performed, erasure is avoided by remaining
close to the guide mark positions, and obtaining magnetic bit information from
areas surrounded by the marks (Fig. E 8.3). The guide marks also allow an image
comparison with high spatial accuracy by overlapping grain images obtained by
TEM with MFM magnetic images. This comparison leads to a better understanding
of how recorded magnetic bits relate to the underlying granular structure of the
magnetic medium.

8.1.4 Head Spacing

In a typical magnetic recording head, there is an inductive write element, and a
magnetoresistive read element. The head is mounted on an air-bearing slider while
flying along the data medium at extremely small heights, about nanometers in
size. The magnetic field distribution as well as the spatial allocation head/medium
controls the recording process. During the writing flight along the medium, the
head stores a sequence of "1" and "0" on the magnetic medium by magnetizing the
material into two possible magnetization directions. When the information is read
back by the head, it is done by detecting the magnetic flux emanating from between
magnetized regions on the medium. Specifically, it is only the fringing fields in
transition areas between these regions that are being detected. As this happens, only
the free-layer magnetization within the head is switched in the low fields originating
from a bit transition written on the magnetic medium. Depending on the orientation
of the detected fields, a "1" or a "0" is detected. Head degradation, saturation, and
overwrite can happen, and therefore measurement techniques were developed to
assess the stability of the recording heads.

In perpendicular recording which is used nowadays, the spacing between the
head air-bearing surface and the soft underlayer, also known as the head keeper
spacing, is a significant parameter that influences the strength of the writing field,
and therefore the playback process. Some assume that perpendicular recording is
not susceptible to head spacing because there is a soft underlayer. However, most

agree that both perpendicular and traditional longitudinal recording have similar sensitivities to head spacing. Additionally, the maximum writing fields provided by the recording write head are also limited by the magnitude of the magnetization of the magnetic materials themselves. Some working parameters such as pressure and temperature further affect how well the head performs and thereby air-bearing design has to account for pressure and temperature variations. Because the head spacing influences both the writing and the read back process, it needs to be carefully monitored. There are methods commonly used for measuring the head keeper spacing, such as for instance determining the playback waveform of a specific recording pattern. In a similar way, the thermal expansion experienced by the magnetic head can be monitored and the information extracted can be used to control the gap between a recording head and the magnetic medium. The following example illustrates a measuring technique that relies on imaging thermal strains and heat distribution in magnetic heads to better control the head spacing.

Example 8.4. Thermal strains and thermal expansion in magnetic heads are important factors for the control of the dynamics of fly height. Comment on how one can determine heating distributions in a flying head for better management of its response to thermal effects.

Answer: The first thought that may come to mind is to record heating distributions using an imaging technique. However, conventional methods may not have the required resolution nor can they capture dynamic variations due to the frequency dependence of the extent and height of the thermal expansion. Nevertheless, the versatile technique with atomic resolution known as scanning probe microscopy (SPM) is likely to have the capability to reveal changes in a gap between the probe tip and the sample surface generated by fine thermal strains in the magnetic head. By using the SPM method, thermal strain images can be obtained at several frequencies.

A group of researchers [4] has undertaken to detect thermal strains generated in the head and image their distribution by using SPM. Their approach is based on the fact that heating generates surface displacements by thermal expansion. The displacements can be quantitatively imaged by heat modulation. This is accomplished by supplying currents to the heater, which are in turn modulated at a frequency, generating cyclic displacements in the surface, visible as surface protrusions. The displacements are measured by an SPM with lock in detection. The lock in amplifier is tuned to the frequency of the heater current. The cantilever vibrates at the current frequency. This lock in method allows simultaneous imaging of the changing surface topography as well as thermal amplitude and phase. As such, amplitude and phase signals given by the vibration of the cantilever are mapped corresponding to the scanning signals. In this case, amplitude images quantitatively show the surface protrusion profiles which correspond to thermal distributions. The phase signals are nearly equal in regions with thermal strain, and thereby illustrate with enough precision the extent of thermal expansion. Thus, the thermal strain distribution will depend on the extent of thermal diffusion which was in this case about $100\,\mu\text{m}$. This means that the surrounding area was not heated; thereby variations in room temperature barely had an influence on the experimental results. Nevertheless, these were static measurements, not accounting for air-bearing pressure due to magnetic hard disk rotation. In any case, the SPM technique employed by the researchers for thermal strain imaging has proven to allow characterization of head spacing. It thereby represents a step forward towards thermal control of fly height of the magnetic heads.

In order to increase the areal density in magnetic recording, the fly height of the head needs to be reduced. Read/write elements must be in close proximity to the recording layer to confine the writing fields to the regions where information is to be recorded, but also to be able to read back the fringing fields. Given the thermally enabled control of fly height, these days head designs include a heater embedded under the magnetic poles of the head, providing thermal protrusions on the air-bearing surface. It should be noted that this design necessitates that the magnetic head be cooled. The gap that requires control is less than 10 nm in size in high-density hard disk recording technologies. When the disk rotates, track changes occur in the gap, making dynamic control of the gap a necessity. Deformations in the air-bearing surface cause the head slider to push back, increasing the gap. The popular all-metallic giant magnetoresistive heads in a current-perpendicular-to-plane geometry have allowed high-density recording sensors due to their superior head design. However, to keep up with rising recording densities, the impedances of these heads have also increased as the cross-sectional area of the sensor was reduced, adding some extra problems to head design. Fortunately, the redesigned improved magnetic heads use nowadays a dual spin valve sensor stack with thin IrMn anti-ferromagnetic top and bottom pinned layers of the type discussed in Chap. 6. These spin valve films also benefit from corrosion resistant configurations by including Ta/NiFe/CoNiFe/CuAu/CoNiFe/PtMn/Ta multilayers built into conventional shield-type giant magnetoresistive heads. Nevertheless, additional challenges await the design of giant magnetoresistive heads in the quest to obtain higher stability, an increased signal to noise ratio, as well as a better resolution. The number of magnetic bits written or read per unit area has increased in recent years at unimaginable rates, and still continues to do so provided improvements in head design allow it, as well as advances in magnetic materials that serve the head, the recording medium, and the spacing between them.

8.2 Bit Patterned Magnetic Media

8.2.1 Advantages and Disadvantages of Prepatterning

Prepatterned magnetic media have been observed over time to satisfy many of the thermal stability requirements. In particular, alloys such as Co/Pd and CoCrPt are again among the materials of choice, in addition to some chemically stable oxide compounds. By patterning the medium, magnetic islands form stable entities which are less likely to switch under thermal fluctuations. Thus, such magnetic

Bit cell

Fig. 8.4 A bit cell in traditional magnetic tapes (*left*) vs. bit cells in a prepatterned magnetic medium (*right*). The grains in the bit cells of conventional tapes are randomly oriented; therefore, statistical averaging over several grains is necessary when recording or reading a bit. In contrast, bit cells of patterned media are formed from single prepatterned entities with a well-defined magnetic easy axis

islands form bit cells, and because these are consequences of prior patterning, the media are known as *bit patterned magnetic media* or *discrete bit media*. This technique of prepatterning the medium is highly favored these days and is likely to continue to be around as a more reliable data storage solution, even as new recording media are being researched. The more dependable discrete bit cells have the potential of maintaining a thermally stable single-domain state, thereby offering several advantages over continuous media. In every bit cell one bit of information is reliably stored, forming more ordered arrays of discrete magnetic entities containing information.

Previously, the conventional magnetic tapes contained in one bit several randomly oriented grains as shown schematically in Fig. 8.4 (left). Nowadays, bits of patterned media are formed from single prepatterned entities, allowing a common and well-defined easy axis of constant orientation with respect to the magnetic head [Fig. 8.4 (right)]. By design, superparamagnetic magnetization reversal is less likely to occur in media with patterned bit cells. Prepatterned magnetic bit cells have, therefore, some distinct advantages over the magnetic grains of conventional media. They allow for instance a larger volume per bit, decreasing the number of magnetic switching volumes per bit to a single entity. This makes the statistical averaging over several grains unnecessary. Because the bit size determines the switching volume, the bit cells are not only thermally stable, but can also be reversed by the magnetic field of the recording head. Furthermore, transitions with magnetic fringing fields across and down the track are now defined by the bit pattern, instead of being determined by the magnetic field of the head.

Like with any technology, not everything is problem free in bit patterned magnetic media. For example, some written-in errors have been noticed that differ from the usual signal to noise ratios. These errors are believed to occur due to statistical fluctuations of the magnetic properties, as well as the locations of the magnetic islands. The size, shape, and location of the patterned magnetic islands could vary from one bit cell to another, most likely due to lithography or self-assembly errors. Under these circumstances, playback heads need to be designed such as to maximize playback signal resolution while minimizing noise sensitivity due to patterning imperfections. The *filling ratio* or *packing density* is a quantity that represents a percentage of the bit area filled with a magnetic material.

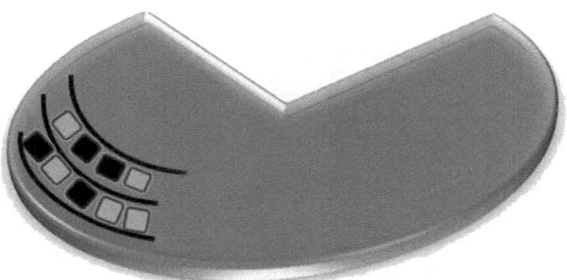

Fig. 8.5 Patterned bits organized in a circular ordered array along patterned discrete tracks at predetermined locations on the magnetic hard disk. The discrete tracks aid in keeping the head on track, relaxing the requirements on head tolerances. The size of the sketched features is exaggerated for clarity. The magnetic easy axis is perpendicular to the film surface, the film forming a perpendicular magnetic recording medium

Unfortunately, it was noticed that the playback signal amplitude experiences a linear loss with filling ratio. Therefore, a head with a differential reading ability allowing bit comparison is likely to perform better with respect to spatial resolution and show increased sensitivity to media imperfections. Nevertheless, bit patterned media allow increased recording densities, provided some challenges specific to prepatterning are overcome, such as imperfections in bit spacings, or size, thickness, and shape irregularities.

Media geometries cannot easily accommodate prepatterning of bit cells. As such, to pattern a longitudinal medium it becomes necessary to induce circumferential magnetic texturing [5], difficult to achieve using polycrystalline films and substrates. This circular symmetry of the magnetic patterns is an unfortunate fabrication challenge for prepatterned media for which Cartesian e-beam lithography developed for mask production in the integrated circuit industry cannot be employed efficiently. For this reason, prepatterned media are being fabricated with the magnetic easy axis normal to the film surface to form a perpendicular recording medium. The patterned bits are then organized in a circular configuration, allowing the head to remain on a single track of units as it goes over a certain data region (Fig. 8.5). By prepatterning tracks on the disk, track locations are physically defined, the head thereby no longer relying solely on the mechanical positioning of the hard drive. Additionally, in order to ensure the head stays on a given path, *servo marks* (shown in Fig. 8.6) are also defined on the disks. During the servo writing process, the head is flown over the entire disk surface while the servo marks are recorded. In traditional recording, these servo marks are written on a plain continuous medium with a recording head, but without predetermined track locations. In patterned media, the locations are predefined which requires the servo marks to be patterned as well.

The thickness of the magnetic medium on which the writing occurs needs to be of the order of the bit spacing to prevent the head from inadvertently writing neighboring areas with the stray fields of the head. Also, fringe fields from adjacent bits can

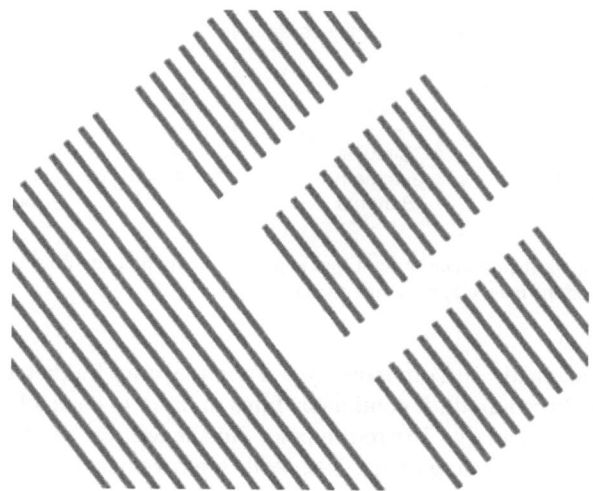

Fig. 8.6 Patterned servo marks at predefined locations on patterned media. The servo marks ensure that the recording head stays on a given path

influence the writing process, and these fields scale with saturation magnetization. Thus, a major difference between discrete bit and traditional media is the need for the synchronization of the head position with the predefined mark locations. In spite of many obstacles, perpendicular discrete bit media have gradually overtaken the magnetic recording market. They have a high-density storage potential given by the larger head write fields, but also superior thermal stability of their single-domain islands. These individual magnetic entities have the added advantage of eliminating the jitter noise of granular longitudinal media that is generally attributed to the grain irregularities in transition regions. Nevertheless, the main source of noise is likely to originate in bit size and bit pitch variations. For this reason, the write field has to be well situated to avoid writing on adjacent bits, while ensuring it writes on the desired bit. Cleanliness and smoothness are also issues, as they can interfere with the ability of the head to fly at the required fly height.

8.2.2 Some Prepatterning Techniques and Their Challenges

While traditional media are very cost efficient, prepatterning on a nanoscale can make disk fabrication economically intolerable. Cost and accessibility are, therefore, important challenges for many fabrication processes currently employed in prepatterned media production. Prepatterning techniques are expected to satisfy the four criteria, summarized in Fig. 8.7. As such, the criteria are: inexpensive fabrication, a capability to produce circular symmetry patterns, have the potential of sub-50 nm resolution, and result in a clean surface for the head to fly over.

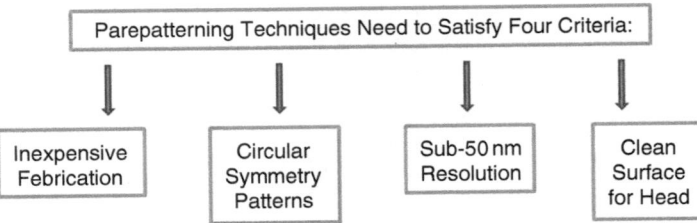

Fig. 8.7 The four most important criteria that magnetic media prepatterning needs to satisfy to stay competitive on the magnetic recording market

Fortunately, a magnetically soft underlayer is not necessary for patterned bit media, as was previously required for continuous films in order to channel flux between a small write pole and a large flux return pole. This underlayer would add unwanted complexity to the patterning of the bits. With improvements obtained nowadays, some types of lithography processes may meet those four criteria; however, they all have their limitations. For instance, optical lithography is limited by its resolution given by $\lambda/(2NA)$, where λ is the light wavelength used and NA is the system numerical aperture. Some resolution enhancement techniques may be possible so that line widths below 90 nm can be achieved. As such, by using immersion lithography, a reduction to 45 nm has been obtained, nevertheless still not sufficient for today's requirements for patterning the recording medium. Only extreme UV light or X-rays can attain the necessary resolution demanded by the market, regrettably these techniques are not commercially viable due to cost issues, yet the situation is improving.

An alternative prepatterning technique gaining more and more recognition is interference lithography, with some reported successful fabrication [6] of magnetic Co nanodots [7], and Co or Ni nanopillars [8]. Figure 8.8a shows a schematic of the fabrication steps that are usually part of the interference lithography process. These patterned dots and pillars have been obtained through successive 90° rotations of the sample between exposures, resulting in patterned areas with sizes depending on the power of the laser beam. Figure 8.8b shows examples of Co nanodots imaged by MFM and obtained by an interference lithography fabrication process. The advantage of interference lithography is that it is maskless due to patterning being done by interference fringes. The fringes are produced by two optical beams giving rise to a pattern period $p = \lambda/2\sin\theta$, where θ is the half-angle between the two beams. In this manner, it was possible to fabricate entire arrays of magnetic nanodots [9], and not just individual dots. Additionally, interference fringes have been successfully used for producing a master for nanoimprinting [10], or for precise, local annealing of previously deposited films [11]. Unfortunately, the sub-50 nm resolution required for discrete bit recording is difficult to achieve by interference lithography because of limitations given by the $\lambda/2$ minimum period requirement. However, as a solution to this problem, a series of diffraction gratings could be used for producing interference minima and maxima. In this manner, achromatic

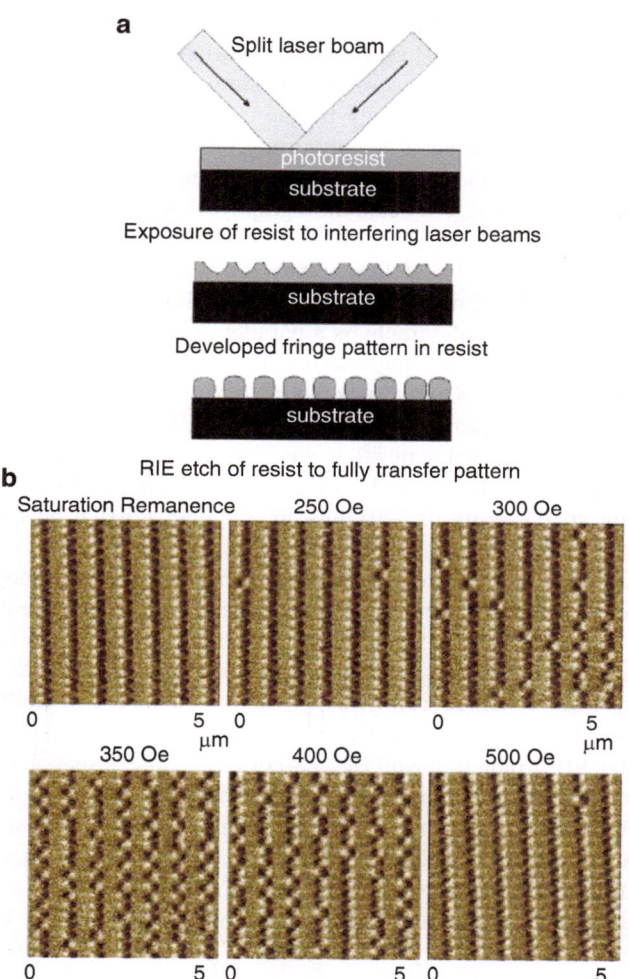

Fig. 8.8 (**a**) Fabrication steps involved in an interference lithography fabrication process. (**b**) Examples of MFM images of 90 nm × 200 nm Co dots fabricated by interference lithography. Reprinted with permission from [6]

interference is obtained that can overcome the wavelength limit of monochromatic interference. Magnetic nanodots have been patterned by achromatic interference with a fringe pitch of 100 nm [12]. Diffraction gratings can also create circular beams not otherwise obtainable with linear fringes of conventional interference lithography. Patterns with circular symmetry have been reported [13], thus making interference lithography a very promising prepatterning technique.

Some prepatterning techniques have been successfully demonstrated in laboratory environments, but have yet to prove themselves for industrial scale fabrication because they are too expensive and too slow for writing processes. Techniques such as ion beam bombardment [14] or electron beam (e-beam) lithography [15] have patterned magnetic nanostructures of sub-100 nm size. Their disadvantage is that the resolution is not determined only by the e-beam diameter but also by resist properties. These are responsible for proximity influences of the overlapping low level exposures that occur due to electron scattering in the resist and substrate [16]. Despite shortcomings, e-beam lithography should not be dismissed as a viable technique for the fabrication of masks used in master/replication methods. Masks fabricated using e-beam lithography are also used in other lithography techniques, specifically X-ray lithography that employs the soft rays of 1–10 nm wavelength. Since X-ray lithography is a parallel process [17], an area as large as the mask can be patterned in a single exposure. Nanoscale features such as 88 nm permalloy dots [18], or 200 nm Co dots [19] have been patterned. Nevertheless, X-ray lithography requires a high-intensity X-ray synchrotron, something not easily available. Even so, patterned features on a nanoscale are challenging to create even when infrastructure challenges and cost issues are resolved. This is because, for example, a bit areal density of 300 Gbits/in^2 requires a 30 nm bit cell. Similarly, 1,500 Gbits/in^2 necessitates a 14 nm cell.

Due to cost and small feature size challenges encountered with lithography, other prepatterning techniques are currently being researched, among them *focused ion beam* (FIB) methods using Ga^+, He^+, or Ar^+ ions [20]. FIB etching with Ga^+ ions has been successfully utilized for patterning servo marks [21] or 100 nm patterned bits [22] for longitudinal recording media. Ga^+ implantation in CoPtCr films can also lead to single-domain remnant states for structures smaller than 70 nm [23]. On the other hand, FIB irradiation with He^+ [24] or Ar^+ [25] can spatially modulate the anisotropy of magnetic systems such as FePt. These are usually ordered films but can be made disordered to some degree and their anisotropy increased over larger areas if they are irradiated and thermally annealed [26]. The interfacial anisotropy of Co/Pt multilayers can also be reduced by FIB [27].

Unfortunately, FIB techniques require radiation resistant masks. As such, resist masks of 600 nm are necessary for He^+ ions of 30 keV, whereas tungsten masks need only to be 200 nm thick [28]. A practicable alternative may be to selectively block ion exposures via contact resist or masks placed directly on the films, such as membrane masks made of etched silicon [29]. A few hundred nanometer thick Si membranes are sufficient for stopping light ions of tens of kiloelectronvolts [30]. Radiation resistance is not the only challenge for FIB masks. The sub-50 nm size requirements for discrete bit recording add further restrictions to these masks. Nevertheless, fabricating membranes for 2.5 inch diameter disks but with very tiny feature sizes may be accomplished with *projection ion beam lithography*, as the diffraction limit of ions does not limit the achievable resolution. With ion wavelengths of the order of 10^{-5} nm, the diffraction limit is only a few nm. A report for SiO_2 masks quoted a resolution limit of 14 nm [31], higher than the *magnetic resolution*. The magnetic resolution defines the magnetic limits due to exchange coupling in magnetic films, and is useful when single domains form. A real concern is too dense magnetic features because in this case projection systems would require masks ten times larger than the disk diameter if patterning is to be done in one exposure [32]. In any case, all prepatterning techniques mentioned above may lack the high throughput necessary for the industrial fabrication of patterned media.

8.2.3   Self-Assembly and Self-Ordering

Natural processes such as self-assembly of templates or self-ordering of nanostructures may offer an appealing and much needed alternative to expensive lithography methods not only because of cost efficiency, but also because of their flexibility. For instance, self-assembled pore templates (shown in Fig. 8.9) distinguish themselves through their capability of forming regular arrays of nanometer sized pores. The pores form on the surface of some materials such as aluminum through a simple anodization process, whereby regular hexagonal arrays of nanometer sized pores (most commonly 10–100 nm) [33] are obtained. These templates can be removed and placed on a different substrate, allowing to be subsequently filled with some material of choice. Their versatility lies in the fact that when employed as masks, self-assembled pore templates [34] can be used for etching or depositing just like an exposed resist. Furthermore, entire regions can be ordered for stretches of over 100 μm, making them similar to grains in traditional magnetic recording.

In some cases of self-assembly, the length scales of self-ordering can be increased by templating the surface before further processing using ion beam exposure or interference lithography. This possibility of combining lithographic processes with natural templates is known as *guided self-assembly* and may solve the problem of coherent long-range ordering. Two categories of techniques can be distinguished as

Fig. 8.9 An example of a self-assembled pore template. Scanning electron microscopy (SEM) image of an alumina mask/membrane self-assembled through a simple anodization process. The alumina membrane displays throughout its thickness a hexagonal array of nanopores similar to a honeycomb structure. [C.-G. Stefanita, S. Bandyopadhyay]

guided self-assembly: *chemical*, where the surface chemistry is selectively modified to allow formation of nanostructures in some areas [35–37], and *topographical*, where the surface geometry is altered [38]. In chemically guided self-assembly, the surface of the substrate is selectively functionalized using for instance poly-dimethylsiloxane so that self-assembled monolayers (SAMs) develop, thus helping the formation of patterned structures [39]. On the other hand, a topographically modified substrate necessitates a combination of masks, stamps, and etching to achieve the desired prepatterning. Packing densities of $450\,Gbits/in^2$ have been reported for magnetic nanodots fabricated in self-assembled templates [40].

Researchers have also been experimenting with other alternatives for data storage media, such as biological self-assembled templates [41] made of proteins. *Ferritin* is a promising protein as it has the capability of self-assembling into a hollow sphere with an $\sim 8\,nm$ cavity. FePt or CoPt nanoparticles can be grown in an *hcp* structure inside the cavity. The arrangement period of the grown structures is influenced only by the ferritin dimensions and not the particle within the ferritin cavity. A problem is the easy axes of the particles which are difficult to orient. However, the major obstacle is the packing fraction which is less than that for conventional recording media [42]. Thus, self-assembly and self-ordering techniques have still a long way to go before they reach a developmental stage that offers a competitive edge for data storage media.

8.2.4   Nanoimprinting

The magnetic recording industry has adopted techniques similar to CD-ROM fabrication to solve, at least for now, its expensive storage problems. CD-ROMs are manufactured by *master/replication* methods where a laser writer produces the master, and its imprint is replicated. Despite rather steep costs, mass manufacture of CD-ROMs by injection moulding from the master is still a viable alternative to other expensive fabrication methods. As such, a similar technique known as nanoimprinting has gained more and more recognition in magnetic recording. The nanoimprinting method uses a rigid or flexible mould/stamper [43] to emboss a relief pattern into a thermosetting or photocuring polymer resist layer [44]. The resist is first patterned with the stamper. Next, the resist is used as an etch mask to further transmit the pattern onto the underlying layer [45]. The advantage of the technique is that the primary stamper can be duplicated using nanoimprinting, while the resulting replica can be used to produce duplicate moulds/stampers.

The patterned resist layer can be made of PMMA heated above its glass transition temperature. A stamper made of Si or SiO_2 is pressed onto it. Self-assembly can also be utilized to pattern the mould or stamper, nevertheless a widespread technique is to use e-beam lithography. A requirement is that the patterned layer be thick enough to shield the unexposed areas underneath. It needs to be at least as thick as the necessary etch depth. The substrate below the SiO_2 film is patterned by reactive ion etching (RIE). This substrate is typically a silicon wafer with a thin metal layer on top that ensures adhesion of the SiO_2 film above it. An obvious advantage of nanoimprinting is that it requires less rigorous lithography constraints. It is known that silicon processing is notorious for its exact alignments, where transfers and repeat exposures delay the process as they add several important steps to fabrication. Severe requirements are enforced during silicon processing which is not the case for nanoimprinting.

Regardless of the fabrication technique, the issue of disk smoothness remains, as the recording head is supported by an air-bearing of sub-10 nm size. Unfortunately, the aerodynamic air-bearing above the disk surface can break down and compromise head performance. The cleanliness and smoothness of the surface are key operation elements for recording heads, as the smallest resist contamination can cause a serious collapse in head performance. Thus, the surface needs to be free of contamination such as resist residue. Only about 0.2–0.4 nm rms surface roughness is allowed, a serious fabrication challenge for those techniques that necessitate comprehensive cleaning steps without damaging the magnetic film. Protective layers usually made of carbon derivatives are used to cover the top. The protective layers have to satisfy the additional requirement of preserving the edges of any patterned features.

RIE is not usually employed for magnetic layers in nanoimprinting processes used for resist patterning. This is because the necessary high aspect ratio required for nanopillars is hard to obtain. Although this problem can be solved, it is still difficult to release the mask due to the high adhesion force in large pillar contact areas. It may be necessary to use a SiO_2 layer on a plating base, that is then etched while a magnetic film is deposited onto the imprinted layer. A followed lift-off process will subsequently expose the nanodots [46]. A reasonable solution for maintaining the required small head disk distances is to pattern the substrate rather than the magnetic film. This is accomplished by creating deep islands and channels in the disk, while etched depressions allow neighbors to decouple from each other. The disk can be easily cleaned after patterning without affecting the magnetic film, however, protective coating is still recommended. The idea of etching and patterning the substrate instead of the magnetic film may, therefore, be the best solution. The magnetic film can then be deposited while the side walls of the trenches provide the necessary domain isolation [47]. Implementing this idea of imprint patterning and etching the substrate has produced 30 nm Co/Pt nanodots with perpendicular anisotropy [48], and discrete tracks in NiP-coated Al disks with CoPtCr films [49].

Example 8.5. Give an example of fabrication of magnetic nanodot arrays by using a current nanoimprinting technique.

Answer: Lithography patterning of a resist layer that is subsequently transferred to a magnetic film by etching/lift-off processes may lead to degradation of shapes or properties when the desired features are of nanoscale dimensions. For this reason, the magnetic recording industry is using other nanoimprinting techniques such as selective exfoliation of sputter-deposited magnetic thin films. Such deposition/exfoliation processes have been previously used for fabrication of either Ni stampers (molds) by electrodeposition, amorphous Si nanoimprint molds by

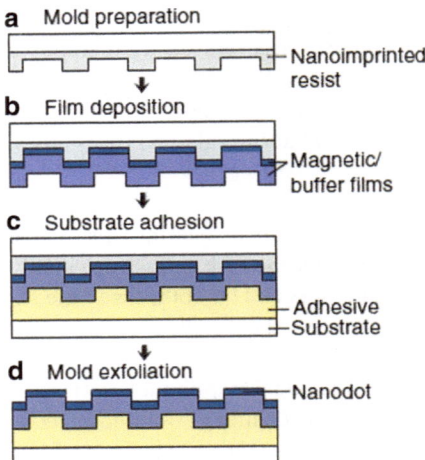

Fig. E 8.5.1 Schematic of the fabrication process depicting: (**a**) preparation of the polymer mold by nanoimprinting. (**b**) Sputter deposition of magnetic and buffer layers. (**c**) Bonding of the glass substrate by UV-cure adhesive. (**d**) Exfoliation of the polymer mold by mechanical separation leaving behind a perpendicular magnetic nanodot array. Reprinted with permission from [50]

plasma enhanced chemical vapor deposition, or Au patterns by evaporation [38]. The advantage of using this fabrication technique is that there is no etching or lift-off which can cause pattern distortion and roughness around the edges. At the same time, the deposited films are firmly bonded to the substrate which would be difficult to obtain if they were simply deposited onto nanoimprinted polymers. The obtained nanostructures can be easily and economically duplicated once the mold is prepared, while it is also possible to deposit relatively thick layers on top, after the initially deposited layer is revealed above the nanostructures. This may be useful if a backing layer is required or additional soft magnetic underlayers are necessary for patterned media.

Figure E 8.5.1 shows a schematic of the most important steps during the fabrication process of the perpendicularly magnetized Co/Pd nanodot arrays discussed in this example [50]. To obtain the nanodots, a Ni mold is replicated by electrodeposition from a master pattern that was previously obtained by e-beam lithography. The mold is then pressed onto the thermoplastic resist to nanoimprint it (Fig. E 8.5.1a). The thermoplastic resist is made of PMMA spin-coated on a Si substrate, and once imprinted it is utilized as a polymer mold for the magnetic nanostructures. The features on the mold are about 60 nm in diameter and 20 nm in depth. Co/Pd multilayer films and a Pd buffer layer are dc sputter deposited onto the polymer mold in an argon atmosphere and at room temperature (Fig. E 8.5.1b). A flat glass substrate is attached to the back surface of the deposited films using a liquid UV-cure adhesive (Fig. E 8.5.1c). After UV curing, the film is exfoliated from the mold so that it almost precisely follows the inverse structure of the mold (Fig. E 8.5.1d). This relatively exact inverse pattern reproduction is possible because of the poor

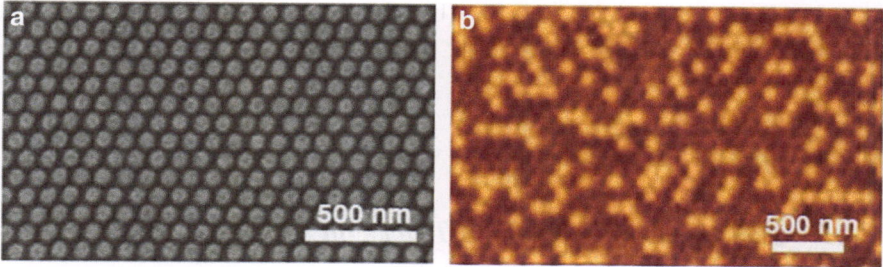

Fig. E 8.5.2 (**a**) SEM image of top view and (**b**) MFM image of exfoliated Co/Pd multilayer handout array with 100 nm pitch. The MFM image is acquired in the remnant state after first applying a negative saturation field and then exposed to a 1 kOe reversal field. Reprinted with permission from [50]

adhesion of the films to the mold which separate mechanically quite easy after bonding the films to the substrate through an adhesive agent. The poor bonding also allows a clean separation without requiring any additional cleaning steps. In the end, perpendicular 60 nm Co/Pd nanodot with sub-100 nm dimensions and arranged hexagonally with a 100 nm pitch are formed over a macroscopically large area of about 18 mm in diameter.

Surface images, topography, and magnetic-domain structure were subsequently obtained by SEM (Fig. E 8.5.2a), AFM, and MFM (Fig. E 8.5.2b) to investigate the samples, and confirm fabrication results. Additionally, magneto-optical Kerr effect (MOKE) as well as superconducting quantum interference device (SQUID) magnetometry investigations were able to give details about the magnetic switching behavior under an applied field. A low moment magnetic probe is used for MFM imaging so not to perturb the domain structures. Each nanodot is in a single-domain state with an either "up" or "down" magnetization. A good exchange interaction exists between the well-isolated dots, as illustrated by the bright spots, proof of individual magnetization reversal.

The procedures mentioned thus far are known as *additive nanopatterning*. This is because a nanopattern is created during an initial step, while the magnetic film is deposited later. During these processes, a resist layer is patterned into an array of depressions which are afterwards exposed to evaporation or sputtering of a magnetic material into those depressions. When resist is used, it is removed at the end of the process by using lift-off. As such, the patterned resist is dissolved in a solvent, while the deposited film on top of the resist is also removed during this process. On the other hand, in *subtractive nanopatterning*, the magnetic films are placed underneath masking layers so that etching of the masks affects the magnetic films as well. The patterned etch mask allows the excess magnetic film to be removed by either ion milling, sputter etching, or wet chemical etching. Unfortunately, ion milling is not the best selective process because different materials have similar etch rates. For this reason, thick polymer masks or hard masks made of metallic films have to be used. In these cases, etch masks obtained by self-assembly may offer an economical alternative to more expensive masks [51].

8.3 Discrete Track Media

8.3.1 Usefulness of Predefined Tracks on Media

An efficient way to increase areal densities may be offered by discrete track media. What this means is that the surface of the magnetic film of the medium is patterned into tracks by etching for instance through an e-beam defined resist mask. In this way, bit transitions can be written into the tracks, allowing certain transition widths between adjacent bits to be defined. Also, head tracing errors can be reduced if tracks are patterned and track locations are physically made out on the disk. Then problems due to mechanical vibrations and imperfect head positioning can be eliminated if misregistration of the head to the track location after each revolution is reduced. A requirement is that regions between tracks should be made nonmagnetic or at least placed at a lower level than the head in order to prevent the read element from sensing any magnetic response from them. It is also possible to manufacture the read element so that it is of comparable width with the write element, thus reducing the effects of side reading. This would improve fabrication yield and lessen requirements on head tolerances. In the end, the linear recording density will depend on the size and uniformity of the tracks, nevertheless, write fringing, erase band, and track-edge noise will still limit the feasible track density [52].

An advantage of having tracks physically defined is that a greater signal to noise ratio is possibly achieved, although writing resolution at the track edges remains an ongoing problem. Even if magnetic isolation of grains improves the write resolution at the track edges, with narrow track recording of $\sim 100\,$nm or less expected for densities of $1.0\,$Tbits$/$in^2 and beyond [53, 54], track edge irregularities are likely to contribute to the track-edge noise. Therefore, at these track widths, e-beam lithography becomes a necessary fabrication step, making the additional processing not cost effective. It is nevertheless possible to use embossing or nanoimprinting to pattern the substrate prior to depositing the film [55]. By etching grooves into the top layer covering the disk, a magnetic film can be deposited on top of the patterned film. If the grooves are deep enough, there is reduced likelihood of a readback amplitude.

A magnetic servo pattern could be used to define the track positions so that data bits can be repeatably written and read by the magnetic head. In this case, the magnetic head is flown over the entire disk surface while track by track servo marks are being recorded during the writing process. When the write field from the recording head delineates the bit length along the track and the bit width across the track, recorded servo marks can thus identify the track locations. A major disadvantage is that this is a slow and expensive process, at least when compared to the stamping or injection molding of optical media such as CDs and DVDs.

With cost and speed issues remaining a significant fabrication challenge, it is up to techniques that meet industrial requirements to demonstrate that discrete track patterning has long-term potential.

8.3.2 Material and Fabrication Requirements for Discrete Tracks

The field gradient from the write head determines down track bit transitions even with lithographically defined track widths. For this reason, material requirements for discrete track media are comparable to those of traditional magnetic media. As such, it is common to use for discrete track media granular CoPtCr-based films [56] deposited onto either prepatterned glass or NiP substrates [57]. At some point, preembossed rigid magnetic disks (PERMs) were used for parallel magnetic data recording, so that the embossed pattern contained track following marks (Fig. 8.10) and magnetic data were recorded on the lands of a land/groove structure [58]. This enabled the simultaneous writing of information on several disks, including formatting the media by inscribing sector headers or position marks. These PERM disks were manufactured by embossing and etching a land/groove pattern into a glass disk. A magnetoresistive device was used to detect an error signal that was representative of the track following on these large grooves. In spite of mechanical complexity, magnetic disk drives are roughly 100 times cheaper to produce than semiconductor memory (DRAM) [59]. The semiconductor memory wafer requires about 10^{10} features, whereas a 5-inch diameter wafer of magnetic heads typically needs only 10^5 features to produce ~40,000 heads. As such, magnetic recording heads are fabricated with fewer processing steps than semiconductor-based memories. Only track widths need to be lithographically defined at the minimum size of ~120 nm, half the track pitch in 40 Gbits/in^2 disks. This is in contrast to the down track spacings between transitions that are defined by the thin film deposition process. In this case, a single head and disk may be able to address 30 GB of data [60].

A magnetic pattern can be transferred by creating a master, then etching a substrate after coating the surface with a soft magnetic film that covers the grooves and lands. An efficient mass replication of content by transferring magnetic patterns onto magnetic disk media is obtained with some types of lithography techniques. These allow improvement of the patterning process of servo information. For instance, *magnetic lithography* [61] uses such a master to transfer the magnetic pattern onto the substrate which is referred to as a slave. The slave requires to be uniformly magnetized along one direction before the mask is placed in closed contact with the disk. Afterwards, the direction of the external magnetic field is reversed while the magnetic mask selectively shields the slave disk from the reverse external field. The magnetic pattern is transferred in the meantime onto the disk [62]. The gap where the substrate is etched prevents magnetic shielding from the grooves, so the pattern can be transferred accurately [63]. Rigid masks with air channel and vacuum [64] can be used to transfer magnetic patterns. Alternatively, rigid Si masks

Fig. 8.10 Patterned discrete tracks. Track widths (shown in *black*) are usually lithographically defined at the minimum size of ~120 nm. However, when preembossed, rigid magnetic disks (PERMs) are manufactured by embossing and etching; groove patterns (shown in *gray*) are embossed into a glass disk. See text for other patterning techniques

with flexible disks such as metal tape [65] can also be used for the transfer. Masters employed in magnetic lithography can be used over one million times to print disks [66]. Even perpendicular recording media with coercivities as high as 10 kOe have benefited from the printing technique [67]. Nevertheless, transferring patterns by magnetic lithography is not likely to reach the tens of nanometers resolution required for high-density recording.

8.4   Alternative Storage Media

8.4.1   The Need for Alternative Storage Media

Many challenges still remain in recording head design and magnetic media fabrication [49] as bit areal density growth has slowed down to 40% per year in recent years. Increases in magnetic storage density have been achieved so far mainly by scaling the components of the disk drive to smaller dimensions, allowing the head to move closer to the disk while enhancing the write and read resolution [68]. Fortunately, improvements in head design have also increased performance so that nowadays high-coercivity media can be written at high data rates. These days, computers use several types of memory for different purposes of storage, such as the already mentioned solid state memories (DRAM) necessary for computational operations. With processors performing 20 million operations while magnetic heads read one bit, more efficient processing of information is attainable with DRAMs which use the electronic state of transistors and capacitors to store data bits. As such, alternative magnetic recording technologies need to be developed in order to achieve higher storage densities while allowing those fast processing speeds of information. It seems that nanotechnology is offering new solutions for increased data densities

[69] as magnetic recording methods discussed so far reach their physical limitations. With magnetic nanostructures becoming the basic units for patterned media, they open new doors for higher storage densities and novel disk drives.

8.4.2   Racetrack Memory

Interesting spintronic chips with *racetrack memory* made of magnetic domains within nanowires have become nowadays one of the latest storage devices to be comprehensively investigated [70]. Nanowires lie horizontally (Fig. 8.11a) or stand vertically (Fig. 8.11b) on silicon substrates while magnetic domains contained in the nanowires store bits of data. The domains are moved rapidly along the tracks by pulses of electric current. Magnetic fields emanating from the domain wall set the magnetization direction in the racetrack's bit. Thus, the domain wall separating two oppositely magnetized domains is used to write a data bit on the racetrack. While the magnetic domains pass a sensor lying underneath the nanowire, the changing magnetization between oppositely magnetized domains is recorded (Fig. 8.11c). Because of this rapid displacement along the track past read/write heads placed at fixed locations, these information storing magnetic nanowires are known as "racetrack memory." The magnetic domains are nonvolatile and rewritable, thereby comparable to HDD. Nevertheless, one of their big advantage over HDD is that the chip has no mechanical moving parts because only electrons read and write bits. This also makes them very fast, while not compromising on reliability.

Each magnetic domain represents a "1" or a "0" of stored data, and retains its data when power is switched off. The sensing layer of the magnetic head

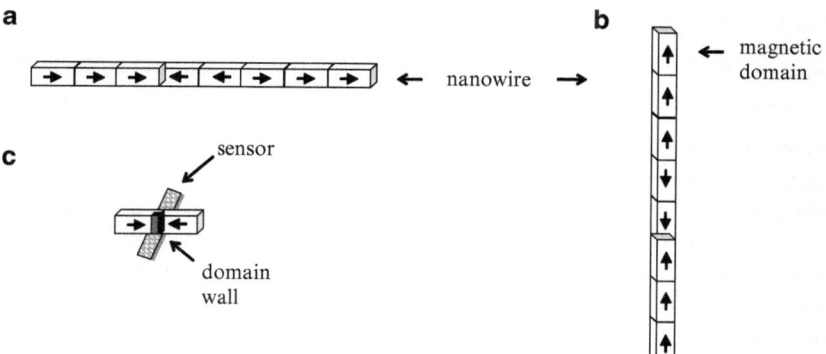

Fig. 8.11 Racetrack memory made of magnetic domains within nanowires: (**a**) lying horizontally and (**b**) standing vertically on silicon substrates. Magnetic domains contained in the nanowires store bits of data, while the domains are displaced rapidly along the tracks by pulses of electric current. (**c**) While the magnetic domains pass a sensor lying underneath the nanowire, the changing magnetization between oppositely magnetized domains is recorded

changes its magnetization back and forth in order to match the field of each domain passing it. Domain walls can move as fast as 150 nm/ns, allowing access millions of times faster than HDDs [70]. With perpendicular columns of nanowires, large data densities become possible on one silicon chip, overcoming many limitations of their predecessor memories. As bits move rapidly along the racetrack, a new generation of data intensive applications is envisaged, especially since the magnetic heads themselves have evolved. From the spin valve sensors relying on GMR to the MTJs exploiting TMR (tunneling magnetoresistance) to achieve a greater sensitivity to small magnetic fields, new spintronic technologies come to the aid of magnetic recording. As such, the racetrack memories benefit from spintronics as well.

Traveling spin-polarized electrons of one magnetic domain will flip as they cross the domain wall into the next domain of opposite magnetization. But spin is a conserved quantity, therefore, if it flips, something else must flip into the state of the initial spin. This something else is an atom on the other side of the domain wall, the domain with opposite magnetization. With spin-polarized electrons flowing through the domain wall, the domain wall itself is moved along the nanowire one atom at a time. Upon reaching a domain with magnetization in the initial direction, the electron spin flips back as it crosses this wall while also moving it a tiny amount along the wire. Now both domain walls are moving simultaneously along the wire, meaning that the bit recorded between these walls moves as well. Research has shown [70] that nanosecond long pulses of spin-polarized current can cause as many as six domain walls to move simultaneously along magnetic nanowires. It is, however, possible for domain walls to drift away, especially when influenced by small stray magnetic fields or variations in the controlling currents. This can be prevented if small grooves are built into the racetrack, at the intended size of the bits. The grooves will then pin the domain walls because the walls would have the least energy at those locations. The following example illustrates part of current research efforts into the development of racetrack memories that use electrical currents for domain wall displacements.

Example 8.6. Describe an experiment for an all electrical control and local detection of multiple magnetic-domain walls in perpendicularly magnetized nanowires.

Answer: It was mentioned in the previous chapter that controlling the direction of magnetization by electrical means is nowadays a highly desired feature for the development of many magnetic devices. Also, we have seen in this chapter that racetrack memory requires ferromagnetic-domain walls to be displaced by injecting an electrical current. For this reason and given the success of nanotechnology, some laboratories are actively investigating a variety of nanowire systems with domain walls that can be moved electrically, systems to be used in magnetic memory applications. These nanowire systems are usually made of Ni/Fe compounds with an in plane magnetization, nevertheless a variety of other types of systems exist. For instance, a group of researchers [71] in Japan is studying perpendicularly magnetized Co/Ni nanowires where a series of domain walls are reproducibly displaced in the same direction with the help of an electric current. Magnetization measurements

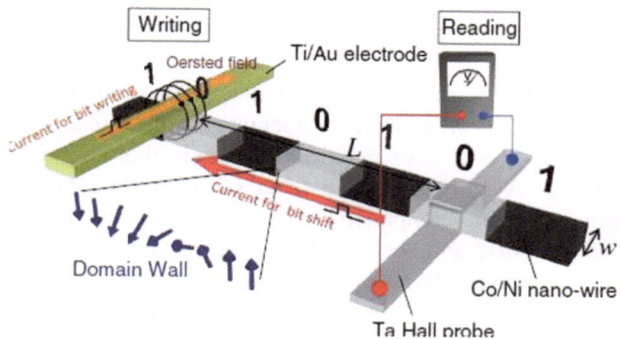

Fig. E 8.6 Co/Ni nanowire device for writing magnetic-domain bits. A Ta Hall probe serves as a local detector for reading the bits. A current flowing through the Ti/Au electrodes controls the domain wall displacement. Reprinted with permission from [71]

are used to confirm that the easy axis of magnetization is perpendicular to the film. All measurements are performed at room temperature (Fig. E 8.6).

The Co/Ni nanowire device is manufactured by sputter depositing a multilayer [Co(0.3 nm)/Ni(0.9 nm)]$_4$/Co(0.3 nm) film through the buffer layers and onto an undoped Si substrate. The nanowires are fabricated from the film by e-beam lithography and Ar ion milling. Two Ti (5 nm)/Au (100 nm) electrodes are formed on the nanowire, as well as a Ta Hall probe. The Hall probe detects the direction of the local magnetization of the wire underneath the probe through the anomalous Hall effect where the Hall resistance is proportional to the perpendicular component of magnetization. An external magnetic field is applied to align the direction of magnetization within the domain. To first create a single-domain wall in the wire, a current pulse is injected into the Ti/Au electrode which creates a pinning field for the domain wall. After the domain wall is created, a series of multiple pulsed currents are injected into the Co/Ni wire. The magnetization direction is checked by using a Hall probe after each pulse is applied. The domain wall is thus moved in the opposite direction of the applied current when the latter is above a threshold value. The wall moves between the two Ti/Au electrodes in a time determined by the interval between the switching of the Hall resistance. From this measurement, the domain wall velocity can be determined which showed a range of 20–50 m/s. The researchers succeeded in keeping the distance between the magnetic domain walls almost the same, while they also managed to shift the walls back and forth by changing the direction of the current. The advantage of having a perpendicularly magnetized system is that it has a much higher domain wall pinning field than that of an in plane magnetized system. This fact renders domain walls a better thermal stability, by avoiding unwanted motion due to thermal fluctuations. Also, the threshold current density for domain wall displacement can be controlled by the wire dimensions.

8.4.3   Memristors

Nanoscale devices that store information without the need for a power source are greatly sought-after for nonvolatile memory applications, even if they are nonmagnetic. As such, we have decided to include this short section about a class of new storage systems that fascinate data storage researchers, yet are based on nonmagnetic electronic components with memory. They are called "memory resistors" or *memristors*, and their properties (in particular, resistance) depend on the state and history of the system. As such, memristors show a "pinched" hysteresis loop in the two constitutive variables that define them, current and voltage (Fig. 8.12). We will discuss very briefly what this means for data storage, as well as a few underlying basic principles of operation. Nevertheless, these types of devices are not the focus of this book since they are not magnetic systems. Hence, we will not devote a lot of time or space to them.

Memristors have been described theoretically nearly 40 years ago [72], but remained a curiosity for a long time, since they were not realized practically. The resistance of the memristor depends on the internal state of the system, and it falls into the same category of two-terminal devices as, for instance, thermistors or some spintronic devices where the resistance depends on spin polarization. The uniqueness of memristors consists in the fact that their behavior cannot be obtained by a circuit built by using only the other three fundamental elements: capacitor, resistor, and inductor. Memristors are, therefore, considered the fourth fundamental element and are believed to be capable of enhancing the performance of digital circuits when employed alongside transistors on a chip, without the necessity of shrinking further the size of these transistors. A two-terminal device, the resistance

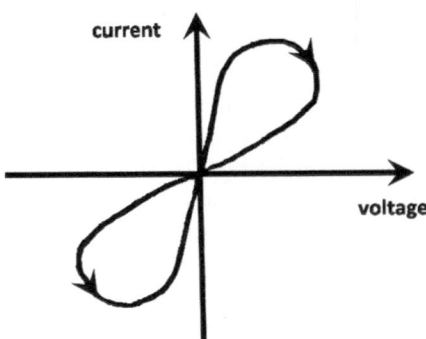

Fig. 8.12 A voltage-controlled memristive system with a "pinched" current–voltage hysteresis representing the switching behavior of the system. As charge flows through the device, the resistance drops and the current increases more rapidly with increasing voltage until the maximum is reached. As the voltage decreases, the current decreases more slowly because charge is flowing through the device and the resistance is still dropping, resulting in an on-switching loop. When the voltage becomes negative, the resistance of the device increases, and an off-switching loop is obtained

of a memristor depends on the magnitude and polarity of the voltage applied to it, as well as the length of time that voltage has been applied. Upon turning off the voltage, the memristor maintains its most recent resistance until the voltage is turned back on, even after prolonged intervals of time. Hence, it is said the memristor "remembers" its history.

Chua [72] described the memristor by associating the charge moving through a circuit with the magnetic flux surrounded by that circuit, similar to Faradays' law, but with an emphasis on the time dependence. As such, a memristor is characterized by a relationship between charge and magnetic flux which can be mathematically written as a time integral of voltage. Chua's relationship couples electric charge to magnetic flux, just like a voltage and current expression in a nonlinear resistor. This means that the resistance of the device varies according to the amount of charge that passes through it, and it remembers the value of the resistance even after the current is switched off. Some researchers noticed in practice this kind of behavior in micrometer scale devices built from polymers and metal oxides [73] and tried to explain it through some other means. Nevertheless, many of these "anomalies" were in fact unrecognized memristor behavior. It seems that memristors are intrinsic to electronic circuits, but this kind of property manifests itself only on very small scales, micro or nanometer in size. The scale on which electronic devices have been built so far has prevented, in most cases, an experimental observation of memristor behavior. Even millimeter scales are too large for any significant memristance effect to be observed [73]. Nevertheless, some research groups have managed to break through these practical barriers, and have reported experiments that illustrate the memristance effect.

With the advent and popularity of nanotechnology, memristors are starting to enter the world of electronic devices, even if sometimes in an unexpected manner. The memristors that have been intentionally built so far seem to depend on the crossbar architecture consisting of an array of perpendicular wires connected by "on" and "off" switches [73] (Fig. 8.13). An open switch represents a zero, and a closed switch represents a one; thereby this crossbar design functions as a storage system. The data are read by probing the switch with a voltage. The ability to store the resistance value when the voltage is turned off relies on the specific design and composition of these switches. Oxygen deficiencies in for instance TiO_{2-x} are repelled or attracted by the voltage applied on the switch. A positive voltage on the switch repels the positive oxygen deficiencies sending them into the insulating layer below. In contrast, a negative voltage on the switch attracts the positive oxygen deficiencies, pulling them out of the insulator. The resistance of the insulator thereby increases, making the switch as a whole more resistive. The more negative voltage is applied, the less conductive the material in the switch becomes. Furthermore, when the voltage is switched off, the oxygen vacancies do not migrate, therefore, maintaining the resistance of the material from the state when the voltage was last applied. Defects are inherent in the nature of these devices, since the position of every atom cannot be completely specified. However, the advantage of the crossbar architecture is its redundancy allowing routing around any parts of the circuit that do not work. Readers who have become further interested in these fascinating

Fig. 8.13 A crossbar architecture consisting of an array of perpendicular wires connected by "on" and "off" switches (where the wires meet). An open switch represents a zero and a closed switch represents a one, thereby this crossbar design functions as a storage system. The data are read by probing the switch with a voltage

nanoscale electronic devices are encouraged to consult the latest literature published on this subject.

8.4.4 Magnetic Random Access Memories and Related Devices

A magnetic random access memory (MRAM) in its basic description has a magnetoresistive structure where the free layer magnetization is the one being used to store data or read it out. TMR type of MRAM are better performing than GMR or other magnetoresistive types, in as much as they get TMR ratio effects higher than 200% at room temperature. Recently, MgO-based TMR materials have been employed in these memory devices and have started to be commercialized given their good performance in cache buffers and configuration storage memories. They represent a particular type of MRAM. In those MRAM based on magnetic tunnel junctions, also known as MTJ cells, the storage element has the familiar trilayer configuration consisting of two ferromagnets separated by a thin tunneling layer. While the magnetization is fixed in one ferromagnet, it is free to move in the other. One bit of information is encoded in the magnetization state of the free layer which is one monodomain. A "1" bit is stored for antiparallel alignment and a "0" bit is stored for parallel alignment of magnetization. The magnetization state of this monodomain can be switched as a whole by the magnetic field generated by the current passing through the "write line" (Fig. 8.14). The bits are then read by measuring the tunnel current through the storage element, a current which depends on the magnetization state. Nevertheless, the trilayer structure is not the only design researchers are experimenting within order to increase writing reliability to a specified cell. The "write line" is in close proximity to the free magnetic layer in the cross-point addressing scheme and may cause writing error. The following

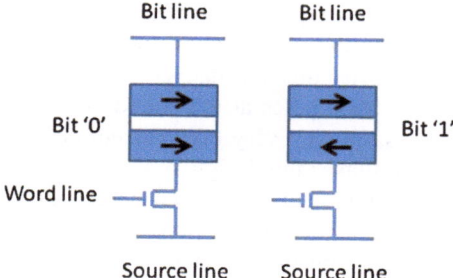

Fig. 8.14 A STTRAM type of structure incorporating an MTJ. The data bit is stored in the MTJ in its magnetization configuration with a bit "0" represented by a parallel state while the programming occurs by sending an electric current from the bit line to the source line. In contrast, a bit "1" is represented by an antiparallel state and is programmed by an electric current from the source line to the bit line

Fig. E 8.7.1 Schematic of a proposed magnetic memory cell based on thin film ferromagnetic nanorings. Reprinted with permission from [74]

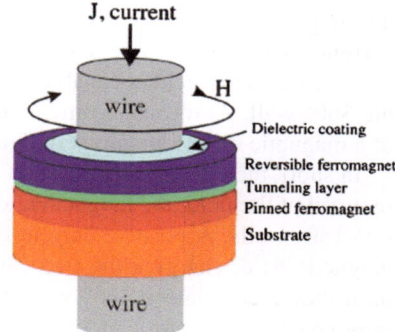

example describes an alternative experimental design for the storage element of an MTJ cell with the intent of improving data writing to a magnetic memory cell.

Example 8.7. Describe an MTJ used in MRAM with an alternative storage element other than the familiar trilayer ferromagnet-insulating barrier-ferromagnet.

Answer: A theoretical design for a magnetic memory cell has been proposed [74] in which thin film nanostructures are fabricated as nanorings that can efficiently store, record, and read out information. The operating principle is based on introducing a stable 360° domain wall into the nanoring so that the information bit is represented by the polarity of the wall. The polarity can be changed by applying a current passing through the wire crossing the ring (Fig. E 8.7.1) which causes splitting of the 360° domain wall into two charged 180° walls. The two charged 180° walls then move to the other end of the ring to recombine into a 360° wall of opposite polarity. This design is an example of current-induced switching, representing a step away from magnetic field driven devices. A challenge for MTJ cells of traditional design is that they need a rather strong current to generate the magnetic fields required for switching, thus necessitating a lot of power.

In similar nanoring designs reported by other research groups, the bit can be encoded by the direction of the vortex magnetization state in the ferromagnetic thin film ring. However, even this mechanism has its difficulties given that it requires creating magnetization poles at the boundaries of the ring, thus needing elaborate writing schemes, or even very high current densities able to switch the magnetization state. What makes the design and operating principle of the device represented in Fig. E 8.7.1 more likely to succeed is that it uses the polarity of strictly localized 360° domain walls in the free ferromagnetic layer (reversible ferromagnet in Fig. E 8.7.1) to encode the bit of information. The existence of these types of domain walls has been previously predicted by several simulations. In the presence of a 360° wall, the magnetization looks like a vortex state in most of the ring, except in that specific area where the ring is localized. However, it is not quite a typical vortex, since inside the wall the magnetization rotates in plane, in the direction opposite to the direction of rotation in the rest of the ring. Therefore, both states representing ±360° walls can be attained by a simple in plane rotation of the magnetization vector, avoiding any out of plane complications. As such, the current carried by the wire passing through the ring configuration and generating a circular magnetic field is capable of easily switching between the two polarities of the 360° wall. At the same time, readout can occur because the nanorings are part of a magnetic tunnel junction configuration.

In short, the magnetization is first saturated by a field along the y-axis, and upon removal of the field, the magnetization settles into a symmetric initial state creating two 180° walls. A positive current through the wire generates a counterclockwise magnetic field, charging the two 180° walls and driving them towards each other until they merge into a 360° wall. It is important to note that once the current is removed, the newly formed 360° wall maintains its integrity and does not break up. However, applying a negative current results in a breakup of the 360° wall into two charged 180° walls which propagate along the ring until they reach the other end where they collide again to form a new 360° wall of opposite polarity. Once more, upon removal of the current, the 360° wall does not break up. Thereby, walls split, move back and forth, and reunite along ring boundaries. The design seems robust in as much as the walls are stable. Furthermore, it promises a space saving alternative because layers can be stacked in the third dimension, freeing space horizontally during fabrication and chip integration (Fig. E 8.7.2).

It is possible to have an MRAM based on the anisotropic magnetoresistance (AMR) effect or on giant magnetoresistance (GMR). Not all random access memories (RAMs) are magnetic. A variety of nonvolatile RAM devices have emerged, believed to be capable of replacing hard disk drives (HDDs) and solving their remaining problems. Among these, it seems that two types of RAM stand out: *spin transfer torque random access memories* (STTRAMs), which are magnetic, and *phase change random access memories* (PCRAMs), which are nonmagnetic. These technologies likely present the best storage alternatives at the moment, due to their small cell size and their ability to store multiple bits per cell.

STTRAMs are related to MRAM devices; however, they rely on a very thin insulating layer between the two ferromagnets of the MTJ structure. The thin

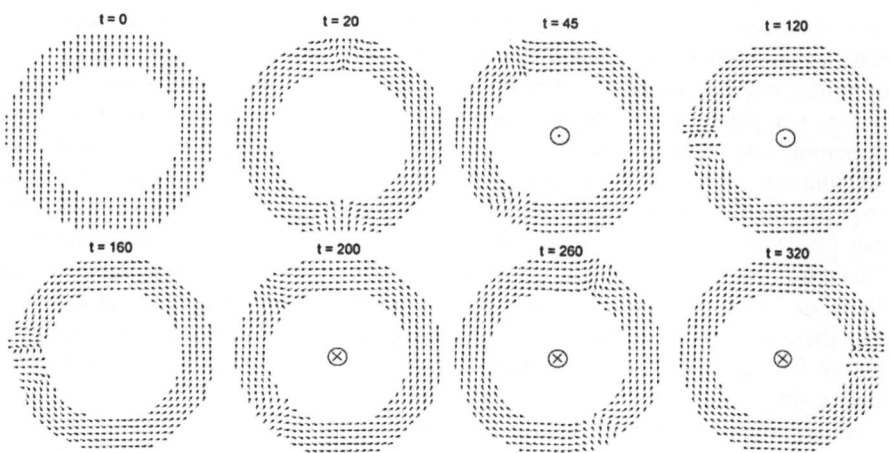

Fig. E 8.7.2 Simulation results representing 360° domain wall splitting, 180° wall propagation and 360° wall formation during magnetization reversal in a ferromagnetic nanoring configuration for a proposed magnetic memory cell. Reprinted with permission from [74]

insulating layer is a MgO-based tunneling barrier of less than 1 nm through which a spin-polarized current tunnels and switches MTJ states from antiparallel "1" to a parallel "0." A schematic is shown in Fig. 8.14. The spin-polarized current transfers angular momentum to the spins in the magnetic free layer, exerting a spin torque and thus switching the magnetization in that layer. In a conventional MRAM, currents are used to generate magnetic fields to switch the free layer magnetization, whereas STTRAMs are programmed by passing electric current through the MTJ elements. Because there is no need for extra current lines in STTRAMs, they are much simpler than field driven MRAM when it comes to building memory array structures and process integration. Nevertheless, MRAM and STTRAM use the same writing technique based on spin-polarized electrons.

STTRAMs are believed by some to have the potential to revolutionize memory technologies because they combine the capacity and cost benefits of DRAM, the fast read and write performance of SRAM, as well as the nonvolatility of flash memories. Because STTRAMs show great promise in terms of read and write speed, power consumption, and scalability, they have become a subject of study for many research groups around the world, in industry and academia. For instance, a group of researchers [75] has decided to study the thermal assist programming scheme in a STTRAM as it is believed to reduce write current and improve scaling. A thermal assisted STTRAM differs from a standard STTRAM only in the memory bit stack. As such, an antiferromagnetic layer is coupled to the free layer of the MTJ stack which allows stabilization of free layer magnetization due to the arising exchange bias. During programming, an electric current is sent through the MTJ causing it to heat up. At a certain blocking temperature, the exchange bias effect vanishes, allowing the free layer to rotate its magnetization. The spin torque of the spin-polarized current is now able to switch the free layer to a parallel of antiparallel state of magnetization, depending on the current polarity. Upon removal of the current, the MTJ temperature drops below the blocking value causing the exchange bias to reappear. The free layer then becomes stabilized in the new magnetization direction. The advantage of such thermal assisted switching in the STTRAM is that the critical switching current is not determined by the spin torque strength required for free layer magnetization switching. Consequently, the write current can be significantly reduced as compared to conventional field driven MRAM. Scalability improves as well in thermally assisted STTRAM because the elliptical shape of the MTJs is no longer required for shape anisotropy and thermal stability. The thermal stability of the bits in the STTRAM is provided by the exchange bias on the free layer. MTJ elements can, therefore, be patterned in simpler and space saving shapes.

In contrast to STTRAM which rely on spin-polarized electrons, PCRAM-based devices use a chalcogenide glass (e.g., $Ge_2Sb_2Te_5$) that changes phase reversibly between amorphous (high-resistance) and crystalline (low resistance), hence the name "phase change." Data in PRAMs are stored in terms of resistance values. The transfer from a high-resistance amorphous phase to a low resistance crystalline phase is called "setting," while the reverse process is known as "resetting." The amorphous-crystalline phase change occurs at a relatively low temperature, where the energy source which triggers the change is the Joule heat generated from the current. A current pulse is applied for a short time to a low resistance cell to increase the temperature of the phase change material to its melting point. This is the resetting phase. The material is then abruptly cooled to become amorphous. A long current pulse is applied to increase the temperature below the melting point but above the crystallization point. This is the setting phase. A slow cooling of the material follows in order to promote crystal growth and decrease the resistance of the device. Thus, it is the current rather than the voltage which controls setting and resetting due to the Joule heat which controls the phase change. The reversible resistance change is similar to the one in GMR, AMR, or TMR structures; nevertheless, PCRAMs are not magnetic structures. Unfortunately, the writing process in PCRAMs takes up to $300\,\mu s$, and the number of rewrites is limited, making them challenging to use in practical applications. There are several other emerging random access memory type of devices, some of them already commercialized, others still in a research stage, however, they are not magnetic so will not be discussed in this chapter. Some memories also use the resistive switching phenomenon; nevertheless, they are based on an oxide thin film placed between two electrodes. Alternatively, a chemical reaction in solid electrolytes can be used for resistive switching, accomplished through organic or inorganic ions moving freely inside the electrolyte. The reader is encouraged to further explore the interesting world of nonvolatile memory devices comprising among others, other types of *resistive RAM (RRAM), carbon nanotube RAM (CNRAM), ferroelectric RAM (FRAM)*, or *copper bridge RAM (CPRAM)* [76].

8.4.5 Concluding Remarks

Nonvolatile memories have become essential for our daily lives, from railway passes and IC cards to cell phones, digital cameras, electric appliances, and cars. For the most part, MOS-type memories such as flash memories, which are semiconductor based, satisfy a lot of these device requirements, especially when large capacity recording is desired. Their widespread usage is due to the low manufacturing cost per bit that is significantly lower than for other kinds of memories. However, they have their drawbacks given their structure with embedded logic that requires high writing voltage, but delivers low writing speeds. In order to keep up with shrinking sizes, flash memories are built by using multilayered and vertical MOS structures; nevertheless, their complexity makes is difficult to further miniaturize, particularly below 32 nm. The main reason why these devices cannot be shrunk more in order to allow increasing memory capacity are the parasitic effects of the interference type resulting from the capacitive coupling with adjacent cells. Furthermore, their limited number of rewrites restricts the type of possible applications.

Ferroelectric RAMs (FeRAMs) are among the first contenders for replacing flash memories, as their embedded logic allows integration in IC cards, event data recorders, and electronic appliances. Their power consumption is lower than for other types of memory, making this advantage particularly attractive for many end user applications. Additionally, these memories have a maximum number of rewrites much larger than for flash memories, as well as much faster writing speeds. Unfortunately, FeRAMs are based on a capacitor type memory cell structure which makes it less suitable for high-density integration. Also, due to their principle of operation, the readout method is destructive because once the readout operation starts, it becomes necessary to rewrite the data after readout. This limits the maximum number of readouts given that bit information is destroyed at the time of readout. However, if current research trends in FeRAM structure allow replacement with transistor type cells enabling a chain structure and high-density integration, FeRAM memory will stay competitive on the market. Furthermore, transistor-type FeRAMs do not require destructive readout, meaning that a much higher rewrite endurance can be achieved.

Many research groups have explored alternative kinds of memories operating on different principles than those of flash memories or the different types of RAM. On these grounds, the development of magnetic memories seems rather timely, especially given the widespread use of electronics in various fields requiring a diversity of memories adapted to each type of application. Intensive studies are still being performed on MRAM and related magnetic devices in order to improve performance and increase memory capacity. In particular, STTRAMs have shown a significant reduction in current density with miniaturization, as well as a reduction in cell area because of a simplified structure. It is believed that in the future, highly integrated and reduced bit cost MRAMs will be increasingly developed along with pioneering applications. MRAM is nonvolatile, has an unlimited read and write endurance, while also showing high-speed read and write operation capabilities.

With a large TMR, a tunable resistance-area product, MRAM with MTJs enables circuit designs with competitive read cycle times.

The key structure at the core of many of these technologies remains the MTJs, in particular those with perpendicular anisotropy. These MTJs allow to scale down the dimensions of spintronic devices to sub-45 nm nodes while keeping the required thermal stability above 60. A challenge remains to overcome the reverse scaling, where the critical current for the switching of free layer in MTJ-based devices needs to be reduced. As such, many solutions are being proposed for better material selection, memory bit design, and not least, programming architecture. Whether using a design with circular pillars instead of elliptical ones like those in p-MTJs, or expanding the variety of magnetic materials employed in their fabrication (CoPt, CoFePd, $L1_0$FePt ordered alloys, and rare earth transition metal ferrimagnets), MTJs have improved their performance significantly. This happened to the point where MRAM based on MTJ bits has proven to be competitive with semiconductor memories with many attributes not found in other types of memory, making MRAM or their spin-offs (e.g., STTRAM) the magnetic storage devices of the future. Therefore, MRAMs are expected to be used as universal memories that combine the functions of both DRAM and SRAM.

Tests, Exercises, and Further Study

P8.1 Describe how perpendicular magnetic anisotropy can be obtained in a Pt/Co/oxide system. Note the thickness of the Co layer. See for instance [77].

P8.2 Give another example of a magnetic system where perpendicular magnetic anisotropy was obtained after annealing. See for instance [78].

P8.3 Give an example of and describe briefly a technique involving a master mold for prepatterning Co/Pt or Co/Pd multilayer magnetic media. See for instance [79] and [80].

P8.4 What type of domain walls can occur in a racetrack memory? See for instance [70].

P8.5 Name a few (e.g., three) types of emerging memories. See for instance [70] and references therein.

P8.6 What are some significant impediments for realizing high-performance crossbar memories of memristor type, and what can be done to overcome them? See for instance [73].

P8.7 Sometimes MRAM fabrication can cause shifted magnetic properties in MTJs and thereby different switching current distributions among multiple cells. What is the influence of MTJ edge concave and convex deformation observed in MRAM fabrication? See for instance [81].

P8.8 Compare a MRAM with a MgO-based MTJ to the one with an AlO_x-based device. How do material attributes for different oxidation processes and free layer alloys influence resistance distributions, bias dependence, free layer magnetic properties, interlayer coupling, breakdown voltage, and thermal endurance? See for instance [82].

P8.9 Describe a stress assisted magnetization reversal method in a perpendicular MRAM (p-MRAM). What are the benefits of such a magnetization reversal method? See for instance [83].

P8.10 Comment on the high-temperature behavior of MTJs. See for instance [84].

P8.11 Comment on factors that influence the quality of a traditional magnetic tape. Give examples from published literature quoting results, the circumstances under which they were obtained and show SEM or TEM images. The questions that should be answered are for instance: What is the role of a crystallographically textured substrate? How does deposition rate influence the quality of a magnetic tape? How do you reduce magnetic grain coupling and lower media noise? What is the most investigated magnetic alloy for the fabrication of magnetic tapes, and what are its advantages? See for instance [85–89].

P8.12 Suggest alternative materials for traditional magnetic tapes, and give examples from published literature showing the properties that were obtained for these materials. See for instance [90–92].

P8.13 What is the role played by the magnetocrystalline and shape anisotropy constants in traditional magnetic tapes? See for instance [93].

P8.14 Why are prepatterned CoPd and CoCrPt alloys more likely to solve the thermal stability problem in traditional magnetic tapes? See for instance [94–96].

P8.15 What makes barium ferrite particles viable candidates for superior traditional magnetic media? See for instance [97–99].

P8.16 What are some advantages of predefined magnetic entities per bit? See [100, 101, 101, 103].

P8.17 Comment on head errors in bit patterned media. See for instance [104].

P8.18 Why cannot MFM be easily used to observe the magnetization of a bit? See for instance [105] and [106].

P8.19 Why does the write head need to be well placed?

P8.20 What are the consequences of the fringes of the head field gradient?

P8.21 FePt is ferromagnetically ordered in the chemical disordered state rather than the chemically ordered state. Various fabrication techniques result in FePt nanoparticles oriented at random and in a disordered fcc crystallographic phase, nevertheless, are found to be magnetic. It is not recommended that some nanoparticles remain in a disordered fcc phase while others attain some degree of ordering. It is always necessary to prevent a large scale spreading of anisotropy in the medium. Comment on the challenges encountered in obtaining ferromagnetically ordered grains in traditional magnetic media. Give examples from published literature.

P8.22 Comment on the usefulness of an artificial antiferromagnet incorporated as a reference layer in an MTJ with perpendicular magnetic anisotropy. The MTJ is used in an MRAM structure. See for instance [107].

Hints and Partial Answers to Problems in Chap. 8

The answers to these problems are in many instances just a hint as to where further information is to be found. They contain the main idea, but the reader needs to find the references that are sometimes recommended for each exercise and complete the answer through further bibliographic reading and study. Reference numbering follows the one used in that particular chapter.

P8.1 A perpendicular magnetic anisotropy was obtained in a Pt/Co/oxide system. The cobalt layers were grown on SiO_2 substrates and were 0.6 nm thick displaying superparamagnetism. A perpendicular magnetic anisotropy is obtained for 1.5 nm films after annealing. Co layers grown on various Al and Mg oxides prepared by sputtering also exhibit perpendicular magnetic anisotropy after annealing. Combined with inverse Pt/Co(CoFeB)/oxide systems, these structures could be employed as tunnel junctions with thicker magnetic electrodes. They show improved thermal stability compared to those based on standard Pt/Co multilayers. These types of systems may be used as magnetic electrodes for magnetic tunnel junctions. *See* [77].

P8.2. A strong perpendicular magnetic anisotropy was observed in systems containing 2 nm thick CoFeB layers. They are believed to be suitable for a MgO-based magnetic tunnel junction. The value of the coercivity is measured under perpendicular applied magnetic fields. It shows a value as high as 1,050 Oe after annealing. Intermixing of Pd with CoFeB, as well as the low saturation magnetization of the Co-rich CoFeB layer, is considered to be responsible for the strong perpendicular magnetic anisotropy. *See* [78].

P8.3 A mechanical rotary stage and a fixed e-beam column ensure patterning of a master mold dot-by-dot while the e-beam resist covered master is rotating. After exposure, the resist is developed and reactive ion etching is used to transfer the pattern into the master mold substrate. A disk substrate is coated with liquid nanoimprint resist, and the previously obtained master mold is pressed against this disk substrate. The liquid resist reflows until the master topography is copied onto the resist, and ultraviolet (UV) light is applied to cure the resist and create a solid replica. The resist pattern is afterwards transferred into the disk substrate using reactive ion etching (RIE). Pillars on the replica correspond to holes on the master mold. After resist stripping and disk substrate cleaning, a magnetic medium is sputter deposited over the patterned disk substrate. *See* [79] and [80] *for further details.*

P8.4 Magnetic-domain walls formed in a racetrack memory can be transverse walls and vortex walls with a twisted pattern of magnetization. *See* [70] *for further details.*

P8.5 Some emerging memory types are resistive random access memory (RRAM), phase-change RAM (PRAM), and spin torque transfer RAM (STT-RAM). Only the latter is a magnetic RAM. *See* [70] *and references therein.*

P8.6 Some significant challenges encountered in the practical realization of memristor memory devices are the small on-to-off resistance ratio of the switches as compared to modern transistors. The ratio can be improved to about 1,000:1 electrical switching if the effective spacing between two wires in the crossbar design is reduced to $\sim$0.3 nm; however, this is very difficult to realize in practice. Molecular type switches may be the answer to this problem. *See* [73] *and references therein.*

P8.7 The edge deformation and the changed switching process generate a magnetization vortex which causes not only a large switching current but also a large switching current distribution of each cell. Reducing the magnetization switching current and its switching distribution is very important for decreasing power consumption. With cell sizes shrinking and edge deformations becoming larger, the necessity arises of controlling MTJ shapes. *See* [81] *for further details.*

P8.8 A TMR greater than 230% is achieved with CoFeB free layers and greater than 85% with NiFe free layers. Both TMR values are much higher than for similar structures with an AlO_x barrier. Bit-to-bit resistance distributions are somewhat wider for MgO barriers. The read access time is reduced in MgO-based MTJs. *See* [82] *for further details.*

P8.9 A mechanical in plane stress of 279 MPa applied to a GMR multilayer structure was shown to change the direction of magnetization of the free layer from perpendicular to in plane. A stress assisted magnetization reversal may reduce the power consumption of p-MRAMs. *See* [83] *for further details.*

P8.10 A challenge is to ensure design margins over a wide range of device temperatures ($-40°C$ to $125°C$), particularly at deeply scaled technology nodes. It is found that antiparallel resistance decreases with increasing temperature while parallel resistance shows a negligible temperature dependence. In a 1T-1MTJ bit cell (i.e., a one transistor-one MTJ), the transistor output current is strongly dependent on MTJ resistance. Lower antiparallel resistances at elevated temperatures can compensate for the degradation of the output current. Because switching voltages are also reduced at high temperatures, the two effects result in a worst temperature state for switching at the "cold" condition, meaning that write margins decrease as the temperature decreases. Simulations show that it is possible to secure enough margins at elevated temperatures. *See* [84] *for further details.*

P8.11 A crystallographically textured substrate can determine the direction of the easy axis of the magnetic layer. Deposition rate is important for the quality of the tape (CoCrPt in a SiO_2 matrix). High deposition rates (e.g., 10 nm/s) may result in faulty microstructure and excessive substrate heating. A SiO_2 matrix can reduce magnetic grain coupling and lower media noise. CoCrPt investigations show differences in microstructure grain size and orientation for varying deposition rates. Slower deposition rate produce better defined SiO_2 network, minimizing coupling between grains. Nonmagnetic SiO_2 as well as oxidation conditions lead to separation of grains. Increased coercivity of Co grains is observed, while media noise performance depends on the oxide type. *See [85–89] for further details.*

P8.12 Fine particles with large coercivity (e.g., iron nitride and Ba ferrite) have been suggested as alternative materials for traditional magnetic tapes. For example, 21 nm Ba ferrite particles can be produced using an advanced precision coating process. They show improved magnetic recording properties. The volume of the Ba particles is roughly 36% smaller than that of metal particles of 45 nm length. However, the coercivity of 160 kA/m is nearly equal to that of metal particles. A thermal stability factor, estimated at 65 is high enough to ensure stability. *See [90–92] for further details.*

P8.13 Pure and Co/Ti-doped Ba ferrite particles show that the magnetocrystalline anisotropy constant dominates the shape anisotropy constant at all temperatures if the Ba ferrite particles are undoped. There is a uniaxial anisotropy along the c-axis. The dominance is reversed for doped particles and at lower temperatures because ionic substitutions weaken the magnetocrystalline anisotropy. With the shape anisotropy becoming larger, the overall magnetic anisotropy is no longer uniaxial, presenting multiple preferred directions for the magnetization. *See [93] for further details.*

P8.14 The enhanced thermal stability comes from the fact that the entire volume of the magnetic islands is the effective switching unit, but only for densities below 10 Tb/in^2. *See [94–96] for further details.*

P8.15 Barium ferrite particles offer a chemical stability common to oxide compounds. High-dispersion technologies for fine Ba ferrite particles are used so that smooth surfaces and uniform particle distributions are obtained in an 80 nm thick magnetic layer. A high signal to noise ratio ($\sim$19 dB) and increased resolution are observed as well. Areal recording densities of 17.5 Gbits/in^2 were demonstrated for Ba ferrite particulate media with GMR read heads. By comparison, densities of 31 Gbits/in^2 were obtained in CoCrPt–SiO$_2$ sputtered media. *See [97–99].*

P8.16 Predefined magnetic entities per bit allow for larger volume magnetic bit-cells, reducing the number of magnetic switching volumes per bit to one entity. This allows about 10^{12} prepatterned magnetic bit-cells per surface to be obtained in patterned media. Estimated size and periodicity of bit-cells are 25 nm and 35 nm, respectively, assuming an initial recording density of 500 Gb/in^2. The bit size determines the switching volume, making bit-cells thermally stable, and yet still reversible by the head technology. At the same time, transitions across and down the track are defined by the bit pattern, and no longer by the magnetic field of the head. *See [100–103] for further details.*

P8.17 The magnetic easy axis of each bit and that of the recording head have the same relative orientation with respect to each other. Because of it, problems arise with written-in errors that differ from the usual signal to noise ratios. The written-in errors are due to statistical fluctuations of the magnetic properties and the locations of the individual dots. The size and shape of the patterned magnetic islands as well as their position may vary from one data cell to another, because of lithography or self-assembly errors in nanoparticle size distribution. Therefore, a magnetoresistive playback head of a certain design must be chosen to maximize playback signal resolution and to minimize sensitivity to noise due to these patterning imperfections. Additionally, the playback signal amplitude experiences a linear loss with filling ratio or packing density, which represents a percentage of the bit area filled with magnetic material. It was observed that differential readers exhibit enhanced performance with respect to spatial resolution and sensitivity to medium imperfections due to their ability to compare bits. *See* [104].

P8.18 Magnetic force microscopy (MFM) was initially used to observe the magnetization states of these patterned bits; however, shortly after it was observed that the tip itself can be used to switch the magnetization of a bit, an undesirable aspect of just observing bits. If a current passes between the tip and the island to be written, or an additional field is applied, some of the bit reversal can be prevented. *See* [105, 106].

P8.19 Because the main source of noise originates in bit size and bit pitch variations, the write field has to be well-placed to write on the desired bit and not on adjacent bits.

P8.20 The thickness of the medium has to be about the size of the bit spacing to prevent the head from inadvertently writing neighboring areas with the fringes of the head field gradient. Stray fields from adjacent bits scale with saturation magnetization, and so does the contribution of the shape anisotropy to the total uniaxial anisotropy.

P8.21 Ion beam irradiation with $700\,keV\,N^+$ ions at a dose of $1 \times 10^6\,sions/cm^2$ destroys the chemical order and induces ferromagnetism in chemically ordered antiferromagnetic $FePt_3$ nanoparticles. Fortunately, nanoparticles as small as 4 nm diameter have sufficient magnetocrystalline anisotropy to remain magnetically stable at these small volumes, avoiding superparamagnetism. Sometimes, annealing at temperatures higher than $600°C$ is required to induce a phase transformation so that the nanoparticles acquire a shared magnetic easy axis. But this can lead to particle agglomeration which can be eliminated by surfactant chemistry optimization, or by hardening the organic matrix with ion beam irradiation. Some studies suggest that ordering depends on particle size. Alloying with a metallic element may also lead to ordering.

P8.22 Artificial antiferromagnets are commonly employed as reference layers in MTJs with perpendicular magnetic anisotropy. Their role is to minimize dipolar interactions between the reference layer and the free layer. By employing such an artificial antiferromagnet, the asymmetry of the free layer reversal is reduced, reversal brought about by stray fields in nanosized MTJs with perpendicular magnetization. *See* [107] *for further details.*

References

1. R. Yamagishi, T. Koike, K. Kodama, N. Hosoito, J. Phys. Soc. Jpn. **79**(9), 094710 (2010)
2. Q.L. Lv, J.W. Cai, H.Y. Pan, B.S. Han, Appl. Phys. Exp. **3**, 093003 (2010)

3. Y. Takahashi, Appl. Phys. Exp. **1**, 037001 (2008)
4. K. Takata, N. Ichimei, M. Okuda, Appl. Phys. Exp. **2**, 126504 (2009)
5. R.M.H. New, R.F.W. Pease, R.L. White, R.M. Osgood, K. Babcock, J. Appl. Phys. **79**, 5851 (1996)
6. Y. Hao, F.J. Castaño, C.A. Ross, B. Vögeli, M.E. Walsh, H.I. Smith, J. Appl. Phys. **91**(10), 7989 (2002)
7. A. Fernandez, P.J. Bedrossian, S.L. Baker, S.P. Vernon, D.R. Kania, IEEE Trans. Magn. **32**, 4472 (1996)
8. M. Farhoud, J. Ferrera, A.J. Lochtefeld, T.E. Murphy, M.L. Schattenburg, J. Carter, C.A. Ross, H.I. Smith, J. Vac. Sci. Technol. B **17**, 3182 (1999)
9. A. Carl, S. Kirsch, J. Lohau, H. Weinforth, E.F. Wassermann, IEEE Trans. Magn. **35**, 3106 (1999)
10. W. Wu, B. Cui, X.Y. Sun, W. Zhang, L. Zhuang, L.S. Kong, S.Y. Chou, J. Vac. Sci. Technol. B **16**, 3825 (1998)
11. L. Gao, S.H. Liou, M. Zheng, R. Skomski, M.L. Yan, D.J. Sellmyer, N. Polushkin, J. Appl. Phys. **91**, 7311 (2002)
12. T.A. Savas, M. Farhoud, H.I. Smith, M. Hwang, C.A. Ross, J. Appl. Phys. **85**, 6160 (1999)
13. H.H. Solak, C. David, J. Vac. Sci. Technol. B **21**, 2883 (2003)
14. Y.J. Chen, J.P. Wang, E.W. Soo, L. Wu, T.C. Chong, J. Appl. Phys. **91**(10), 7323 (2002)
15. C.A. Ross, Annu. Rev. Mater. Sci. **31**, 203 (2001)
16. J.C. Lodder, J. Magn. Magn. Mater. **272–276**, 1692 (2004)
17. H.I. Smith, J. Vac. Sci. Technol. B **13**, 2323 (1995)
18. C. Miramond, C. Fermon, F. Rousseaux, D. Decanini, F. Carcenac, J. Magn. Magn. Mater. **165**, 500 (1997)
19. F. Rousseaux, D. Decanini, F. Carcenac, E. Cambril, M.F. Ravet, C. Chappert, N. Bardou, B. Bartenlian, P. Veillet, J. Vac. Sci. Technol. B **13**, 2787 (1995)
20. C.T. Rettner, M.E. Best, B.D. Terris, IEEE Trans. Magn. **37**, 1649 (2001)
21. X.D. Lin, J.G. Zhu, W. Messner, IEEE Trans. Magn. **36**, 2999 (2000)
22. J.G. Zhu, X.D. Lin, L.J. Guan, W. Messner, IEEE Trans. Magn. **36**, 23 (2000)
23. C.T. Rettner, S. Anders, T. Thomson, M. Albrecht, Y. Ikeda, M.E. Best, B.D. Terris, IEEE Trans. Magn. **38**, 1725 (2002)
24. W.H. Bruenger, C. Dzionk, R. Berger, H. Grimm, A. Dietzel, F. Letzkus, R. Springer, Microelectron. Eng. **61–62**, 295 (2002)
25. A. Dietzel, R. Berger, H. Loeschner, E. Platzgummer, G. Stengl, W.H. Bruenger, F. Letzkus, Adv. Mater. **15**, 1152 (2003)
26. D. Ravelosona, T. Devolder, H. Bernas, C. Chappert, V. Mathet, D. Halley, Y. Samson, B. Gilles, A. Marty, IEEE Trans. Magn. **37**, 1643 (2001)
27. C.T. Rettner, S. Anders, J.E.E. Baglin, T. Thomson, B.D. Terris, Appl. Phys. Lett. **80**, 279 (2002)
28. J. Fassbender, D. Ravelosona, A. Samson, J. Phys. D.: Appl. Phys. **37**, R179 (2004)
29. P. Warin, R. Hyndman, J. Glerak, J.N. Chapman, J. Ferre, J.P. Jamet, V. Mathet, C. Chappert, J. Appl. Phys. **90**, 3850 (2001)
30. B.D. Terris, D. Weller, L. Folks, J.E.E. Baglin, A.J. Kellock, H. Rothuizen, P. Vettiger, J. Appl. Phys. **87**, 7004 (2000)
31. T. Devolder, C. Chappert, Y. Chen, E. Cambril, H. Bernas, J.P. Jamet, J. Ferre, Appl. Phys. Lett. **74**, 3383 (1999)
32. J. Melngailis, A.A. Mondelli, I.L. Berry, R. Mohondro, J. Vac. Sci. Technol. B **16**, 927 (1998)
33. C.G. Stefanita, F. Yun, M. Namkung, H. Morkoc, S. Bandyopadhyay, Electrochemical self-assembly as a route to nanodevice processing, in *Handbook of Electrochemical Nanotechnology*, ed. by H.S. Nalwa (American Scientific, 2006)
34. R. Zhu, Y.T. Pang, Y.S. Feng, G.H. Fu, Y. Li, L.D. Zhang, Chem. Phys. Lett. **368**, 696 (2003)
35. S.O. Kim, H.H. Solak, M.P. Stoykovich, N.J. Ferrier, J.J. de Pablo, P.F. Nealey, Nature **424**, 411 (2003)
36. M. Kimura, M.J. Misner, T. Xu, S.H. Kim, T.P. Russell, Langmuir **19**, 9910 (2003)

37. L. Rockford, S.G.J. Mochrie, T.P. Russell, Macromolecules **34**, 1487 (2001)
38. A.A. Zhukov, M.A. Ghanem, A. Goncharov, P.A.J. de Groot, I.S. El-Hallag, P.N. Bartlett, R. Boardman, H. Fangohr, J. Magn. Magn. Mater. **272–276**, 1621 (2004)
39. S. Palacin, P.C. Hidber, J.P. Bourgoin, C. Miramond, C. Fermon, G.M. Whitesides, Chem. Mater. **8**, 1316 (1996)
40. Z.B. Zhang, D. Gekhtman, M.S. Dresselhaus, J.Y. Ying, Chem. Mater. **11**, 1659 (1999)
41. B.D. Reiss, C.B. Mao, D.J. Solis, K.S. Ryan, T. Thomson, A.M. Belcher, Nano Lett. **4**, 1127 (2004)
42. E. Mayes, A. Bewick, D. Gleeson, J. Hoinville, R. Jones, O. Kasyutich, A. Nartowski, B. Warne, J. Wiggins, K.K.W. Wong, IEEE Trans. Magn. **39**, 624 (2003)
43. M. Colburn, A. Grot, B.J. Choi, M. Amistoso, T. Bailey, S.V. Sreenivasan, J.G. Ekerdt, C.G. Willson, J. Vac. Sci. Technol. B **19**, 2162 (2001)
44. S.Y. Chou, P.R. Krauss, W. Zhang, L.J. Guo, L. Zhuang, J. Vac. Sci. Technol. B **15**, 2897 (1997)
45. S.Y. Chou, P.R. Krauss, L.S. Kong, J. Appl. Phys. **79**, 6101 (1996)
46. L.S. Kong, L. Zhuang, M.T. Li, B. Cui, S.Y. Chou, Jpn. J. Appl. Phys. **37**, 5973 (1998)
47. J. Moritz, B. Dieny, J.P. Nozieres, S. Landis, A. Lebib, Y. Chen, J. Appl. Phys. **91**, 7314(2002)
48. J. Moritz, S. Landis, J.C. Toussaint, P. Bayle-Guillemaud, B. Rodmacq, G. Casali, A. Lebib, Y. Chen, J.P. Nozieres, B. Dieny, IEEE Trans. Magn. **38**, 1731 (2002)
49. D. Wachenschwanz, W. Jiang, E. Roddick, A. Homola, P. Dorsey, B. Harper, D. Treves, C. Bajorek, IEEE Trans. Magn. **41**, 670 (2005)
50. H. Oshima, H. Kikuchi, H. Nakao, K. Itoh, T. Morikawa, H. Tamura, K. Nishio, H. Masuda, Appl. Phys. Exp. **1**, 054001 (2008)
51. C. Ross, Annu. Rev. Mater. Res. **31**, 203 (2001)
52. K. Miura, D. Sudo, M. Hashimoto, H. Muraoka, H. Aoi, Y. Nakamura, IEEE Trans. Magn. **42**(10), 2261 (2006)
53. R. Wood, IEEE Trans. Magn. **36**(1), 36 (2000)
54. M. Mallary, A. Torabi, M. Benaki, IEEE Trans. Magn. **38**(4), 1719 (2002)
55. T. Ishida, O. Morita, M. Noda, S. Seko, S. Tanaka, H. Ishioka, IEEE Trans. Fund. Electron. Commun. Comput. Sci. **E76A**, 1161 (1993)
56. H.J. Richter, IEEE Trans. Magn. **35**, 2790 (1999)
57. D. Weller, M.F. Doerner, Annu. Rev. Mater. Sci. **30**, 611 (2000)
58. K. Watanabe, T. Takeda, K. Okada, H. Takino, IEEE Trans. Magn. **29**, 4030 (1993)
59. http://www.reed-electronics.com/semiconductor
60. M. Plummer et al. (ed.), *The Physics of High Density Recording* (Springer, Berlin, 2001))
61. S.A. Nikitov, L. Presmanes, P. Tailhades, D.E. Balabanov, J. Magn. Magn. Mater. **241**, 124 (2002)
62. Z.Z. Bandic, H. Xu, Y.M. Hsu, T.R. Albrecht, IEEE Trans. Magn. **39**, 2231 (2003)
63. R. Sugita, T. Muranoi, M. Nishikawa, M. Nagao, J. Appl. Phys. **93**, 7008 (2003)
64. T. Ishida, K. Miyata, T. Hamada, H. Hashi, Y.S. Ban, K. Taniguchi, A. Saito, IEEE Trans. Magn. **39**, 628 (2003)
65. R. Sugita, O. Saito, T. Muranoi, M. Nishikawa, M. Nagao, J. Appl. Phys. **91**, 8694 (2002)
66. T. Ishida, K. Miyata, T. Hamada, K. Tohma, IEEE Trans. Magn. **37**, 1875 (2001)
67. A. Saito, T. Ono, S.J. Takenoiri, IEEE Trans. Magn. **39**, 2234 (2003)
68. E. Grochowski, D.A. Thompson, IEEE Trans. Magn. **30**, 3797 (1994)
69. http://public.itrs.net
70. S.S.P. Parkin, Sci. Am. **300**, 76 (2009)
71. D. Chiba, G. Yamada, T. Koyama, K. Ueda, H. Tanigawa, S. Fukami, T. Suzuki, N. Ohshima, N. Ishiwata, Y. Nakatani, T. Ono, Appl. Phys. Exp. **3**, 073004 (2010)
72. L.O. Chua, IEEE Trans. Circuit Theory **CT-18**, 507 (1971)
73. R.S. Williams, IEEE Spectrum **29** (2008)
74. C.B. Muratov, V.V. Osipov, IEEE Trans. Magn. **45**(8), 3207 (2009)
75. H. Xi, J. Stricklin, H. Li, Y. Chen, X. Wang, Y. Zheng, Z. Gao, M.X. Tang, IEEE Trans. Magn. **46**(3), 860 (2010)

76. M.H. Kryder, C.S. Kim, IEEE Trans. Magn. **45**(10), 3406 (2009)
77. L.E. Nistor, B. Rodmacq, S. Auffret, B. Dieny, Appl. Phys. Lett. **94**, 012512 (2009)
78. J.H. Jung, S.H. Lim, S.R. Lee, Appl. Phys. Lett. **96**, 042503 (2010)
79. T. Suzuki, Scr. Metall. Mater. **33**(10–11), 1609 (1995)
80. T.C. Ulbrich, D. Makarov, G. Hu, I.L. Guhr, D. Suess, T. Schrefl, M. Albrecht, Phys. Rev. Lett. **96**, 077202 (2006)
81. Y. Katoh, S. Saito, H. Honjo, R. Nebashi, N. Sakimura, T. Suzuki, S. Miura, T. Sugibayashi, IEEE Trans. Magn. **45**(10), 3804 (2009)
82. R.W. Dave, G. Steiner, J.M. Slaughter, J.J. Sun, B. Craigo, S. Pietambaram, K. Smith, G. Grynkewich, M. DeHerrera, J. Akerman, S. Tehrani, IEEE Trans. Magn. **42**(8), 1935(2006)
83. K. Jimbo, N. Saito, S. Nakagawa, IEEE Trans. Magn. **46**(6), 1649 (2010)
84. K. Lee, S.H. Kang, IEEE Trans. Magn. **46**(6), 1537 (2010)
85. J.J. Miles, B.K. Middleton, IEEE Trans. Magn. **26**, 2137(1990)
86. T. Oikawa, M. Nakamura, H. Uwazumi, T. Shimats, H. Muraoka, Y. Nakamura, IEEE Trans. Magn. **38**(5), 1976 (2002)
87. L. Wang, H.S. Lee, Y. Qin, J.A. Bain, D.E. Laughlin, IEEE Trans. Magn. **42**(10), 2306 (1006)
88. J. Ariake, T. Chiba, N. Honda, IEEE Trans. Magn. **41**, 3142 (2005)
89. G. Choe, A. Roy, Z. Yang, B.R. Acharya, E.N. Abarra, IEEE Trans. Magn. **42**(10), 2327 (2006)
90. Y. Sasaki, N. Usuki, K. Matsuo, M. Kishimoto, IEEE Trans. Magn. **41**(10), 3241(2005)
91. S. Saito, H. Noguchi, Y. Endo, K. Ejiri, T. Mandai, T. Sugizaki, Fujifilm Res. Develop. **48**, 71 (2003)
92. T. Nagata, T. Harasawa, M. Oyanagi, N. Abe, S. Saito, IEEE Trans. Magn. **42**(10), 2312 (2006)
93. G. Bottoni, IEEE Trans. Magn. **42**(10), 2309 (2006)
94. P.F. Carcia, A.D. Meinhaldt, A. Suna, Appl. Phys. Lett. **47**(2), 178 (1985)
95. J.H. Judy, J. Magn. Magn. Mater. **287**, 16 (2005)
96. E. Miyashita, K. Kuga, R. Taguchi, T. Tamaki, H. Okuda, J. Magn. Magn. Mater. **235**, 413 (2001)
97. S. Saitoh, R. Inaba, A. Kashiwagi, IEEE Trans. Magn. **31**(6), 2859 (1995)
98. A. Matsumoto, Y. Endo, H. Noguchi, IEEE Trans. Magn. **42**(10), 2315 (2006)
99. R.M. Palmer, M.D. Thornley, H. Noguchi, K. Usuki, IEEE Trans. Magn. **42**(10), 2318 (2006)
100. J. Lohau, A. Moser, C.T. Rettner, M.E. Best, B.D. Terris, Appl. Phys. Lett. **78**, 990 (2001)
101. Z.Z. Bandic, E.A. Dobisz, T.W. Wu, T.R. Albrecht, Solid State Technol. **S7** (2006)
102. M. Ichida, K. Yoshida, IEEE Trans. Magn. **42**(10), 2291 (2006)
103. A.F. Torabi, D. Bai, P. Luo, J. Wang, M. Novid, IEEE Trans. Magn. **42**(10), 2288 (2006)
104. R. Fischer, T. Leineweber, H. Kronmüller, Phys. Rev. B **57**(10), 723 (1998)
105. G.A. Gibson, S. Schultz, J. Appl. Phys. **73**, 4516 (1993)
106. M. Kleiber, F. Kummerlen, M. Lohndorf, A. Wadas, D. Weiss, R. Wiesendanger, Phys. Rev. B **58**, 5563 (1998)
107. S. Bandiera, R.C. Sousa, Y. Dahmane, C. Ducruet, C. Portemont, V. Baltz, S. Auffret, I.L. Prejbeanu, B. Dieny, IEEE Magn. Lett. **1**, 3000204 (2010)

Index